P E E X A M P R E P

MECHANICAL ENGINEERING

PE LICENSE REVIEW

Seventh Edition

Jerry H. Hamelink, EdD, PE & John D. Constance, PE

KAPLAN AEC EDUCATION

President: Roy Lipner
Vice President & General Manager: David Dufresne
Vice President of Product Development and Publishing: Evan M. Butterfield
Editorial Project Manager: Laurie McGuire
Director of Production: Daniel Frey
Creative Director: Lucy Jenkins

Published by Kaplan AEC Education
30 South Wacker Drive
Chicago, IL 60606-7481
(312) 836-4400
www.kaplanaecengineering.com

Printed in the United States of America.

08 09 10 10 9 8 7 6 5 4 3 2 1

CONTENTS

Introduction

OUTLINE

HOW TO USE THIS BOOK

Mechanical Engineering PE License Review and its companion texts, *Mechanical Engineering PE Problems & Solutions* and *Mechanical Engineering PE Sample Exam*, form a three-part approach to preparing for the Principles and Practice of Mechanical Engineering exam:

- *Mechanical Engineering PE License Review* contains the conceptual review of mechanical engineering topics for the exam. Solved examples illustrate how to apply the equations and analytical methods discussed in the text.
- *Mechanical Engineering PE Problems & Solutions* provides problems for you to solve in order to test your understanding of concepts and techniques. Ideally, you should solve these problems after completing your conceptual review. Then, compare your solution to the detailed solutions provided, to get a sense of how well you have mastered the content and what topics you may want to review further.
- *Mechanical Engineering PE Sample Exam* provides complete morning and afternoon exam sections so that you can simulate the experience of taking the PE test within its actual time constraints and with questions that match the test format. Take the sample exam after you're satisfied with your review of concepts and problem-solving techniques, to test your readiness for the real exam.

BECOMING A PROFESSIONAL ENGINEER

To achieve registration as a professional engineer there are four distinct steps: (1) education, (2) the Fundamentals of Engineering/Engineer-In-Training (FE/EIT) exam, (3) professional experience, and (4) the professional engineer (PE) exam, more formally known as the Principles and Practice of Engineering Exam. These steps are described in the following sections.

Education

The obvious appropriate education is a B.S. degree in mechanical engineering from an accredited college or university. This is not an absolute requirement. Alternative, but less acceptable, education is a B.S. degree in something other than mechanical engineering, or a degree from a non-accredited institution, or four years of education but no degree.

Fundamentals of Engineering (FE/EIT) Exam

Most people are required to take and pass this eight-hour multiple-choice examination. Different states call it by different names (Fundamentals of Engineering, E.I.T., or Intern Engineer), but the exam is the same in all states. It is prepared and graded by the National Council of Examiners for Engineering and Surveying (NCEES). Review materials for this exam are found in other Kaplan AEC books such as *Fundamentals of Engineering: FE/EIT Exam Preparation.*

Experience

Typically one must have four years of acceptable experience before being permitted to take the Professional Engineer exam (California requires only two years). Both the length and character of the experience will be examined. It may, of course, take more than four years to acquire four years of acceptable experience.

Professional Engineer Exam

The second national exam is called Principles and Practice of Engineering by NCEES, but just about everyone else calls it the Professional Engineer or P.E. exam. All states, plus Guam, the District of Columbia, and Puerto Rico, use the same NCEES exam.

MECHANICAL ENGINEERING PROFESSIONAL ENGINEER EXAM

The reason for passing laws regulating the practice of mechanical engineering is to protect the public from incompetent practitioners. Most states require engineers working on projects involving public safety to be registered, or to work under the supervision of a registered engineer. In addition, many private companies encourage or require engineers in their employ to pursue registration as a matter of professional development. Engineers in private practice, who wish to consult or serve as expert witnesses, typically also must be registered. There is no national registration law; registration is based on individual state laws and is administered by boards of registration in each of the states. You can find a list of contact information for and links to the various state boards of registration at the Kaplan AEC Web site: *www.kaplanaecengineering.com.* This list also shows the exam registration deadline for each state.

Examination Development

Initially the states wrote their own examinations, but beginning in 1966 the NCEES took over the task for some of the states. Now the NCEES exams are used by all states. This greatly eases the ability of an engineer to move from one state to another and achieve registration in the new state.

The development of the engineering exams is the responsibility of the NCEES Committee on Examinations for Professional Engineers. The committee is composed of people from industry, consulting, and education, plus consultants and subject matter experts. The starting point for the exam is a task analysis survey, which NCEES does at roughly 5- to 10-year intervals. People in industry, consulting, and education are surveyed to determine what mechanical engineers do and what knowledge is needed. From this NCEES develops what it calls a "matrix of knowledge" that forms the basis for the exam structure described in the next section.

The actual exam questions are prepared by the NCEES committee members, subject matter experts, and other volunteers. All people participating must hold professional registration. Using workshop meetings and correspondence by mail, the questions are written and circulated for review. Although based on an understanding of engineering fundamentals, the problems require the application of practical professional judgment and insight.

Examination Structure

The exam is organized into breadth and depth sections.

The morning breadth exam consists of 40 multiple-choice questions covering the following areas of mechanical engineering (relative exam weight for each topic is shown in parentheses):

- General knowledge, codes and standards (30%)
- Machine design and materials (17%)
- Hydraulics and fluids (17%)
- Energy conversion and power systems (18%)
- HVAC and refrigeration (18%)

You will have four hours to complete the breadth exam.

The afternoon depth exam is actually three exams; you choose the depth exam you wish to take. The depth exams are

- HVAC and Refrigeration
- Machine Design
- Thermal and Fluids Systems

Clearly, you should choose the exam that best matches your training and professional practice. You will have four hours to answer the 40 multiple-choice questions that make up the depth exam.

Both the breadth and depth questions include four possible answers (A, B, C, D) and are objectively scored by computer.

For more information on the topics and subtopics and their relative weights on the breadth and depth portions, visit the NCEES Web site at www.ncees.org.

Exam Dates

The National Council of Examiners for Engineering and Surveying (NCEES) prepares Mechanical Engineering Professional Engineer exams for use on a Friday in April and October of each year. Some state boards administer the exam twice a year in their state, whereas others offer the exam once a year. The scheduled exam dates for the next ten years can be found on the NCEES Web site (*www.ncees.org/exams/schedules/*).

People seeking to take a particular exam must apply to their state board several months in advance.

Exam Procedure

Before the morning four-hour session begins, proctors will pass out an exam booklet, answer sheet, and mechanical pencil to each examinee. The provided pencil is the only writing instrument you are permitted to use during the exam. If you need an additional pencil during the exam, a proctor will supply one.

Fill in the answer bubbles neatly and completely. Questions with two or more bubbles filled in will be marked as incorrect, so if you decide to change an answer, be sure to erase your original answer completely.

The afternoon session will begin following a one-hour lunch break.

In both the morning and afternoon sessions, if you finish more than 15 minutes early you may turn in your booklet and answer sheet and leave. In the last 15 minutes, however, you must remain to the end of the exam in order to ensure a quiet environment for those still working and an orderly collection of materials.

Exam-Taking Suggestions

People familiar with the psychology of exam taking have several suggestions for people as they prepare to take an exam.

1. Exam taking involves really, two skills. One is the skill of illustrating knowledge that you know. The other is the skill of exam taking. The first may be enhanced by a systematic review of the technical material. Exam-taking skills, on the other hand, may be improved by practice with similar problems presented in the exam format.

2. Since there is no deduction for guessing on the multiple choice problems, answers should be given for all of them. Even when one is going to guess, a logical approach is to attempt to first eliminate one or two of the four alternatives. If this can be done, the chance of selecting a correct answer obviously improves from 1 in 4 to 1 in 3 or 1 in 2.

3. Plan ahead with a strategy. Which is your strongest area? Can you expect to see several problems in this area? What about your second strongest area? What will you do if you still must find problems in other areas?

4. Plan ahead with a time allocation. Compute how much time you will allow for each of the subject areas in the breadth exam and the relevant topics in the depth exam. You might allocate a little less time per problem for those areas in which you are most proficient, leaving a little more time in subjects

that are difficult for you. Your time plan should include a reserve block for especially difficult problems, for checking your scoring sheet, and to make last-minute guesses on problems you did not work. Your strategy might also include time allotments for two passes through the exam—the first to work all problems for which answers are obvious to you, and the second to return to the more complex, time-consuming problems and the ones at which you might need to guess. A time plan gives you the confidence of being in control and keeps you from making the serious mistake of misallocation of time in the exam.

5. Read all four multiple-choice answers before making a selection. An answer in a multiple-choice question is sometimes a plausible decoy—not the best answer.
6. Do not change an answer unless you are absolutely certain you have made a mistake. Your first reaction is likely to be correct.
7. Do not sit next to a friend, a window, or other potential distractions.

Exam Day Preparations

The exam day will be a stressful and tiring one. This will be no day to have unpleasant surprises. For this reason we suggest that an advance visit be made to the examination site. Try to determine such items as

1. How much time should I allow for travel to the exam on that day? Plan to arrive about 15 minutes early. That way you will have ample time, but not too much time. Arriving too early, and mingling with others who also are anxious, will increase your anxiety and nervousness.
2. Where will I park?
3. How does the exam site look? Will I have ample workspace? Where will I stack my reference materials? Will it be overly bright (sunglasses), cold (sweater), or noisy (earplugs)? Would a cushion make the chair more comfortable?
4. Where are the drinking fountains and lavatory facilities?
5. What about food? Should I take something along for energy in the exam? A bag lunch during the break probably makes sense.

What to Take to the Exam

The NCEES guidelines say you may bring only the following reference materials and aids into the examination room for your personal use:

1. Handbooks and textbooks, including the applicable design standards.
2. Bound reference materials, provided the materials remain bound during the entire examination. The NCEES defines "bound" as books or materials fastened securely in their covers by fasteners that penetrate all papers. Examples are ring binders, spiral binders and notebooks, plastic snap binders, brads, screw posts, and so on.
3. A battery-operated, silent, nonprinting, noncommunicating calculator from the NCEES list of approved calculators. For the most current list, see the NCEES Web site (*www.ncees.org*). You also need to determine whether or not your

state permits preprogrammed calculators. Bring extra batteries for your calculator just in case; many people feel that bringing a second calculator is also a very good idea.

At one time NCEES had a rule barring "review publications directed principally toward sample questions and their solutions" in the exam room. This set the stage for restricting some kinds of publications from the exam. *State boards may adopt the NCEES guidelines, or adopt either more or less restrictive rules.* Thus an important step in preparing for the exam is to know what will—and will not—be permitted. We suggest that if possible you obtain a written copy of your state's policy for the specific exam you will be taking. Occasionally there has been confusion at individual examination sites, so a copy of the exact applicable policy will not only allow you to carefully and correctly prepare your materials, but will also ensure that the exam proctors will allow all proper materials that you bring to the exam.

As a general rule we recommend that you plan well in advance what books and materials you want to take to the exam. Then they should be obtained promptly so you use the same materials in your review that you will have in the exam.

License Review Books

The review books you use to prepare for the exam are good choices to bring to the exam itself. After weeks or months of studying, you will be very familiar with their organization and content, so you'll be able to quickly locate the material you want to reference during the exam. Keep in mind the caveat just discussed—some state boards will not permit you to bring in review books that consist largely of sample questions and answers.

Textbooks

If you still have your university textbooks, they are the ones you should use in the exam, unless they are too out of date. To a great extent the books will be like old friends with familiar notation.

Bound Reference Materials

The NCEES guidelines suggest that you can take any reference materials you wish, so long as you prepare them properly. You could, for example, prepare several volumes of bound reference materials, with each volume intended to cover a particular category of problem. Maybe the most efficient way to use this book would be to cut it up and insert portions of it in your individually prepared bound materials. Use tabs so that specific material can be located quickly. If you do a careful and systematic review of civil engineering, and prepare a lot of well-organized materials, you just may find that you are so well prepared that you will not have left anything of value at home.

Other Items

In addition to the reference materials just mentioned, you should consider bringing the following to the exam:

- *Clock*—You must have a time plan and a clock or wristwatch.
- *Exam assignment paperwork*—Take along the letter assigning you to the exam at the specified location. To prove you are the correct person, also bring something with your name and picture.

- *Items suggested by advance visit*—If you visit the exam site, you probably will discover an item or two that you need to add to your list.
- *Clothes*—Plan to wear comfortable clothes. You probably will do better if you are slightly cool.
- *Box for everything*—You need to be able to carry all your materials to the exam and have them conveniently organized at your side. Probably a cardboard box is the answer.

Examination Scoring and Results

The questions are machine-scored by scanning. The answers sheets are checked for errors by computer. Marking two answers to a question, for example, will be detected and no credit will be given.

Your state board will notify you whether you have passed or failed roughly three months after the exam. Candidates who do not pass the exam the first time may take it again. If you do not pass you will receive a report listing the percentages of questions you answered correctly for each topic area. This information can help focus the review efforts of candidates who need to retake the exam.

The PE exam is challenging, but analysis of previous pass rates shows that the majority of candidates do pass it the first time. By reviewing appropriate concepts and practicing with exam-style problems, you can be in that majority. Good luck!

Acknowledgments

Several reviewers provided very helpful guidance and accurancy review in the preparation of this edition. The author and publisher are grateful to:

Y. I. Sharaf-Eldeen, PhD, PE
Florida Institute of Technology

Forrest G. Lowe, PhD, PE
University of Missouri—Kansas City

Ghanashyam Joshi, PhD, PE
Southern University and A & M College

Gerald J. Micklow, PhD, PE
East Carolina University

Abraham Tchako, PhD
Union College

Frank Wicks, PhD, PE
Union College

CHAPTER 1

Mechanics of Materials

OUTLINE

Mechanics of materials is normally divided into two topics: statics and dynamics. Statics deals with the effects of forces on rigid and nonrigid non-moving members or bodies. Dynamics deals with the corresponding actions of motion caused by forces on members or bodies that are free to move.

STATICS

Statics uses the equations of equilibrium for analysis of systems. These equations are

$$\sum F_x = 0 \quad \sum M_x = 0$$
$$\sum F_y = 0 \quad \sum M_y = 0$$
$$\sum F_z = 0 \quad \sum M_z = 0$$

Example 1.1

Consider the trolley braking system in Exhibit 1, with the dimensions in feet indicated. A force of 85 lbf is applied to the handle of the brake at B. What are the horizontal and vertical reactions at the brake drum "O?"

Solution

First make a free body diagram of the parts (Exhibit 2).

$$\sum M_A = 0$$
$$4(85) + F_{Bx}(2) + F_{By}(2) = 0$$
$$\sum M_O = 0 = -M + F_{By}(0.5)$$
$$F_{By} = 60/0.5 = 120 \text{ lb}$$
$$\text{Therefore, } F_{Oy} = 120 \text{ lb}\uparrow$$
$$\text{Reaction } F_{Oy} = 120 \text{ lb}\downarrow$$

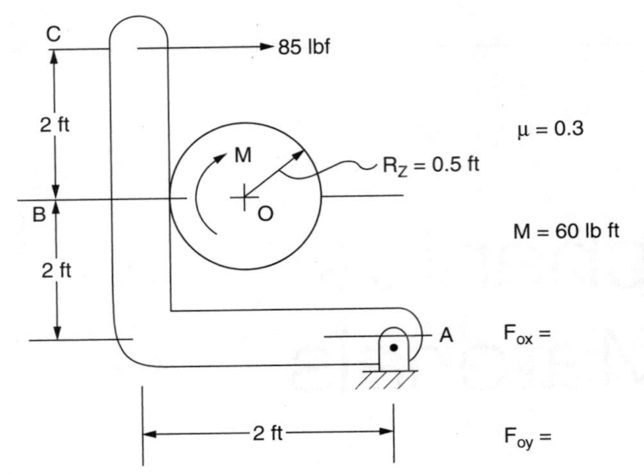

Exhibit 1

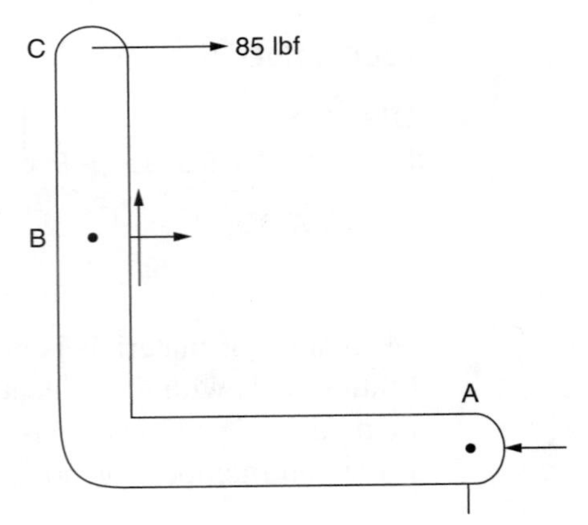

Exhibit 2

From the previous equation,

$$4(85) + 2F_{Bx} + 2F_{By} = 0$$
$$340 + 2F_{Bx} + 2(120) = 0$$
$$F_{Bx} = -290 \text{ lb or,}$$
$$F_{Ox} = 290\rightarrow$$
$$\text{Reaction } F_{Ox} = 290\leftarrow$$

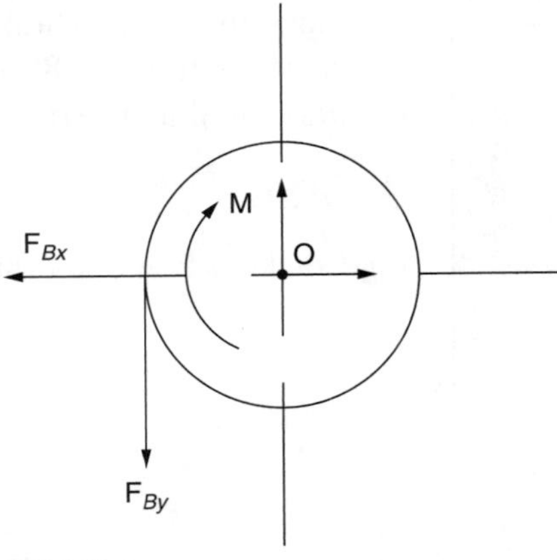

Exhibit 3

Frames and Trusses

Frames and trusses are simple structures in which the basic equations of equilibrium may be used to great advantage. One basic assumption is that these structures are held together by frictionless pins. Another usual assumption is that the members are rigid and weightless. The analysis follows the general procedure shown in Examples 1.2 and 1.3.

Example 1.2

Find the force in truss members BD and BE in Exhibit 4.

$$\Sigma M_A = 0 = -8.66(5000) - 10{,}000(12) + F_H(24)$$
$$F_H = 6804\uparrow$$

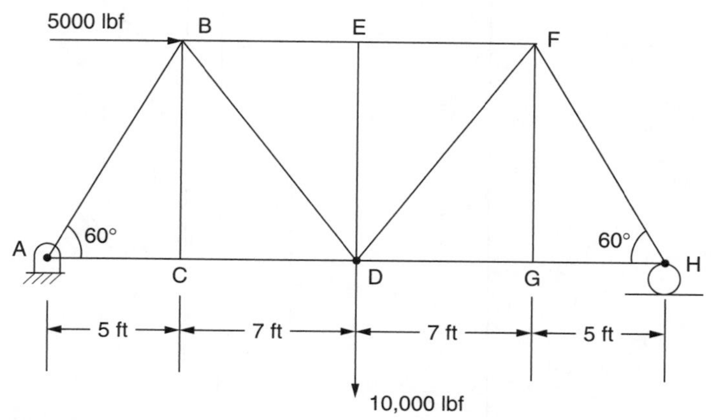

Exhibit 4

Solution

Draw a free body diagram of the section (Exhibit 5).

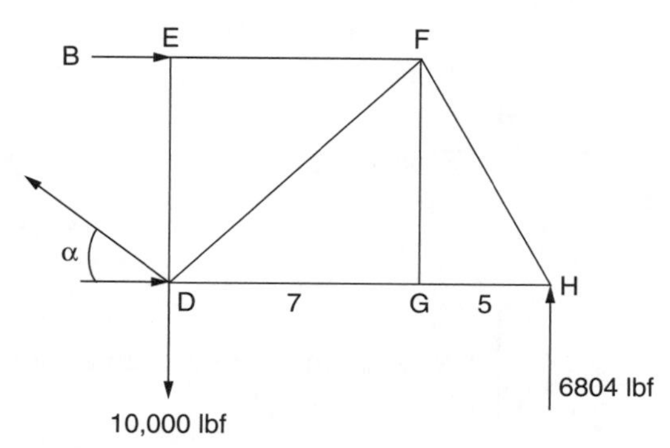

Exhibit 5

$$\Sigma M_D = 0 = 6804(12) - F_E(8.66)$$
$$F_E = 9428\rightarrow$$
Therefore, $F_{BE} = 9428$ lbf compression
$$\tan\alpha = 8.66/7$$
$$\alpha = 51.1^\circ$$
$$\Sigma F_y = 0 = F_{BD}\sin\alpha - 10{,}000 + 6804$$
$$F_{BD}\sin 51.1^\circ = 3196$$
$$F_{BD} = 4107 \text{ lbf tension}$$

Example 1.3

The frame in Exhibit 6 is holding up a 200lbf sign. Determine (a) the moment at "o" and (b) the horizontal and vertical components of the forces on pins A and B.

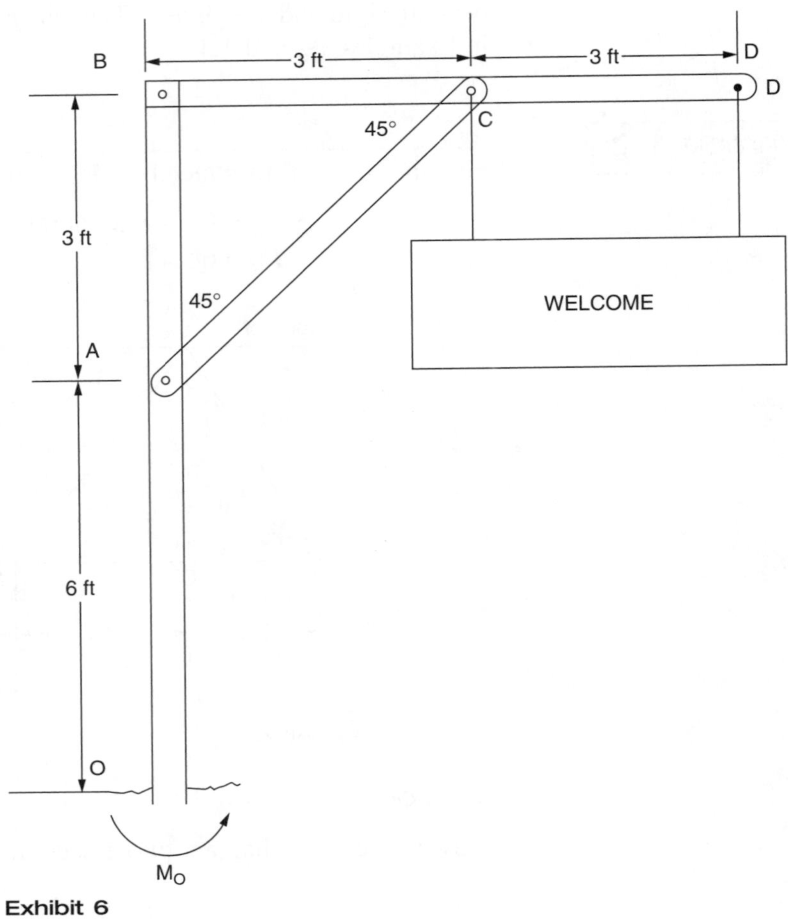

Exhibit 6

Solution

$$\Sigma M_O = M_O - 200 \text{ lbf}(4.5 \text{ ft})$$
$$M_O = 9000 \text{ lb ft}$$

Now dismantle the frame into various free body diagrams (Exhibits 7 and 8).

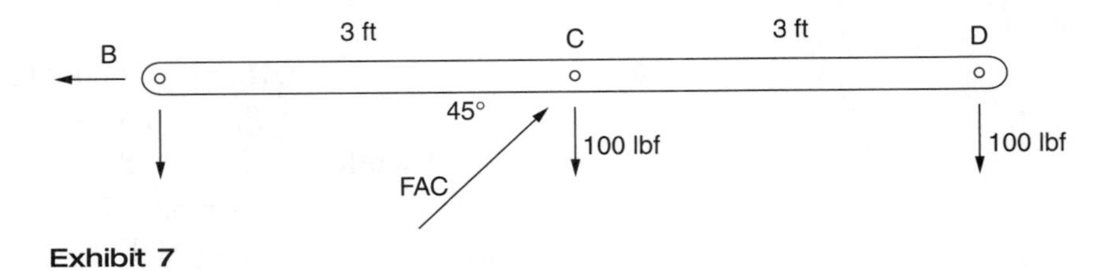

Exhibit 7

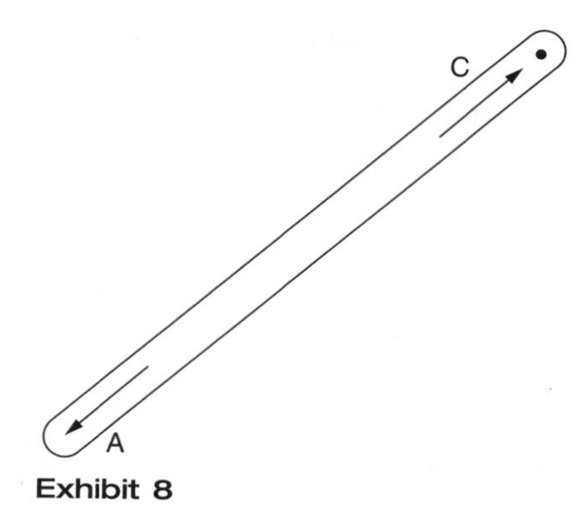

Exhibit 8

$$\Sigma M_C = 0 = -(3\text{ft})(100 \text{ lbf}) + 3\text{ft}(F_{BY})$$
$$F_{BY} = 100 \text{ lbf}\downarrow$$
$$\Sigma F_Y = 0 = F_{BY} + F_{CY} + F_{DY} + F_{AC} \sin 45°$$
$$F_{AC} = 300/\sin 45° = 424 \text{ lbf}$$
$$\Sigma F_X = 0 = -F_{BX} + 424 \cos 45°$$
$$F_{BX} = 300 \text{ lbf}\leftarrow$$
$$F_{AY} = 300 \text{ lbf}\downarrow$$
$$F_{AX} = 300 \text{ lbf}\leftarrow$$

Friction

Friction is the force that impedes an object from moving relative to another object. It uses an average coefficient of friction that varies depending on the types of material of the mating surfaces, environmental conditions, and whether there is relative motion between the surfaces.

Example 1.4

Consider the wedges with the forces and coefficients of friction indicated in Exhibit 9. Determine how much the spring compresses before Block B slips on Block A and determine the force F required to cause the action.

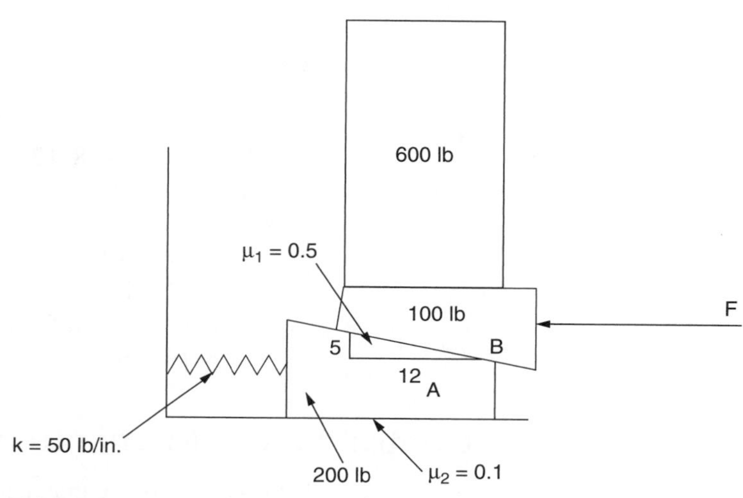

Exhibit 9

Solution

Draw a free body diagram of wedge B (Exhibit 10).

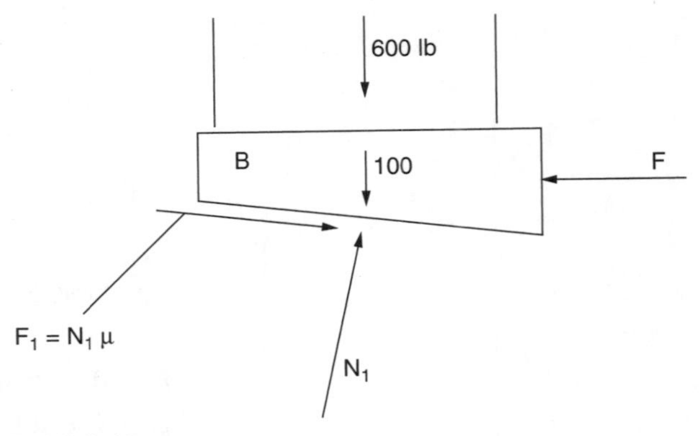

Exhibit 10

$$\Sigma F_Y = 0 = -600 - 100 - F_1(5/13) + N_1(12/13)$$
$$-5/13(N_1)(\mu_1) + 12/13N_1 = 700$$
$$-5/13(N_1)(0.5) + 12/13N_1 = 700$$
$$N_1 = 958 \text{ lbf}$$
$$\Sigma F_X = 0 = F_1(12/13) + N_1(5/13) - F$$
$$N_1(0.5)(12/13) + 5/13(N_1) = F$$
$$F = 811 \text{ lbf}$$

Now draw a free body diagram of wedge A (Exhibit 11).

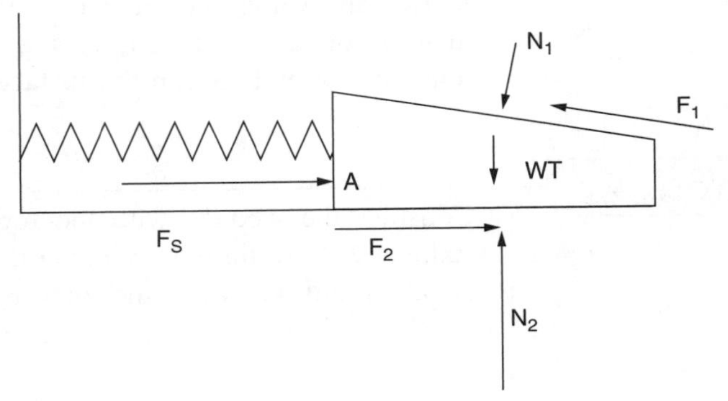

Exhibit 11

$$\Sigma F_Y = 0 = N_2 - 200 - N_1(12/13) + F_1(5/13)$$
$$N_2 = 200 + 958(12/13) - 958(0.5)(5/13)$$
$$N_2 = 900 \text{ lb}$$
$$\Sigma F_X = 0 = F_S + F_2 - N_1(5/13) - F_1(12/13)$$
$$F_S = -N_2(0.1) + (12/13)(958)(0.5) + 958(5/13)$$
$$F_S = 720 \text{ lb}$$
$$F_S = K_S = (50 \text{ lbf/in})(x \text{ in}) = 720 \text{ lbf}$$
$$x = 14.4 \text{ in}$$

Centroids and Moments of Inertia

A centroid is that point where we may generally assume the center of mass is located in a non-homogeneous body. The moment of inertia is a function of the force required to move a body about a given axis. It is defined as the second moment of area.

Example 1.5

A plate with a hole in it is to be rotated about the x-axis (Exhibit 12). Determine the y distance to the centroid and the moment of inertia about the x-axis.

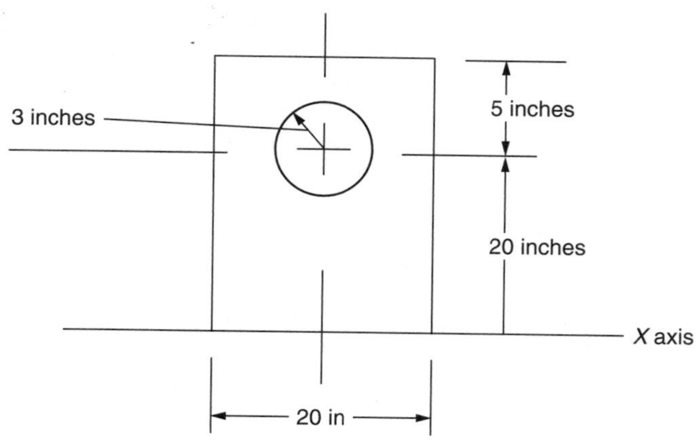

Exhibit 12

Solution

Break the body down into two parts, the plate and the hole. Knowing $\bar{y} = \Sigma yA/\Sigma A$

Body	Area	y	yA
Plate	500	12.5	6250
Hole	−28.27	20	−565.4
Σ	471.73		5684.6

$$\bar{y} = \Sigma yA/\Sigma A = 5{,}684.6/471.73 = 12.05 \text{ inches}$$
$$I_x = I_x + Ady^2$$

Rectangle

$$I_x = 1/12\, bh^3 + Ady^2$$
$$I_x = (1/12)(20)(25)^3 + (500)(12.5)^2$$
$$I_x = 104{,}167 \text{ inches}^4$$

Hole

$$I_x = 1/4\Pi r^4 + Ady^2$$
$$I_x = (1/4)(\Pi)(3)^4 + 28.3(20)^2$$
$$I_x = 11{,}384 \text{ inches}^4$$
$$I_{x\text{Total}} = I_{x\text{Rect}} - I_{x\text{Hole}}$$
$$I_{x\text{Total}} = 104{,}167 - 11{,}384 = 92{,}783 \text{ inches}^4$$

DYNAMICS

As indicated at the beginning of this chapter, dynamics deals with bodies in motion and the forces required to change their direction or speed of motion. Some types of situations include motions of bodies which are dependent on each other. This dependence may occur because the particles are interconnected by a cable, as in Example 1.6.

Example 1.6

Determine the velocity of pulley 1 in Exhibit 13.

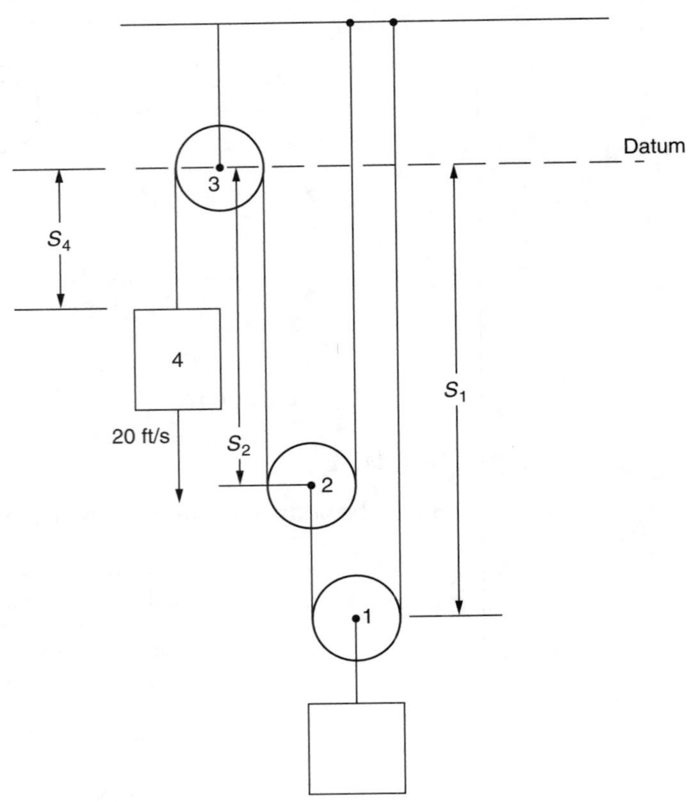

Exhibit 13

Solution

In working a problem of this type, make use of the following:

- Ropes are constant length.
- Select a datum from which to work.

Lengths of cables

(a) $$L_1 = s_4 + 2s_2 \quad \text{or} \quad s_2 = (L_1 - s_4)/2$$

(b) $$L_2 = s_1 + (s_1 - s_2)$$

Substitute into Equation (b).

$$L_2 = s_1 + s_1 - (L_1 - s_4)/2$$
$$2L_2 = 4s_1 - L_1 + s_4$$
$$4s_1 + s_4 = 2L_2 + L_1$$

Taking d/dt and noting L_1 and L_2 are constant,

$$4v_1 + v_4 = 0$$
$$v_1 = 20 \text{ ft/s}/4\uparrow = 5 \text{ ft/s}\uparrow$$

Many dynamics problems may be solved using the energy balance approach, that is, a form of the first law of thermodynamics.

Example 1.7

In Exhibit 14, a 500 lbm block is held at position 1, in which the spring is compressed 10 inches. The block is then released and allowed to move. The coefficient of friction between the block and the surface is $\mu = 0.3$. Determine the velocity when the block has moved 12 inches.

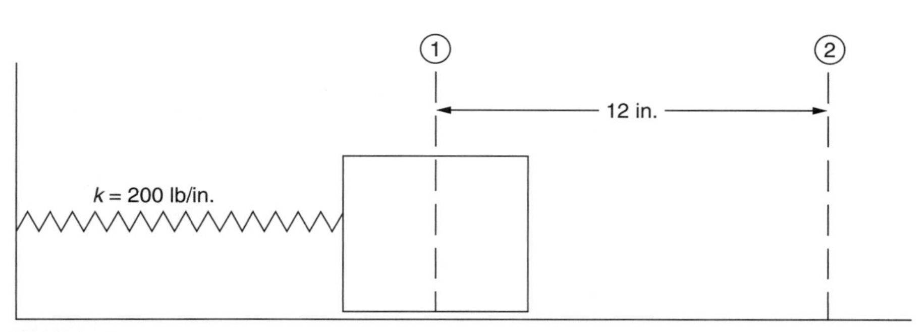

Exhibit 14

Solution

$$\sum \text{Energy (1)} = \sum \text{Energy (2)} + \text{Work}$$

$$\frac{1}{2}kx^2 = \frac{1}{2}mV^2 + \text{Fds}$$

$$\frac{1}{2}\left(200\frac{\text{lb}}{\text{in}}\right)\left(12\frac{\text{in}}{\text{ft}}\right)\left(\frac{100\text{ in}^2}{144\frac{\text{in}^2}{\text{ft}^2}}\right) = \frac{\frac{1}{2}500\text{ lbm}}{32.2\frac{\text{ft lbm}}{\text{lbf } s^2}}\left(V\frac{\text{ft}}{s}\right)^2 + (500\text{ lbf})\mu(x_2 - x_1)$$

$$833\text{ ft lbf} = 7.8(V)^2\text{ft lbf} + 500\text{ lbf}\quad(0.3)\left(\frac{12\text{in}}{12\text{in/ft}}\right)$$

$$\text{V} = 9.4\text{ ft/s}$$

Often, in dynamics, motion of bodies includes impact. Because impact is difficult to analyze where the energy is dissipated, summation of momentum is used.

Example 1.8

A $\frac{1}{2}$ ounce bullet traveling at 3000 ft/s impacts a 40 lb log that has a $\mu = 0.5$. Determine the velocity of the log after impact and how far the log will travel after impact.

Solution

Velocity has to be determined by momentum balance. Exhibit 15 shows the scenario.

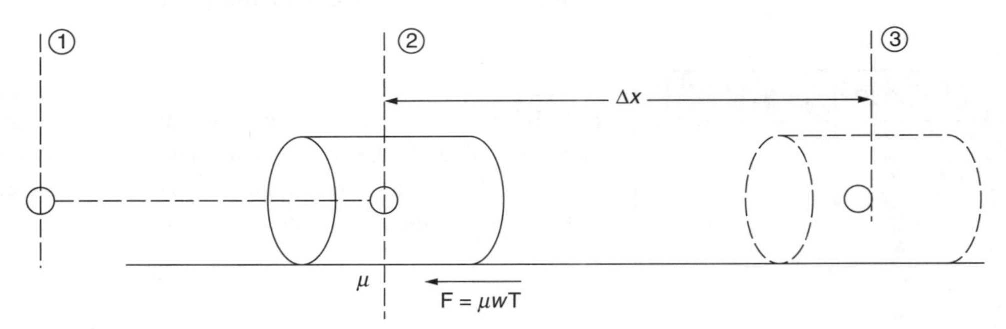

Exhibit 15

$$\Sigma mV_1 = \Sigma mV_2$$

$$\frac{\left(\frac{1}{32}\text{ lbm}\right)\left(3000\,\frac{\text{ft}}{s}\right)}{32.2\,\frac{\text{ft lbm}}{\text{lbf } s^2}} = \frac{\left(\frac{1}{32}+40\right)\text{lbm}\left(V_2\right)}{32.2}$$

$$V_2 = 2.34\,\frac{\text{ft}}{s}$$

$$\Sigma mV_1 = \Sigma mV_3 + w \quad \text{and} \quad V_3 = 0$$

$$\frac{\left(\frac{1}{32}\right)(3000)}{32.2} = \frac{\left(40+\frac{1}{32}\right)}{32.2}(\mu)\Delta x$$

$$2.91 = 1.24(0.5)\Delta x$$

$$\Delta x = 4.7\text{ ft}$$

Example 1.9

An elevator weighs 8000 lb and moves downward at a constant velocity of 300 ft per min. At the instant when the length of cable from elevator to cable drum is 50 ft, an accident occurs that causes the drum to stop instantaneously. Assume a hemp-filled steel cable of six strands, with each strand made up of 19 wires having a metal area of 2 sq in. Neglecting the weight of the cable, find: (a) total stretch of the cable, (b) maximum stretch induced in the cable, and (c) frequency of oscillation.

Solution

(a) Refer to Exhibit 16. δ_d = dynamic deflection due to dynamic force.

Total stretch of cable after accident = $\delta_{static} + \delta_{dynamic} = \delta_s + \delta_d$

$$\delta_s = \frac{PL}{AE} = \frac{WL^*}{AE} = \frac{8{,}000(50\times 12)}{2\times 12\times 10^6} = 0.2\text{ in.}$$

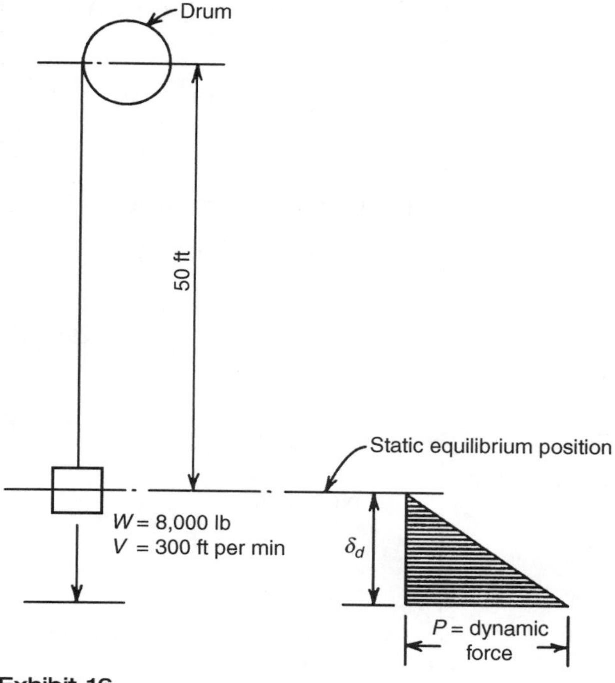

Exhibit 16

Strain energy u in wire rope due to dynamic load is

$$u = 1/2 \times \delta_d \times P$$

From the elastic behavior of wire rope,

$$\delta_d = \frac{PL}{AE}.$$

$$\text{Therefore,} \quad u = \frac{AE\delta_d^2}{2L}.$$

$$\text{Kinetic energy at accident} = \frac{W}{g} \times \frac{v^2}{2}$$

Because of conservation of energy, $E_{total} = 0 = \Delta u$. Then strain energy = kinetic energy. Thus

$$\frac{AE\delta_d^2}{2L} = \frac{Wv^2}{2g}.$$

By rearrangement,

$$\delta_d^2 = \frac{Wv^2L}{AEg}$$

$$\delta_d = v\sqrt{\frac{W}{g} \times \frac{L}{AE}} = \frac{300 \times 12}{60}\sqrt{\frac{8{,}000}{386} \times \frac{50 \times 12}{2 \times 12 \times 10^6}} = 1.366 \text{ in.}$$

Total stretch = $\delta_s + \delta_d = 0.2 + 1.366 = 1.566$ in.

(b) Max stress = $S_s + S_d$. Also, $S = P/A = E\delta/L$.

$$S = \frac{E}{L}(\delta_s + \delta_d) = \frac{12\times10^6}{50\times12}\times0.2 + \frac{12\times10^6}{50\times12}\times1.366$$

$$S = 4000 + 27{,}300 = 31{,}300 \text{ psi}$$

(c) Oscillation frequency $= f$, and t (time of oscillation) $= 1/f$.

Refer to Exhibit 17. Then

$$f = \frac{1}{2\pi}\sqrt{\frac{K}{M}} \text{ cycles per sec (Hz)} = \frac{1}{2\pi}\sqrt{\frac{W}{\delta_s}\Big/\frac{W}{g}} = \frac{1}{2\pi}\sqrt{\frac{g}{\delta_s}}$$

$$f = \frac{1}{2\pi}\sqrt{\frac{386}{0.2}} = 7 \text{ cycles per sec.}$$

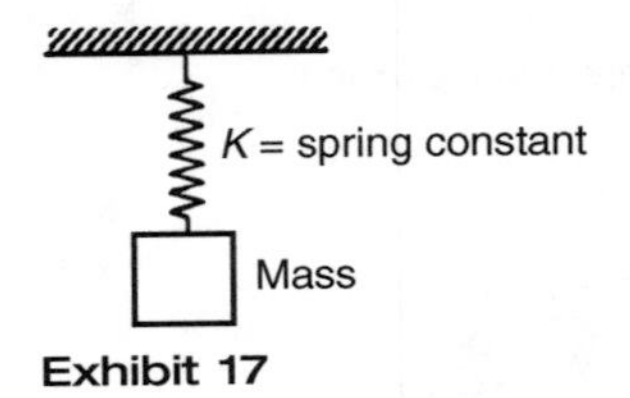

Exhibit 17

Note that

$$W/\delta_s = AE/L$$
$$= K.$$

Also, 386 = in. per sec per sec.*

Example 1.10

The bars in Exhibit 18 have the same cross-sectional area.

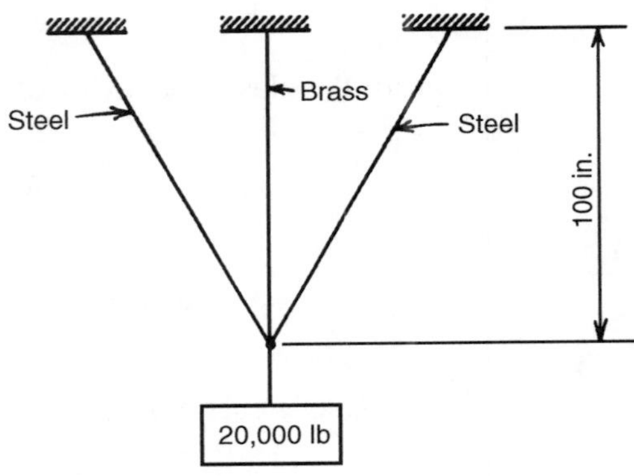

Exhibit 18

There is no stress in the bars before the load is applied. Each bar is 0.5 in. square. Outer bars hang at 60° from horizontal.

(a) Find the force in each bar.

(b) Find the force in each bar if the temperature drops 100°F.

Solution

(a) $\delta = Pl/AE$, and by rearrangement, $P = \delta AE/l$. Now for the steel,

$$P_s = \frac{0.866\delta A\times30{,}000{,}000}{115.47} = 225{,}000\delta A$$

See Exhibit 19.

* See Marks Handbook for wire rope.

Vertical component:

$$(P_s)_v = 0.866\delta \times 225{,}000A = 194{,}850\delta A$$

For brass:

$$P_b = \frac{(\delta A \times 15 \times 10^6)}{(100)} = 150{,}000\delta A$$

$$\text{Total: } \delta A(2 \times 194{,}850 + 150{,}000) = 20{,}000$$

$$\delta A = \frac{20{,}000}{539{,}700} \quad \text{and} \quad P_s = 225{,}000 \times \frac{20{,}000}{539{,}700} = 8338 \text{ lb}$$

$$P_b = 150{,}000 \times \frac{20{,}000}{539{,}700} = 5559 \text{ lb}$$

(b) Change of length due to temperature

For steel, $\delta = 115.47 \times 0.0000065 \times 100 = 0.075056$ in. shortening
For brass, $\delta = 100 \times 0.0000102 \times 100 = 0.102000$ in. shortening

Stretch of brass to load: See Exhibit 20.

$$\delta_b = \frac{P_b \times 100 \times 2}{15 \times 10^6}$$

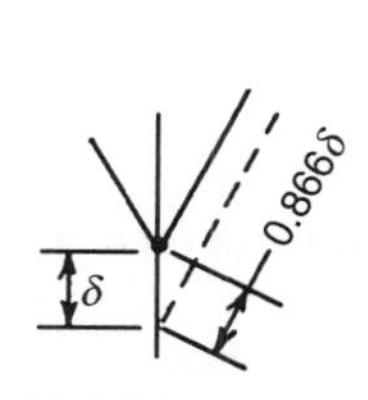

Exhibit 19

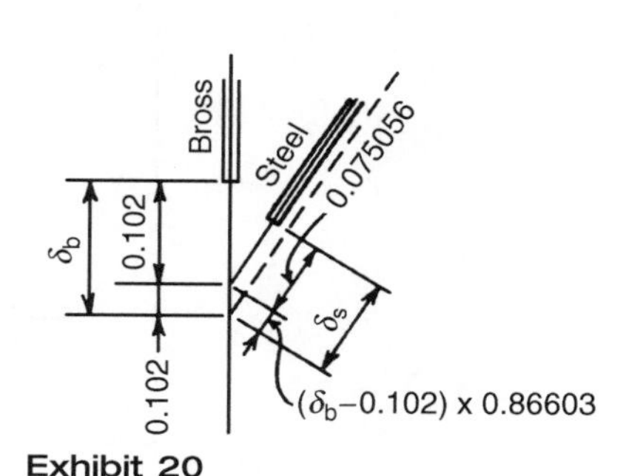

Exhibit 20

Stretch of steel due to load:

$$\delta_s = (\delta_b - 0.102) \times 0.86603 + 0.075056 = 0.86603\delta_b - 0.013279$$

$$2(P_s)_v = 20{,}000 - P_b$$

Force in steel:

$$P_s = \frac{20{,}000 - P_b}{2 \times 0.86603} = 11{,}547 - 0.57735P_b$$

$$\delta_s = 0.86603\delta_b - 0.013278 = \frac{P_s \times 115.47 \times 4}{30 \times 10^6}$$

$$0.86603 \times \frac{P_b \times 400}{15 \times 10^6} - 0.013278 = \frac{(11{,}547 - 0.57735P_b)(115.47)}{0.25 \times 30 \times 10^6}$$

$$0.000023094\,P_b - 0.013278 = 0.177778 - 0.000008889\,P_b$$

$$0.000031983P_b = 0.191057$$

$$P_b = 5974 \text{ lb}$$

$$P_s = 11{,}547 - 0.57735 \times 5974$$

$$= 11{,}547 - 3449 = 8098 \text{ lb}$$

Example 1.11

Exhibit 21 is a view of two beams, both of which have the same EI of 33,333 lb in.^{2}

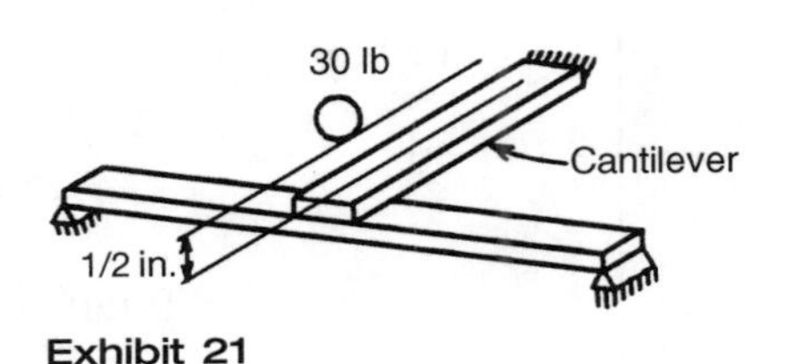

Exhibit 21

The end of the 10-in. cantilever beam rests on the midpoint of the 20-in. simply supported beam. A 30-lb weight falls freely through a distance of $\frac{1}{2}$ in. and strikes the free end of the cantilever beam. Find the maximum deflection that the cantilever beam undergoes.

Solution

Consider each beam as a spring that can store the potential energy of the weight $W = 30$ lb in the form of strain energy (see Exhibit 22). For the cantilever beam alone, shown in Exhibit 23,

$$\Delta = \frac{PL^3}{3EI} \qquad K_1 = \frac{P}{\Delta} = \text{lb/in.} = \frac{3EI}{L^3}$$

$$= \frac{3(33,333)}{10^3} = 100 \text{ lb per in.}$$

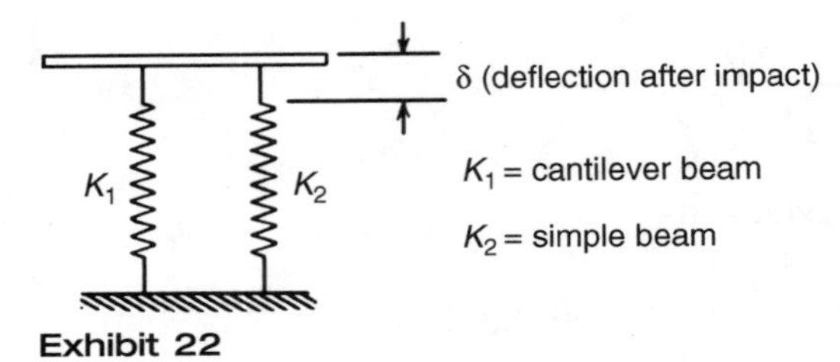

Exhibit 22

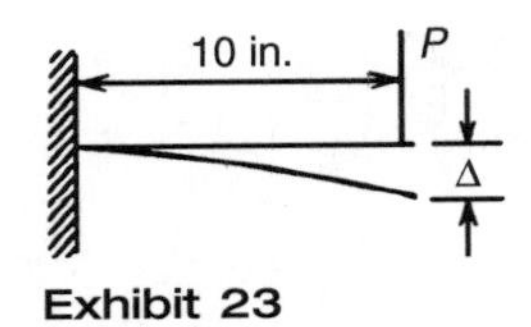

Exhibit 23

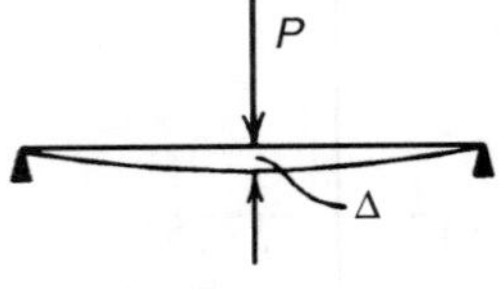

Exhibit 24

For the simple beam alone, shown in Exhibit 24,

$$\Delta = \frac{PL^3}{48\,EI} \qquad K_2 = \frac{P}{\Delta} = \frac{48\,EI}{L^3} = 200 \text{ lb per in.}$$

$$K(\text{effective}) = K_1 + K_2 = 100 + 200 = 300 \text{ lb per in.}$$

The strain energy stored in the beams for deflection δ is

$$u = \frac{1}{2}K_e\delta^2 = \frac{1}{2}(300)(\delta^2) = 150\delta^2.$$

$$\text{Potential energy of weight} = E = Wy = 30(0.5 + \delta)$$

Conservation of energy principle

$$u = E \quad \text{and} \quad 150\delta^2 = 30(0.5 + \delta)$$

Then by rearrangement, $150\delta^2 - 30\delta - 15 = 0$ and by application of the binomial theorem,

$$\delta = (30 \pm 100)/300 = 0.43 \text{ in.}$$

Example **1.12**

A punch punches a 1-in.-diameter hole in a steel plate $\frac{3}{4}$ -in. thick every 10 sec. The actual punching takes 1 sec. The ultimate shear strength of the plate is 60,000 psi. The flywheel of the punch press has a mass moment of inertia of 500 in.-lb-sec^2 and rotates at a mean speed of 150 rpm.

(a) What is the horsepower required for the punch operation?

(b) What is the total speed fluctuation of the flywheel in revolutions per minute?

Solution

(a) The force required for the actual punching of the metal is

$$F = \pi d S t = \pi \times 1 \times 60{,}000 \times 0.75 = 141{,}372 \text{ lb.}$$

Assume that the force required during half the thickness of punch is

$$\text{Energy} = \frac{1}{2} \times 141{,}372 \times 0.75 \times 1/12 = 4419 \text{ ft-lb} \approx 4420$$

Then the horsepower required to punch is $4420/(1 \times 550) = 8.04$ hp.

(b) The speed fluctuation may be found from

$$\left(\frac{I_0}{2}\right)(w_2 - w_1)(w_2 + w_1)$$

where $(w_2 - w_1)$ is the total fluctuation in angular velocity and $(w_2 + w_1)$ is equal to twice the average angular velocity $= 2(150/60)(2\pi) = 2 \times 15.70$.

$$(w_2 - w_1) = (4420 \times 12)/(500 \times 15.7) = 6.76 \text{ radians per sec}$$

Expressed as revolutions per minute,

$$(6.76)(60/2\pi) = 64.55 \text{ rpm}$$

Example **1.13**

The valve push rod for an overhead valve engine is $\frac{1}{4}$ in. in diameter and 14 in. long. Find the critical load when the rod is considered as a column with round ends.

Solution

$$I = \frac{\pi d^4}{64} = \frac{\pi}{4^4 \times 64} = \frac{\pi}{256 \times 64}$$

$$P_{cr} = \frac{\pi^2 EI}{l^2} = \frac{\pi^2 \times 30 \times 10^6 \pi}{14^2 \times 256 \times 64} = 290 \text{ lb}$$

Example 1.14

A structural steel rod extends through an aluminum tube as shown in Exhibit 25. The cross-sectional area of the rod is 0.8 sq in., and its upper end has 20 threads per inch. The aluminum tube is 20 in. long and has a cross-sectional area of 1.8 sq in. Determine the stresses in the rod and in the tube due to a quarter turn of the nut on the bolt. Assume

$$E_s = 30 \times 10^6 \text{ psi} \quad \text{and} \quad E_a = 10 \times 10^6 \text{ psi}.$$

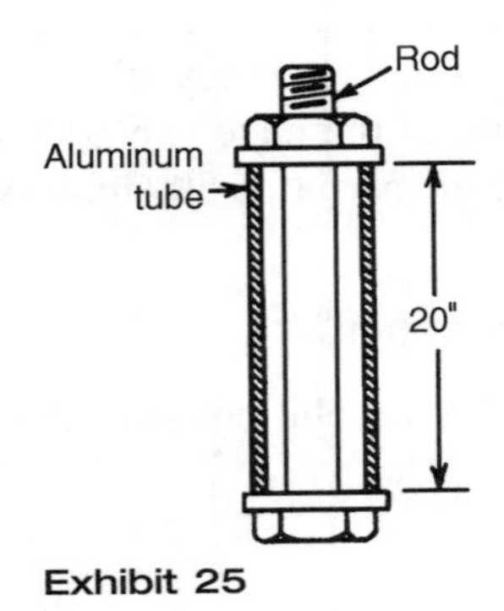

Exhibit 25

Solution

The same force P must exist in both the tube and the bolt. Also, the one-quarter-turn deformation will be taken up by a shortening of the tube and lengthening of the bolt. Accordingly,

$$\Delta = \Delta_{\text{bolt}} + \Delta_{\text{tube}} = \left(\frac{P \times L}{A \times E}\right)_{\text{bolt}} + \left(\frac{P \times L}{A \times E}\right)_{\text{tube}}$$

$$\Delta = \frac{(P)(20)}{(0.8)(30 \times 10^6)} + \frac{(P)(20)}{(1.8)(10 \times 10^6)} = \left(\frac{1}{4}\right)\left(\frac{1}{20}\right)$$

$$(8.33 \times 10^{-7})P + (1.111 \times 10^{-6})P = 0.0125$$

from which P is found to equal 6430 lb.

$$\sigma_s = \frac{P}{A} = \frac{6430}{0.8} = 8040 \text{ psi for the steel}$$

$$\sigma_a = \frac{P}{A} = \frac{6430}{1.8} = 3570 \text{ psi for the aluminum}$$

Example 1.15

A steel specimen is undergoing destructive testing. Data supplied on the specimen include: total length is 19 in.; sections at each end of the specimen are 1 by 1 in. by 6 in. long; center section is $\frac{7}{16}$ in. in diameter by 7 in. long with center section concentric with the square sections at each end. Center 7-in. section has a 2-in. gauge length marked. The specimen under test broke when the load reached a maximum of 11,274.75 lb, and the break occurred through the original $\frac{7}{16}$-in. diameter. But the diameter was now $\frac{1}{4}$ in. The cross-sectional area of a $\frac{7}{16}$-in.-diameter bar is 0.15033 in.^{2}. The length of the 2-in. gauge section had stretched to 2.55 in.

(a) Determine the ultimate strength of the material.

(b) What is the ultimate strength in pounds per square inch of the square section at each end?

(c) Determine the percentage elongation of the 2-in. gauge section.

Solution

(a) Refer to Exhibit 26. The break occurred in the $\frac{7}{16}$-in. section, so that the stress S is $S = F/a$, where a is original cross-sectional area.

$$S = \frac{11{,}274.75}{0.15033} = 75{,}000 \text{ lb/in.}^2 \text{ ultimate strength}$$

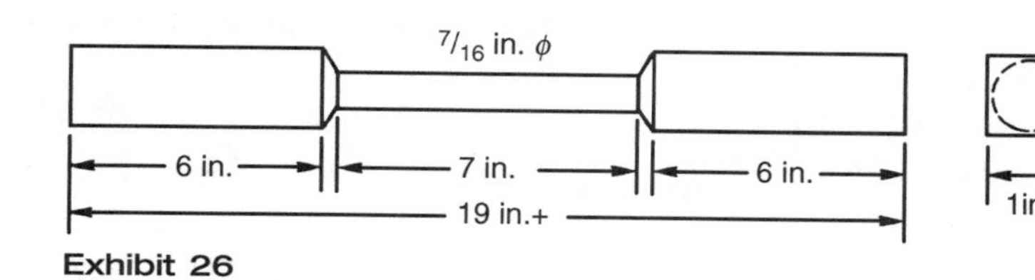

Exhibit 26

(b) The ultimate strength at the square section at each end is the same (75,000 lb/in.^{2}) because it is still the same material. It did not break at this section because the cross-sectional area is larger than at the $\frac{7}{16}$-in.-diameter section.

(c) $$\text{Percentage elongation} = \frac{(\text{final length} - \text{original length}) \times 100}{\text{original length}}$$

$$= (2.55 - 2)/2 \times 100$$
$$= 27.5 \text{ percent}$$

Example **1.16**

A steel ring having an internal diameter of 8.99 in. and a thickness of $\frac{1}{4}$ in. is heated and allowed to shrink over an aluminum cylinder having an external diameter of 9.00 in. and a thickness of $\frac{1}{2}$ in. After the steel cools, the cylinder is subjected to an internal pressure of 800 lb/in.^{2}. Find the stresses in the two materials. For aluminum, $E = 10 \times 10^6$ lb/in.^{2}.

Solution

(a) Compute the radial pressure caused by prestressing.

$$p = \frac{2\Delta D}{D^2(1/t_a E_a + 1/t_s E_s)}$$

where

p = radial pressure resulting from prestressing, lb/in^2.
D = internal diameter of cylinder, in.
t_a = cylinder wall thickness for aluminum, in.
t_s = cylinder wall thickness for steel, in.

$$p = \frac{2(0.01)}{9^2[(1/0.5 \times 10 \times 10^6) + 1/0.25 \times 30 \times 10^6]} = 741 \text{ lb/in}^2$$

(b) Compute the corresponding prestresses. Using the subscripts 1 and 2, denote the stresses caused by prestressing and internal pressure, respectively, $s_{a1} = pD/2t_a$. Thus,

$$s_{a1} = \frac{741(9)}{2(0.5)} = 6670 \text{ lb/in}^2 \text{ compression}$$

$$s_{s1} = \frac{741(9)}{2(0.25)} = 13,340 \text{ lb/in}^2 \text{ tension}$$

(c) Compute the stresses caused by internal pressure. From the relation

$$\frac{S_{s2}}{S_{a2}} = \frac{E_s}{E_a} = \frac{30 \times 10^6}{10 \times 10^6} = 3 \qquad \text{Therefore } S_{s2} = 3S_{a2}$$

$$\text{using } t_a s_{a2} + t_s S_{s2} = \frac{PD}{2} = (800\text{lb/m}^2)\left(\frac{9}{2}\text{in}\right) = 3600\text{lb/in}$$

$$(0.5\text{in})S_{a2} + 0.25\text{in}(3S_{a2}) = 3600 \text{ lb/in}$$

$$S_{a2} = 2880\text{lb/in}^2 \quad \text{tension}$$

$$S_{s2} = 3(2880) = 8640\text{lb/in}^2 \quad \text{tension}$$

(d) Compute the final stresses.

$$S_{a3} = S_{a1} + S_{a2} = 6670 \text{ comp} + 2880 \text{ tension}$$

$$\text{Therefore } S_{a3} = 6670 - 2880 = 3790\text{lb/in}^2 \text{ comp.}$$

$$S_{s3} = S_{a1} + S_{a2} = 13{,}340 \text{ tension} + 8640 \text{ tension}$$

$$\text{so } S_{s3} = 21{,}980 \text{ lb/in}^2 \text{ tension}$$

CHAPTER 2

Thermodynamics

Thermodynamics is the study of heat and the useful work that can be obtained through the transfer of heat. This heat is transferred to a pure substance such as water, air or other gases, or HFC 134a. These pure substances can be classified as ideal gases or liquid/vapor pure substances.

An ideal gas is a superheated vapor that operates in a region where it obeys the ideal gas laws. When operating in this region, the following equations are true.

The change of internal energy and/or enthalpy can be found by using Cv or Cp by:

$$du = Cv\,dT \quad \text{internal energy}$$
$$dh = CP\,dT \quad \text{enthalpy.}$$

In addition, the ideal equation of state can be used to calculate various properties:

$$PV = mRT.$$

Example 2.1

If the specific heat for a certain process is given by the equation $c = 0.2 + 0.00005T$ Btu per lb-deg, how much heat should be transferred to raise the temperature of 1 lb from 500°R to 2000°R?

Solution

$$Q = w\int_{500}^{2,000} c dT = \int_{500}^{2,000} (0.2 + 0.00005T)\,dT$$
$$= 0.2T + \frac{0.00005T^2}{2}\Bigg]_{500}^{2,000} = 394 \text{ Btu per lb}$$

Example 2.2

A water-jacketed air compressor receives 300 cfm of air at 14 psia and 70°F and discharges the air at 70 psia and 280°F. Cooling water enters the jacket at 60°F and leaves at 70°F, or a 10°F rise for a flow rate of 24 lb min. Determine the horsepower required for the driving motor for this reciprocating compressor. In your calculations assume reasonable efficiencies.

Solution

Conditions given in problem

$$p_1 = 14 \text{ psia}$$
$$T_1 = 530°\text{R}$$
$$p_2 = 70 \text{ psia}$$
$$T_2 = 740°\text{R}$$

Using the equation of state for an ideal gas,

$$PV = mRT,$$
$$(14\text{lbf/in.}^2 \ \ 144 \text{ in.}^2/\text{ft}^2)(300\text{ft}^3/\text{min}) = (\text{m lbm/min})(53.3\text{ft lbf/lbm R})(530 \text{ R})$$
$$\text{mass flow rate} = 21.4 \text{ lbm/min.}$$
$$\text{Using } Q = \dot{m}C\Delta T$$
$$\text{Compression heat to be removed} = 21.4*0.24*(740\text{-}530) = 1{,}079 \text{ Btu/min}$$
$$\text{Heat absorbed by cooling water} = 24*1*(70\text{-}60) = 240 \text{ Btu/min}$$

Total power = 1079 + 240 = 1319 Btu/min, and converting to horsepower is given by 1319/(2545/60) = 31.1hp.

Note: It appears that not all the friction losses in the system are absorbed by the cooling water. Technical literature indicates that bearing losses and windage can account for as much as 5 percent of the total power, so that the actual horsepower requirements are greater than those calculated in this example.

If the vapor is reasonably close to the saturation line or if the problem deals with a liquid and vapors, the ideal gas relationships cannot be used. Rather, the steam tables or HFC 134a tables are to be used.

Example 2.3

Estimate the steam flow rate required to drive a boiler feedpump, which requires 100 bhp at 5000 rpm. The turbine is a velocity-compounded impulse turbine with two rows of moving blades. Steam is available at 500 psi and a total steam temperature of 560°F. The turbine back pressure is 60 psi.

Solution

Given the following efficiencies:

Turbine mechanical efficiency = 95 percent

Stage efficiency = 80 percent

Throttle conditions:

Enthalpy = 1273.4 Btu per lb

Ideal back pressure conditions H_{2s} = 1095 Btu per lb. isentropic expansion

Ideal enthalpy change $H_1 - H_{2s} = 1273.4 - 1095 = 178$ Btu per lb

Actual enthalpy change = $(H_1 - H_2) = 0.80 \times 178 = 142.4$ Btu per lb

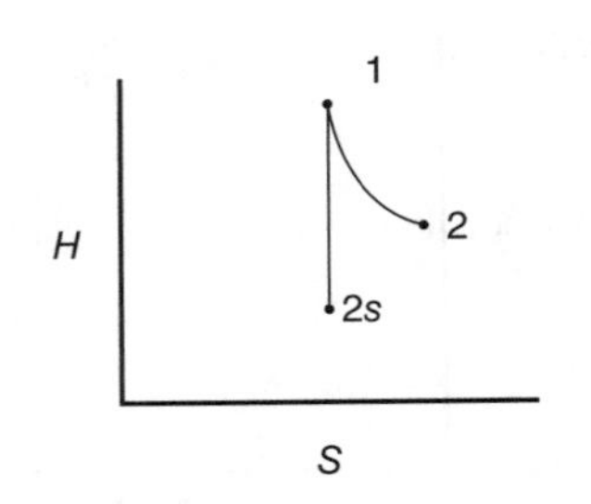

Exhibit 1

This is the energy delivered to the turbine shaft.

$$\text{Power required at the turbine shaft} = 100 \times 2545/0.95 = 267{,}894 \text{ Btu}$$

Finally,

$$\text{Steam required} = 267{,}894/142.4 = 1881 \text{ lb per hr}$$

Example 2.4

If 4670 cfm of air saturated at 95°F enters a four-stage adiabatic compressor, having a compression ratio of 2.33 to 1, at atmospheric pressure, how much heat must be removed in the first-stage intercooler?

Solution

Inlet air = 4670 cfm and, using steam tables,

Saturation partial pressure of water vapor at 95°F = 0.8153 psia

Saturation specific volume of water vapor at 95°F = 404.3 cu ft per lb

The air and the water both occupy the same volume at their respective partial pressures. The pounds of water entering with the air per hour = 4670 × 60/404.3 = 692 lb per hr.

First Stage. After the 2.33 to 1 compression ratio, $P_2 = 14.7 \times 2.33 = 34.2$ psi. Then, using $\gamma = 1.4$,

$$\left(\frac{T_2}{T_1}\right)_{\text{abs}} = \left(\frac{p_2}{p_1}\right)^{(\gamma-1/\gamma)}$$

$$(T_{2\text{abs}})/(460+95) = 2.33^{(1.4-1/1.4)}$$

and

$$T_{2\text{abs}} = 707°\text{R} \quad \text{or} \quad 247°\text{R}.$$

Intercooler.

$$\text{Final gas volume} = 4670 \times 60 \times 14.7/34.2 = 120{,}000 \text{ cu ft per hr}$$
$$\text{Water remaining in air} = 12{,}000/404.3 = 297 \text{ lb per hr}$$
$$\text{Condensation in intercooler} = 692 - 297 = 395 \text{ lb per hr}$$

$$\text{Specific volume of atmospheric air} = (359/29)(555/492)(14.7/14.7 - 0.8153)$$
$$= 14.8 \text{ cu ft per lb}$$
$$\text{Air in inlet gas} = 4670 \times 60/14.8 = 18{,}900 \text{ lb per hr}$$

Heat load. (245°F to 95°F)

Sensible heat

$$Q_{\text{air}} = \dot{m}Cp\Delta T$$

$$Q_{\text{air}} = 18{,}900 \times 0.25(247 - 95) = 708{,}000 \text{ Btuh}$$
$$Q_{\text{water}} = 692 \times 0.45(247 - 95) = 46{,}700 \text{ Btuh}$$

Latent heat

$$Q_{\text{water}} = 395 \times 1040.1 = 411{,}000 \text{ Btuh}$$
$$\text{Total} \qquad 1{,}165{,}700 \text{ Btuh}$$

If condensation had not been accounted for, an error of 33 percent would have resulted. Note that over half of the water condenses in the first-stage intercooler.

Example 2.5

A cylinder of nitrogen has been returned to the filling plant for recharging. It had originally contained the gas in the filled condition at 80°F and a pressure of 2000 psig. When weighed for water content, it was found to hold 96 lb of water at 70°F. How many cubic feet of "free" gas will the cylinder discharge as its pressure returns to atmospheric? Compressibility factor for nitrogen at 2,000 psig and 80°F may be taken as 1.125.

Solution

First find the number of moles contained in the nitrogen bottle. But in order to do this the volume of the bottle must be calculated. Density of water at 70°F is 62.22 lb per cu ft. Then the volume of the bottle is simply found by $V = 96/62.22 =$ 1.54 cu ft. Proceeding,

$$N = \frac{pV}{ZKT} = \frac{(2000+14.7)\times 1.54}{1.125\times 10.71\times (460+80)}$$
$$= 0.476 \text{ lb mole of nitrogen.}$$

This quantity of gas will occupy the following volume at 70°F and 14.7 psia (standard for the compressed gas industry)

$$\left(0.476\times 379\frac{460+70}{460+60}\right) - 1.54 = 183.83 - 1.54 = 182.29 \text{ cu ft.}$$

Note that we have subtracted the volume of the bottle because this is the amount remaining within the bottle itself after internal pressure has dropped from 2000 psig to atmospheric 14.7 psia. There is no longer a differential of pressure to drive the remainder out.

Example 2.6

Calculate the volume occupied by 30 lb of chlorine at a pressure of 743 mm of mercury (Hg) and a temperature of 70°F. Molecular weight of chlorine may be taken as 71.

Solution

One lb mole of chlorine will occupy 379 cu ft volume at standard conditions (60°F and 14.7 psia). This pressure of 14.7 psia is also equivalent to 760 mm Hg. We can say, in effect, that the molal volume (379 cu ft) will vary directly as the ratio of the absolute temperatures and indirectly as the ratio of the absolute pressures. Thus,

$$V_2 = V_1\left(\frac{m}{M}\right)\left(\frac{P_1}{P_2}\right)\left(\frac{T_2}{T_1}\right)$$

$$V_2 = 379\times\frac{30}{71}\times\frac{760}{743}\times\frac{460+70}{460+60} = 167 \text{ cu ft.}$$

Whenever there is a decrease in pressure compared to standard conditions, the pressure ratio is greater than unity; for an increase, the ratio becomes less than unity. As for change in temperature, its effect on volume is to increase it with increase in temperature or decrease volume at lowered temperatures.

Example 2.7

A 1000-gal tank filled with acetylene at 15 psia is supercharged with 4.3 lb of the gas. What would be the final pressure? Assume no change in temperature (60°F) takes place during the charging period. Volume of 1 lb mol of gas at STP = 379 ft^3.

Solution

The 1000-gal tank has the equivalent volume of 1000/7.48 = 133.6 cu ft. The weight of gas originally in the tank is given by

$$\frac{(V\text{ft}^3)(\text{mol wt lbm/lb mol})}{(\text{ft}^3/\text{lb mol})}$$

$$(133.6)\frac{26}{379} = 9.5 \text{ lb.}$$

Thus, the total weight of the gas under the new pressured conditions is 9.5 + 4.3 = 13.8 lb. We then proceed to say, "What happens to the original 15 psia when we add the additional weight of gas?"

$$P_f = P_i\left(\frac{\text{Vol at STP}}{\text{Vol}}\right) \times \left(\frac{\text{total mass}}{\text{mol wt}}\right)$$

$$P_f = 15 \times \frac{379}{133.6} \times \frac{13.8}{26} = 15 \times 2.83 \times 0.53 = 22.6 \text{ psia}$$

The above equation is to say, if one mole is involved and compressed into a smaller volume, what would be the final pressure?

Example 2.8

A mixture of gases has the following composition by volume: oxygen, 6.3 percent; sulfur dioxide, 14.6 percent; nitrogen, 79.1 percent. Calculate the composition by weight of this mixture. Assume gases will not react chemically.

Solution

Basis: 1 lb mole of the mixture. Now, because volume percent equals mole percent or mole fraction, and mole fraction multiplied by molecular weight gives weight in pounds, we can set up the following table.

Gas	Mole Fraction	×	Molecular Weight M	=	Pounds
Oxygen	0.063		32		2.02
Sulfur dioxide	0.146		64		9.34
Nitrogen	0.791		28		22.15
Total weight	—		—		33.51

Because this is the weight of 1 lb mole, it is the average molecular weight. Composition by weight is found simply.

$$\% = \left(\frac{\text{wt of constituent}}{\text{total wt}}\right) \times 100$$

For oxygen 2.02/33.5, or 6 percent by weight

For sulfur dioxide 9.34/33.5, or 27.9 percent by weight

For nitrogen 22.15/33.51, or 66.1 percent by weight

Example 2.9

A mixture of gases has the following composition by weight: oxygen, 10.7 percent; carbon monoxide, 0.9 percent; nitrogen, 88.4 percent. Calculate the composition by volume (volumetric analysis) of this mixture.

Solution

Use as a basis 1 lb weight of the mixture, setting up the following table:

Gas	Wt. Fraction	÷	M	=	Lb mole
Oxygen	0.107		32		0.00335
Carbon monoxide	0.009		28		0.000321
Nitrogen	0.884		28		0.0315
Total					0.03517

Now, because mole percent is equal to volume percent, all we need to do is to calculate mole percent.

$$\text{Vol\%} = \frac{\text{lb mol}}{\text{total lb mol}}$$

For oxygen: 0.00335/0.03517, or 9.5 percent by volume

For carbon dioxide: 0.000321/0.03517, or 0.9 percent by volume

For nitrogen: 0.0315/0.03517, or 89.6 percent by volume

This same treatment may be given to any composition of gases as long as no chemical reaction takes place.

Example 2.10

Air having a total volume of 490.3 cu ft at 212°F and 1 atm pressure contains

0.20 mole of oxygen

0.78 mole of nitrogen

0.02 mole of water vapor.

(a) What is the average molecular weight of the gas?

(b) What is the weight percent of the constituents?

(c) What are the partial pressures of the constituents?

(d) If the constituents were separated, what volume would they occupy at the same conditions of temperature and pressure?

Solution

(a) Assume one lb mole of the mixture. Then set up the table as in Example 2.8.

Gas	Mole Fraction	×	M	=	Pounds	
Oxygen	0.20		32		6.4	
Nitrogen	0.78		28		21.84	
Water vapor	0.02		18		0.36	
Total					28.60	Average molecular weight

(b) As in Example 2.8, the weight percent is

$$\% = \left(\frac{\text{wt const.}}{\text{wt total}}\right) \times (100)$$

For oxygen 6.4/28.6, or 22.8 percent by weight

For nitrogen 21.84/28.60, or 76.4 percent by weight

For water vapor 0.36/28.60, or 0.8 percent by weight.

(c) The partial pressures are

$$\text{partial pressure} = (P_{\text{tot}})(\text{mol fraction})$$

For oxygen $14.7 \times 0.20 = 2.94$ psia

For nitrogen $14.7 \times 0.78 = 11.47$ psia

For water vapor $14.7 \times 0.02 = 0.294$ psia.

This total equals 14.704 psia. Merely drop the last digit.

(d) For oxygen: $379 \times 0.20 \times \dfrac{212+460}{60+460} = 98$ cu ft

$$\text{For nitrogen: } 379 \times 0.78 \times \frac{212+460}{60+460} = 382 \text{ cu ft}$$

$$\text{For water vapor: } 379 \times 0.02 \times \frac{212+460}{60+460} = 9.8 \text{ cu ft}$$

Example **2.11**

A boiler flue-gas analysis shows, after converting to percent composition by weight, CO_2, 0.1 percent; O_2, 6.2 percent; CO, 0.1 percent; N_2, 74.2 percent. Find the instantaneous specific heat at constant pressure at 500°F if c_p (specific heat at constant pressure) for CO_2 at 500°F is 0.235, for CO is 0.251, for O_2 is 0.22, and for N_2 is 0.256.

Solution

Basis: 100 lb of flue gas.

$$\begin{aligned} CO_2 &= (19.5/100)\,0.235 = 0.0458 \\ CO &= (0.1/100)\,0.251 = 0.000251 \\ O_2 &= (6.2/100)\,0.22 = 0.01364 \\ N_2 &= (74.2/100)\,0.256 = 0.1864 \\ \text{Total} &= 0.2461 \end{aligned}$$

The study of thermodynamics involves the study of the first and second laws of thermodynamics. The first law of thermodynamics is a conservation of energy law, of changing from one form of energy to another without any net loss.

$$\frac{v_1^2}{2gJ} + \frac{y_1}{J} + H_1 + Q = \frac{W}{J} + \frac{v_2^2}{2gJ} + \frac{y_2}{J} + H_2$$

where

v_1 = velocity entering system, fps
J = 778 ft-lb per Btu
y_1 = potential distance above a certain datum, ft
H_1 = enthalpy of working substance entering system, Btu
Q = heat transferred into or out of system, Btu
W = flow work, ft-lb
v_2 = velocity leaving system, fps
y_2 = potential distance above a certain datum, ft
H_2 = enthalpy of working substance leaving system, Btu

Example **2.12**

A steam turbine receives 3600 lb of steam per hour at 110 fps velocity and 1525 Btu per lb enthalpy. The steam leaves at 810 fps and 1300 Btu per lb enthalpy. What is the horsepower output?

Solution

Let the basis be 1 lb steam flow per second.

$$\frac{v_1^2}{2gJ}+\frac{y_1}{J}+H_1+Q=\frac{W}{J}+\frac{v_2^2}{2gJ}+\frac{y_2}{J}+H_2$$

Assume centerline of nozzle passes through datum horizontally so that $y_1 = y_2$. Also $Q = 0$, because adiabatic process is assumed. Then

$$\frac{v_2^2-v_1^2}{2gJ}+(H_1-H_2)=\frac{-W}{J}$$

By rearranging for W, we obtain

$$-W=\frac{v_2^2-v_1^2}{2g}-J(\Delta H)=\frac{810^2-110^2}{64.4}+778(1300-1525)$$

$$=-165{,}000 \text{ ft-lb} \quad \text{or} \quad W=165{,}000 \text{ ft-lb/lb}$$

$$\text{hp}=(165{,}000)(1 \text{ lb steam per sec})/550=300 \text{ hp.}$$

Example 2.13

A vessel with a capacity of 5 cu ft is filled with air at a pressure of 125 psia when at a temperature of 600°F. It is desirable to lower the pressure to 60 psia. (a) What amount of heat will have to be extracted and what will be the final temperature of the gas, assuming that the vessel does not change in size with change in temperature and that the air may be treated as an ideal gas? (b) Find the gain in entropy during this process.

Solution

As the problem indicates, this is a constant-volume process involving a gas.

a)
$$\frac{P_1V_1}{T_1}=\frac{P_2V_2}{T_2}$$

$$V_1=V_2$$

$$\frac{(125)\text{psia}}{1060°\text{R}}=\frac{60\text{psia}}{T_2}$$

$$T_2=508.8°\text{R heat extracted}$$

Constant volume process:

$$Q=\Delta u = wCv\Delta T$$

$$Pv = wRT$$

$$w=\frac{(125)(144)(5)}{(53.3)(1060)}=1.59\text{lbm}$$

$$\therefore Q=(1.59)(0.171)(1060-508.8)$$

$$Q=149.9\text{Btu out.}$$

b) $$\Delta s = wCv \ln\left(\frac{T_2}{T_1}\right)$$

$$\Delta s = (1.59)(0.171) \ln\frac{508.8}{1060}$$

$$\Delta s = -0.20 \text{ Btu/lb°R}$$

Example 2.14

A mole of a gas at a pressure of 500 psia undergoes a constant pressure nonflow process with the temperature changing from 1000 to 1100°R. What is the maximum amount of work that can be obtained from this expansion process?

Solution

The maximum work is the reversible work

$$W = P(V_2 - V_1) \text{ ft-lb per lb.}$$

At any state

$$pV = RT.$$

Upon substitution in the above

$$W = R(T_2 - T_1)$$
$$W = (1.986 \text{ Btu/mole°R})(1100 - 1000)\text{°R}$$
$$W = 198.6 \text{ Btu/mole} = 154{,}500 \text{ ft-lb per mole.}$$

Note that gas constant was in molal dimensions. Another form would be as previously shown to be $R = 1544$ molecular weight, approximately.

Example 2.15

A volume of gas having an initial entropy of 2800 Btu per °F is heated at constant temperature of 1000°F until the entropy is 4300. How much heat is added and how much work is done during the process?

Solution

Heat added is the area beneath the *TS* curve for the process.

$$Q_{1-2} = T(S_2 - S_1) = (460 + 1000)(4300 - 2800) = 2.19 \times 10^6 \text{ Btu}$$

Work done. Because work of an isothermal process is exactly compensated for by transferred heat, then

$$W_{1-2} = (2.19 \times 10^6 \text{ Btu})(778 \text{ ft-lb/Btu}) = 1.704 \times 10^9 \text{ ft-lb}$$

Example 2.16

(a) How much work will be required to compress 2 lb of an ideal gas from an initial volume of 25 cu ft and an initial pressure of 13.5 psia to a final pressure of 75 psia, according to the equation $PV^{1.35}$ equals a constant? (b) What will be the final temperature if the initial temperature is 55°F?

Solution

(a)

$$W = 144\frac{(75\times V_2)-(13.5\times 25)}{1-1.35} = 78{,}700 \text{ ft-lb}$$

The preceding equation was solved with V_2 first determined from the treatment below and then inserted.

$$\frac{V_2}{V_1} = \left(\frac{P_1}{P_2}\right)^{1/1.35} = \frac{V_2}{25} = \left(\frac{13.5}{75}\right)^{1/1.35} = 0.2808$$

from which

$$V_2 = 0.2808 \times 25 = 7.02 \text{ cu ft.}$$

(b)

$$\frac{T_1}{T_2} = \left(\frac{P_1}{P_2}\right)^{(1.35-1)/1.35} = \frac{55+460}{T_2} = \left(\frac{13.5}{75}\right)^{0.35/1.35} = 0.6414$$

from which

$$T_2 = 515/0.6414 = 803°\text{R} \quad \text{or} \quad 803-460 = 343°\text{F}.$$

Example 2.17

Air drawn into a compressor is at 61°F and 14.7 psia. Flash point of the lubricating oil used is 350°F. (a) If compression is a reversible adiabatic (isentropic), what pressure could be attained in the compressor if the maximum allowable temperature is 50°F below the flash point of the oil? (b) If compression follows the law PV^n with n equal to 1.3, what is the maximum allowable pressure?

Solution

(a) After first finding T_2 and letting $k = 1.4$ for air,

$$T_2 = (350 - 50) + 460 = 760°\text{R}$$

$$P_2 = P_1\left(\frac{T_2}{T_1}\right)^{\frac{k}{k-1}} = 14.7\left(\frac{760}{521}\right)^{\frac{1.4}{1.4-1}}$$

from which

$$P_2 = 14.7/0.267 = 55 \text{ psia.}$$

(b) For air during a reversible polytropic process of compression (polytropic)

$$\frac{P_1}{P_2} = \left(\frac{T_1}{T_2}\right)^{n/(n-1)} = \left(\frac{521}{760}\right)^{1.3/0.3} = 0.195$$

$$P_2 = 14.7/0.195 = 75.4 \text{ psia.}$$

Example 2.18

It is desired to have air delivered from a nozzle at a velocity of 1800 fps, a pressure of 15 psia, and 40°F. The nozzle coefficient is 0.98. Neglecting the velocity of approach, assume air to be an ideal gas and find (a) the initial temperature of the air and (b) the initial pressure. Assume constant c_p.

Solution

(a) $$v_2 = 0.98\sqrt{(2c_p)(T_1 - T_2)g_c J}$$

$$v_2 = 223.8\sqrt{0.240(T_1 - 460 + 40)} \times 0.98 = 1800$$

Solving for T_1, this is found to be equal to 770°R, or 310°F.

(b) Assuming adiabatic process where k equals 1.4 for air and $k/(k-1)$ is equal to 3.5

$$\left(\frac{T_1}{T_2}\right)^{3.5} = \frac{P_1}{P_2} = \left(\frac{770}{500}\right)^{3.5} = 1.54^{3.5} = 4.532 = \frac{P_1}{15}$$

Thus $P_1 = 15 \times 4.532 = 68.5$ psia.

Example 2.19

Steam at a pressure of 170 psia arrives at a steam engine with a quality of 97 per cent. For 1 lb, what are (a) the enthalpy, (b) the volume, (c) the entropy, (d) the internal energy, and (e) the temperature of this steam?

Solution

Refer to Exhibit 2. With the use of steam tables at 170 psia

(a)
$$\begin{aligned} H_1 &= H_f + 0.97(H_{fg}) \\ &= 341.11 + 0.97(855.2) \\ &= 1171.11 \text{ Btu per lb} \end{aligned}$$

(b)
$$\begin{aligned} \bar{v}_1 &= \bar{v}_f + 0.97\bar{v}_{fg} \\ &= 0.01821 + 0.97(2.656) \\ &= 2.5945 \text{ cu ft per lb} \end{aligned}$$

(c)
$$\begin{aligned} S_1 &= S_f + 0.97(S_{fg}) \\ &= 0.5266 + 0.97(1.0327) \\ &= 1.5283 \text{ Btu/lbm°R} \end{aligned}$$

(d)
$$u_1 = u_f + xu_{fg}$$
$$u_1 = 340.75 + 0.97(772.0)$$
$$u_1 = 1089.1 \text{ Btu/lb}$$

(e) From the steam tables, the temperature is 368.42°F.

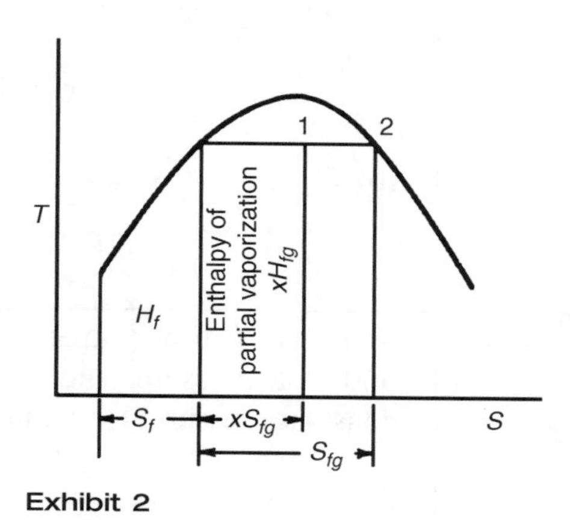

Exhibit 2

Example **2.20**

Two boilers discharge equal amounts of steam into the same main. The steam from one is at 200 psia and 420°F, and from the other, at 200 psia and 95 percent quality. (a) What is the equilibrium condition after mixing? (b) What is the loss of entropy by the higher-temperature steam? Assume no pressure drop in the pipeline.

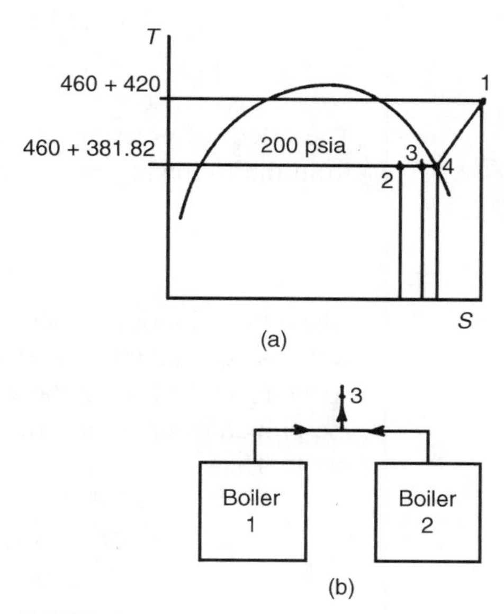

Exhibit 3

Solution

Refer to Exhibit 3 and use the Mollier diagram.

(a) $$H_3 = \frac{H_1 + H_2}{2} = \frac{1225 + 1164}{2} = 1194.5 \text{ Btu per lb}$$

At 200 psia, steam tables give data to determine the quality of the mixture.

$$1194.5 = H_f + x_3 H_{fg} = 355.4 + x_3(843.3)$$

Solving for x_3,

$$x_3 = 0.995, \quad \text{or} \quad 99.5 \text{ percent quality}$$

(b) $$S_1 - S_3 = 1.575 - 1.541 = 0.034 \text{ Btu/lb°R}$$

Example 2.21

Water at 70°F and atmospheric pressure is pumped into a boiler, evaporated at 300 psia pressure, and superheated to a total steam temperature of 600°F at constant pressure. How much heat must be supplied per pound of water?

Solution

Except for the step of pumping the water into the boiler, the entire process occurs at constant pressure. The effect of pressure on the enthalpy of the liquid is very small (and will be calculated), but ordinarily we may assume that the enthalpy of the liquid water at 70°F and atmospheric saturation pressure is the same as the heat at 70°F and 300 psia. Now, from the steam tables and the Mollier diagram

$$H_1 = 38.05 \text{ Btu per lb} \quad \text{or} \quad (70 - 32 = 38)(\text{enthalpy at } 70°\text{F})$$
$$H_2 = 1314.7 \text{ Btu per lb (enthalpy of superheated vapor)}$$
$$H_2 - H_1 = 1314.7 - 38.05 = 1276.65 \text{ Btu per lb water vaporized.}$$

The effect of pressure on the liquid to get it into the boiler may be determined from the following expression:

$$H_1 = H_a + A(P_1 - P_a)\bar{v}_a$$

where the subscript a refers to the saturated liquid at the temperature t_1 at which the water is pumped into the boiler, i.e., 70°F in this case. The saturation pressure corresponding to 70°F from the steam tables is 0.3628 psia. From the steam tables giving the properties of the saturated liquid, H_a = 38.05 Btu per lb and $\bar{v}_a$ = 0.01605 cu ft per lb. Hence,

$$H_1 = 38.05 + 144/778(300 - 0.3628)(0.01605)$$
$$= 38.05 + 0.88 = 38.93 \text{ Btu per lb.}$$

From the tables for superheated steam, with steam pressure at 300 psia and temperature at 600°F, the enthalpy is 1314.7 Btu per lb. Then the heat added to each pound of water becomes

$$q = (H_2 - H_1) = 1314.7 - 38.93 = 1275.77 \text{ Btu.}$$

By comparison the difference between both methods is very small. Note that enthalpies were obtained from the Mollier diagram and not directly from the steam tables for greater precision. However, the use of the Mollier diagram is acceptable for the written examination.

Example 2.22

Saturated steam at 200 psia expands continuously through a throttle to atmosphere (14.7 psia). What are the state of expanded steam and the change in entropy in the process if no heat is lost (constant enthalpy) to the surroundings, and if all kinetic energy due to any high-velocity jets is dissipated?

Solution

A throttling process is one occurring at constant enthalpy. The Mollier diagram is the handiest tool to use for the solution of this problem. Locate the point on the Mollier diagram where the 200-psia pressure line crosses the saturation line. This is the initial state, and from the coordinates of the diagram we obtain

$$H_1 = 1198.4 \text{ Btu per lb}$$
$$S_1 = 1.545 \text{ Btu/(lb)(°F)}.$$

Now proceed horizontally to the right or left as the case may be at constant enthalpy to the intersection with the 14.7 psia line. This is the final state. Read

$$H_2 = 1189.4 \text{ Btu per lb}$$
$$S_2 = 1.823 \text{ Btu/(lb)(°R)}$$
$$t_2 = 311.5\text{°F}.$$

Change in entropy $S_2 - S_1 = 1.823 - 1.545 = 0.278$ Btu/lb°R. Degrees superheat are 99°F.

Example 2.23

Five pounds of steam expand isentropically (nonflow) from p_1 equal to 300 psia and t_1 equal to 700°F to t_2 equal to 200°F. Find x_2 and the work.

Solution

From the superheat steam tables, we find S_1 = 1.6751 Btu/lb°R, H_1 = 13,683.3 Btu/lb, and $\overline{v}_1 = 2.227$ ft^3/lb. From the saturated steam tables, we find $S_{f2} = 0.2938$, $H_{f2} = 167.99$, $S_{fg2} = 1.4824$, $H_{fg2} = 977.9$, $v_{g2} = 33.64$, $p_2 = 11.526$. With $S_1 = S_2$ we see that

$$s_2 = 1.6751 = 0.2938 + 1.4824x_2,$$

from which $x_2 \equiv 93.2$ percent. Then we can proceed to find

$$H_2 = H_f + xH_{fg}$$

$$H_2 = 167.99 + 0.932(977.9) = 1079 \text{ Btu per lb}$$
$$\bar{v}_2 = 0.932 \times 33.64 = 31.3 \text{ cu ft}$$
$$U_2 = H_2 - \frac{P_2V_2}{J} = 1{,}079 - \frac{(11.526 \text{ lb/in}^2)(144 \text{ in}^2/\text{ft}^2)(31.3 \text{ ft}^3)}{778 \text{ ft-lb/Btu}}$$
$$= 1012.2 \text{ Btu per lb}$$
$$U_1 = 1{,}368.3 - \frac{(300)(144)(2.227)}{778} = 1244.6 \text{ Btu per lb}$$
$$W = \Delta U = 1{,}244.6 - 1{,}012.2 = 232.4 \text{ Btu per lb}$$
$$W = 5 \times 232.4 = 1162 \text{ Btu for 5 lb of steam.}$$

Example **2.24**

A vapor with a quality of 100 percent and a temperature of 100°F is supplied to suitable heat-transfer coils for special heating or, as the case may be, cooling purposes. The condensate leaves the coils as a saturated liquid at the same pressure. The heat load is 1 million Btu per hr. Find the weight of vapor to be supplied to the coils in pounds per hour if the vapor used is steam, ammonia, or sulfur dioxide. The enthalpy of vaporization or condensation for steam (from steam tables) is 1036.4 Btu per lb; for ammonia (from ammonia tables), 477.79 Btu per lb; for sulfur dioxide (from sulfur dioxide tables), 140.8 Btu per lb.

Solution

For steam: $\frac{1 \times 10^6}{1036.4} = 965$ lb per hr

For ammonia: $\frac{1 \times 10^6}{477.79} = 2093$ lb per hr

For sulfur dioxide: $\frac{1x10^6}{140.8} = 7100$ lb per hr

Example **2.25**

In determining the quality of steam in a main with a throttling calorimeter, the following readings were taken: pressure in main 110 psig, barometer 30.6 in. Hg, manometer reading 2.04 in. Hg, and thermometer after stem correction 220°F. Find the quality of the steam in the main.

Solution

Use the Mollier chart. Convert barometer to psia reading,

$$(30.6 \text{ in Hg}) \times (0.4912 \frac{\text{lb/in}^2}{\text{in Hg}}) = 15.04 \text{ psia.}$$

Main pressure is

$$110 + 15.04 = 125.04 \text{ psia} = P_1.$$

Manometer pressure is

$$(2.04 \times 0.4912) + 15.04 = 16.04 \text{ psia} = P_2.$$

From the Mollier diagram with steam at 16.04 psia and a total steam temperature of 220°F, the enthalpy is equal to 1163.4 Btu per lb. This is also the enthalpy existing in the main (constant enthalpy process). Now on the Mollier diagram move horizontally left or right (depending on the diagram used) along the enthalpy line of 1163.4 and where this line intersects the pressure line (interpolation) of 125.04 psia read constant moisture content of 4.2 percent. The quality is 100 − 4.2 = 95.8 percent.

Example **2.26**

Water from a boiler operating at 150 psig is blown down to a flash tank held at a pressure of 25 psig. Neglecting pressure drop in the blowdown line and assuming insulation of line and flash tank, how many pounds of water are flashed into vapor per pound of water fed to the tank? Use 15 psia as atmospheric pressure.

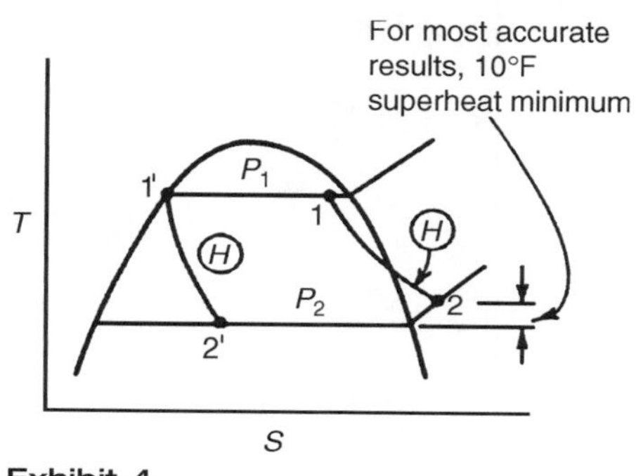

Exhibit 4

Solution

This is a throttling process at constant enthalpy taking place on the left side of the *TS* diagram. One method of solution is to use a *TS* diagram for steam, beginning at a point on the saturated liquid line, and to expand the liquid into the area under the dome at constant enthalpy (see Exhibit 4). Where the expansion meets the horizontal 25 psig (40 psia) line, read quality from the x lines. However, we shall calculate the results, and the candidate should check using the *TS* diagram. On the basis of an energy balance

Enthalpy of saturated water = enthalpy of flashed vapor
+ enthalpy of saturated liquid at final state.

From the steam tables,

Enthalpy of saturated water at 150 psig	338.53 Btu per lb = H_1
Enthalpy of saturated water at 25 psig	236.02 Btu per lb = H_2
Enthalpy of saturated vapor at 25 psig	1169.7 Btu per lb = H_3

Let x equal the weight of water flashed into saturated vapor at the flash tank pressure of 25 psig; then the enthalpy balance is

$$338.53 \times 1 = 1169.7(x) + [236.02(1 - x)].$$

Solving for x, we find it to be equal to 0.107 lb water flashed into saturated vapor. Then, $1 - 0.107 = 0.893$ lb of water remains as saturated liquid at 25 psig. Finally,

$$\text{lb water/lb vapor} = 0.893/0.107 = 8.34 \text{ lb}$$

Note that this treatment may be applied to all other pure fluids: ammonia, carbon dioxide, methyl chloride, Freon-12, etc.

Example 2.27

A carnot heat pump is used to heat a house and maintain it at 72°F. An analysis of the house indicates the heat loss fram the house is:

$$Q = \frac{T_1}{T_2}[2000(T_1 - T_2)]$$

where T is in °R.
If the outside temperature is 0°F find:
a) heat loss in Btu/hr
b) C.O.P.
c) Input power required

Solution

a)

$$Q = \frac{(460+72)}{460}[2000(460+72)-460]$$

$$Q = 166{,}500 \text{ Btu/hr}$$

b)

$$\text{C.O.P.} = \frac{Q_H}{Q_H - O_L} = \frac{T_H}{T_H - T_L} = \frac{(420+72)}{72} = 7.39$$

c)

$$\text{Input power} = (Q_H - Q_L) = \frac{Q_H}{\text{C.O.P.}}$$

$$\text{Power} = \frac{166{,}500 \text{ Btu/hr}}{7.39} = 22{,}530 \text{ Btu/hr}$$

$$\text{Power} = \frac{22{,}530 \text{ Btu/hr}}{2545 \text{ Btu/hr}} = 8.85 \text{ hp}$$

Example 2.28

Compute the work per pound of steam and the steam consumption in pounds per indicated horsepower-hour when steam is used in a Carnot cycle between zero and a quality of unity and 150 psia and 3 psia.

Solution

Refer to Exhibit 5 and the Mollier diagram.

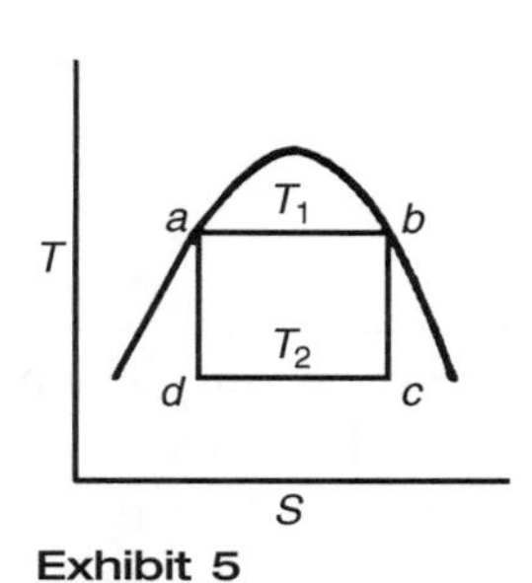

Exhibit 5

$$\text{Net work (area of rectangle } abcd) = (H_b - H_a)JE_t$$

$$\text{Steam rate} = \frac{2545}{(H_b - H_a)(1/E_t)} \text{ lb per hp-hr}$$

Then $P_a = 150$ psia
$t_a = 358.43°F$
$H_a = 330.53$ Btu per lb
$P_b = 150$ psia
$t_b = 358.43°F$
$H_b = 1194.4$ Btu per lb
$P_d = 3$ psia
$t_d = 141.5°F$

From this we can obtain the thermal efficiency of the cycle as follows:

$$E_t = (358.43 - 141.5)/(358.43 + 460) = 0.265$$
$$\text{Net work} = (1194.4 - 330.53)778 \times 0.265 = 178{,}000 \text{ ft-lb per lb}$$
$$\text{Water rate} = \frac{2545}{(1194.4 - 330.53)(1/0.265)} = 11.1 \text{ lb per hp-hr.}$$

Example **2.29**

An actual engine supplied with steam at 115 psia and containing 1 percent moisture was found to use 21 lb of steam per indicated horsepower-hour. This occurred when the steam in the condenser was 140°F. Compute (a) engine efficiency, (b) heat rate, and (c) actual thermal efficiency of the engine.

Solution

Refer to Exhibit 6 and use the Mollier diagram and steam tables.

(a) Engine efficiency

From Mollier diagram

$H_1 = 1180$ Btu/lb

$H_2 = 936$ Btu/lb

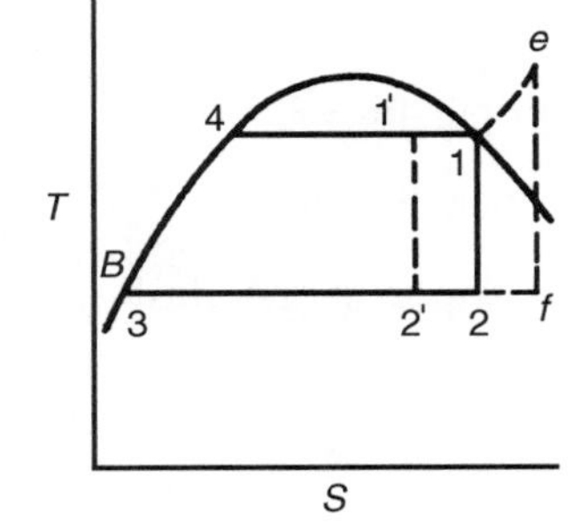

Exhibit 6

$$E_e = \frac{2545}{21 \times (1180 - 936)} \times 100 = 0.495 \times 100 = 49.5 \text{ percent}$$

(b) Heat rate

$$Q = \dot{m}(H_1 - H_3)$$

$$Q = 21 \times (1180 - 107.9) = 22{,}500 \text{ Btu per hp-hr}$$

(c) Actual thermal efficiency

$$E_{ea} = \frac{2545}{(1180 - 107.9) \times 21} \times 100 = 0.113 \times 100 = 11.3 \text{ percent}$$

Example 2.30

Process heat exchangers are supplied with 290 psig saturated steam, which is trapped and flows to a flash tank maintained at 20 psig. The flash tank is close by the trapping system. Flash steam is carried off as flash steam to a 20-psig steam system, and the remaining condensate is pumped back to the power plant. The steam load for heating is 29,000 lb per hr. Assume no subcooling of condensate after trapping. Determine in pounds per hour how much 20-psig steam is produced and how much condensate is returned to the power plant as condensate return.

Solution

Refer to Exhibit 7. Assume a steady flow rate and no flashing in the pipeline to the flash tank. On the basis of an energy balance,

Enthalpy of saturated water as condensate leaving exchangers
= enthalpy of flashed saturated vapor leaving trap
+ enthalpy of saturated liquid at final state.

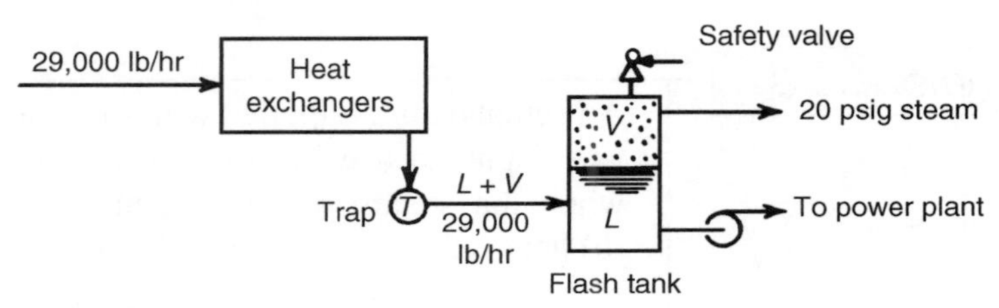

Exhibit 7

From steam tables,

Enthalpy of saturated water at 290 psig 395.49 Btu per lb = H_1

Enthalpy of saturated water at 20 psig 227.82 Btu per lb = H_2

Enthalpy of saturated vapor at 20 psig 1166.7 Btu per lb = H_3

Let x equal the weight of water flashed into saturated vapor at the flash-tank pressure of 20 psig. Then the enthalpy balance is

$$H_1 = xH_3 + (1 - x)H_2$$

$$395.49 \times 1 = 1166.7(x) + [227.82(1 - x)].$$

Solving for x, we find it to be equal to 0.18 lb water flashed into saturated vapor. Then, 1 – 0.18 = 0.82 lb of water remaining as saturated liquid at 20 psig. Finally,

20-psig steam to process = 0.18 × 29,000 = 5220 lb per hr
Condensate returned to boiler house = 0.82 × 29,000 = 23,780 lb per hr.

Example 2.31

A heat exchanger that is well insulated is used to heat water from 50°F to 150°F using a brine solution from an underground hot spring. The hot spring supplies brine with a specific heat of 1.21 Btu/lb°F and a temperature of 185°F which exits at 120°F.

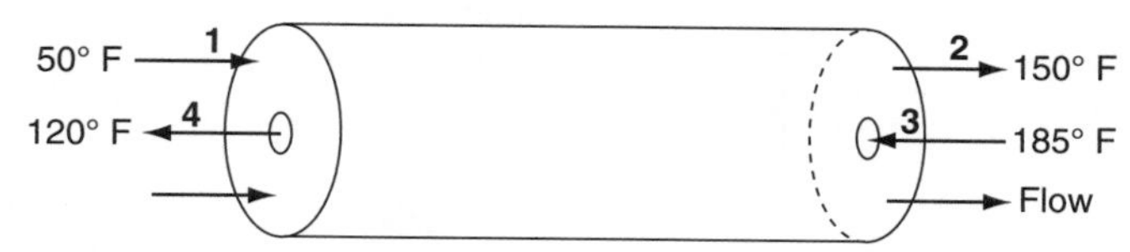

Exhibit 8

Brine is available at 30 lb/s.

a) Determine flow rate of water that can be heated.

b) Determine rate of entropy produced.

Solution

a) Energy transfer to water = energy from brine

$$W_{H_2O}C_{WATER}(T_2 - T_1) = W_{BRINE}\,C_{BRINE}(T_3 - T_4)$$

$$W_{H_2O}(1\,\text{Btu/lb°F})(150 - 50)\text{°F} = (30\text{lb/s})(1.21\,\text{Btu/lb°F})(185 - 120)\text{°F}$$

$$W_{H_2O} = \frac{2359.5}{100} = 23.6\text{ lb/s}$$

b) $\dot{S}_{gen} = W_{H_2O}C\ln\frac{T_2}{T_1} + W_{BRINE}\,C\ln\frac{T_4}{T_3}$

$$\dot{S}_{gen} = (23.6\text{lb/s})(1.0)\text{Btu/lb°F})\ \ln\frac{610}{510} + (3\text{lb/s})(1.21\,\text{Btu/lb°F})\ln\frac{580}{645}$$

$$\dot{S}_{gen} = 4.22 + (-\,0.386)$$

$$\dot{S}_{gen} = 3.83\text{ Btu/s°F}$$

Example 2.32

A real incompressible fluid enters a machine at areas A_1 and A_2 and leaves at area A_3. The temperature is constant. The mass density of the fluid is 80.5 lbm per cu-ft. Consider all openings to be at the same elevation. Calculate the shaft horsepower and indicate whether the horsepower is in or out.

$$\begin{array}{lll} V_1 = 30\text{ fps} & p_1 = 115\text{ psia} & A_1 = 0.2\text{ sq ft} \\ V_2 = 40\text{ fps} & p_2 = 75\text{ psia} & A_2 = 0.1\text{ sq ft} \\ V_3 = 65\text{ fps} & p_3 = 40\text{ psia} & \end{array}$$

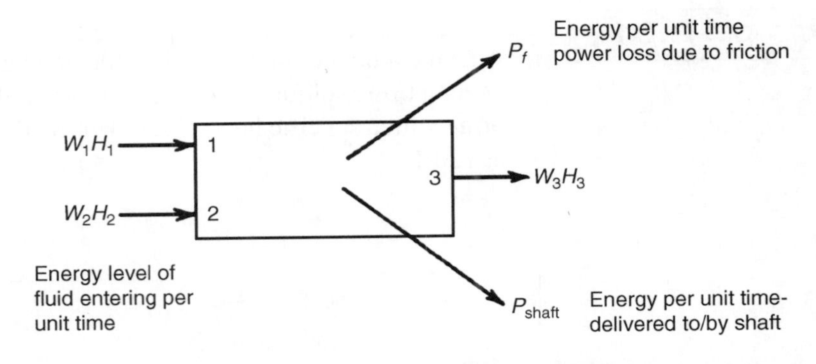

ρ = mass density of fluid, lb per ft³
slug = 1lb force per ft per sec²

P = power
Hd = head, ft
W = mass flow rate, lb per sec

Exhibit 9

Solution

Refer to Exhibit 9.

$$Hd = \frac{P}{\rho} + \frac{V^2}{2g} + Z_{\text{elevation}} = \text{total head, ft}$$

$$W = \rho A V$$

where

ρ = fluid density, lb per cu ft
A = area, sq ft
V = velocity, fps

$$Hd_1 = \frac{p_1}{\rho_1} + \frac{V_1^2}{2g}$$

$$p_1 = 115 \times 144 = 16{,}560 \text{ lb per sq ft}$$

$$Hd_1 = (16{,}560/80.5) + (30^2/64.4) + 0 = 219.7 \text{ ft}$$

$$Hd_2 = (75 \times 144)/(80.5) + (40^2/64.4) + 0 = 159 \text{ ft}$$

$$Hd_3 = (40 \times 144)/(80.5) + (65^2/64.4) + 0 = 137 \text{ ft}$$

$$W_1 = \rho_1 A_1 V_1 = 80.5 \times 0.2 \times 30 = 483 \text{ lb per sec}$$

$$W_2 = \rho_2 A_2 V_2 = 80.5 \times 0.1 \times 40 = 322 \text{ lb per sec}$$

$$W_1 + W_2 = W_3 = 805 \text{ lb per sec}$$

Then

$$W_1 H_1 = 483 \times 219 = 106{,}115 \text{ lb-ft per sec}$$

$$W_2 H_2 = 322 \times 159 = 51{,}200 \text{ lb-ft per sec}$$

$$W_3 H_3 = 805 \times 137 = 110{,}285 \text{ lb-ft per sec}$$

Thus

$$\text{Fluid power in} = W_1H_1 + W_2H_2 = 157{,}300 \text{ lb-ft per sec}$$
$$\text{Fluid power out} = W_3H_3 = 110{,}285 \text{ lb-ft per sec}$$
$$\text{Net power in} = 157{,}300 - 110{,}285 \text{ lb-ft per sec} = 47{,}000 \text{ lb-ft per sec}$$

Assume a reasonable efficiency (mechanical and hydraulic) = 70 percent.

$$\text{Efficiency} = \frac{\text{output}}{\text{input}}$$

Therefore,

$$\text{Output} = \text{shaft power} = 0.70 \times 47{,}000 = 32{,}900 \text{ lb-ft per sec}$$
$$\text{Shaft horsepower} = 32{,}900/550 = 59.8 \text{ shaft horsepower}$$

This is a turbine because power is out.

Example **2.33**

Superheated steam at 200 psia and a total steam temperature of 500°F enters an ideally insulated nozzle with negligible velocity and expands to a back pressure of 20 psia with a quality of 98 percent.

(a) Calculate the velocity leaving the nozzle in feet per second.

(b) Calculate the nozzle efficiency percent and the nozzle velocity coefficient.

Solution

Refer to Exhibit 10 and an *h-S* diagram for steam.

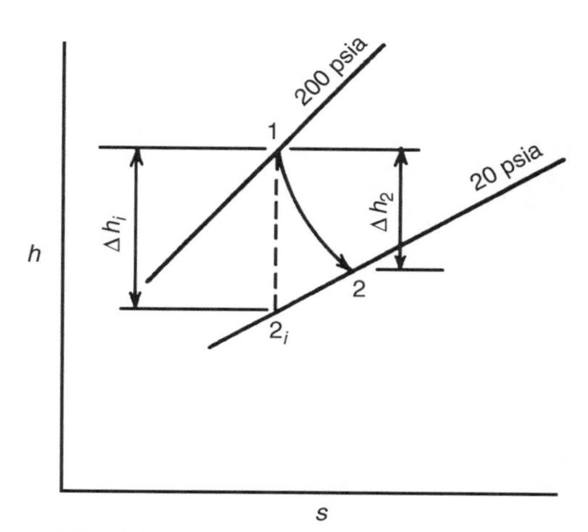

Exhibit 10

$$p_1 = 200 \text{ psia}$$
$$t_1 = 500°\text{F}$$
$$x_2 = 0.98$$
$$h_1 = 1268.9 \text{ Btu per lb}$$

$$S_1 = 1.624$$
$$S_1 = S_{2i} = 1.624$$

$$S_{f2} + x_2 \times S_{fg2} = S_2$$
$$S_2 = 0.3356 + 0.98 \times 1.3962 = 1.7038$$

There is an attendant entropy increase.

$$h_2 = h_{f2} + x_2 \times h_{fg2} = 196.16 + 0.98 \times 960.1 = 1137.1 \text{ Btu/lb}$$

$$\text{Actual heat drop} = \Delta h_a = h_1 - h_2 = 1268.9 - 1137.1 = 131.8 \text{ Btu per lb}$$

$$S_1 = S_{2i} = 1.624 = S_{f2i} + x_{2i} \times S_{fg2i} = 0.3356 + x_{2i} \times 1.3962,$$

from which

$$x_{2i} = 0.923$$

Then $h_{2i} = 196.16 + 0.923 \times 960.1 = 1{,}082.33$ Btu per lb

$$\text{Ideal heat drop} = \Delta h_i = 1268.9 - 1082.33 = 186.57 \text{ Btu per lb}$$

(a) V_2 exit nozzle velocity $= 223.84(h_1 - h_2)^{1/2} = 2570$ fps

(b) Nozzle efficiency = actual heat drop/ideal heat drop

$= 131.8/186.57 - 0.7066$ or 70.66%

Nozzle velocity coefficient $= 0.7066^{1/2} = 0.84$

CHAPTER 3

Fluid Mechanics

OUTLINE

Fluid mechanics is the branch of engineering that deals with the actions of forces on and by fluids. The study of fluids may be divided into two branches: hydrostatics (fluids at rest) and hydrodynamics (fluids in motion).

HYDROSTATICS

Fluid mechanics involves the study of liquids, vapors, and gases (highly superheated vapors). Vapors and gases usually are assumed to exert constant pressure throughout the contained volume. This of course does not hold true for our atmosphere because, as we know, the barometric pressure decreases as we go higher into the atmosphere. This fact explains the need for pressurization of the inside of aircraft.

Hydrostatics deals with the action of static fluids on bodies or within bodies. There is no motion of the fluid relative to stationary bodies, therefore the forces on the bodies are caused by normal pressure forces.

Example 3.1

A cylindrical log is 1 ft in diameter and 20 ft long. What weight of iron must be tied to one end of the log to keep it floating in an upright position in seawater with 18 ft of the log submerged? Specific gravity of the log is 0.7, of iron 7.9, and of seawater 1.03.

Solution

The basic equation is given by

$$\text{wt of log} + \text{wt of iron} = \text{buoyant force log} + \text{buoyant force of iron}$$

$$[(\pi D^2/4)](L)(62.4)(0.7) + \text{wt of iron} = [(\pi D^2/4)](62.4)(1.03)(18) + (\text{vol of iron})(62.4)(1.03)$$

$$686 \text{ lb} + \text{wt iron} = 909 \text{ lb} + 64.3(\text{vol of iron})$$

and wt of iron = (S.G. iron)(wt of water lb/ft^3)(vol of iron)

wt iron = (7.9)(64.3)(vol iron) = 508(vol iron)

Therefore, 508 vol iron − 64.3 vol iron = 223

Vol iron = 0.503ft^3 and wt iron = (0.503)(7.9)(64.3) = 256 lb

Forces on Submerged Areas

Pressure is a non-directional force and acts perpendicularly on a surface. This force varies directly with the depth or the distance down from the surface. The vertical distance from the surface of the liquid to the centroid of the body is labeled h_c. The force acting on the body is equal to

$$F = \gamma h_c A.$$

Where

F = total force acting upon the body
γ = specific weight of the liquid
h_c = distance from the surface of the liquid to the centroid of the body
A = area of the body
h_p = distance from the surface of the liquid to the center of pressure of the body
I_c = centroidal moment of inertia

That is, $h_p = h_c + I_c/h_c A$.

Example 3.2

A rectangular sluice gate is 6 ft wide and 9 ft high. It is immersed vertically in water with the 6-ft edges horizontal. Water stands on one side of the gate at a depth level with its upper edge and on the other side at a depth of 4.5 ft below the upper edge. The gate is hinged at the top edge and is held in equilibrium by a horizontal force applied at its lower edge. Calculate this force.

Solution

$$h_p = h_c + \frac{I_c}{h_c A} = 4.5 + \frac{\frac{1}{12}69^3}{(4.5)(6\times 9)} = 6 \text{ ft}$$

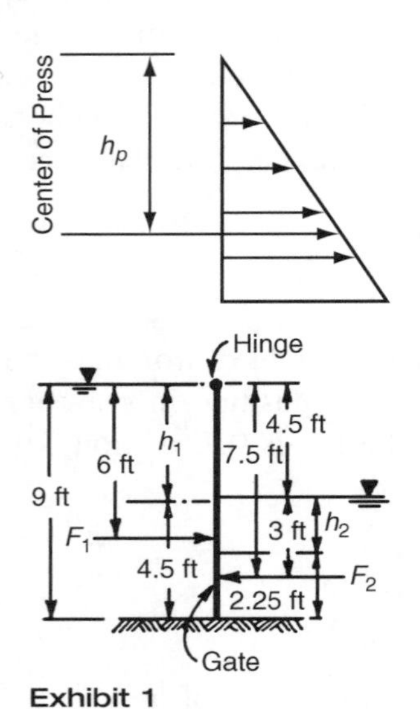

Exhibit 1

Refer to Exhibit 1. The center of pressure acts two-thirds of the distance down from the top (at water level). Then

$$F_1 = wh_1A_1 = 62.4\times 4.5\times(9\times 6) = 15{,}200 \text{ lb}$$
$$F_2 = wh_2A_2 = 62.4\times 2.25\times(4.5\times 6) = 3800 \text{ lb}$$

Now take the summation of moments about hinge and set equal to zero. Counterclockwise moments equal clockwise moments. Now set up this relationship.

$$F_1\times 6 = (F_2\times 7.5) + (F_0\times 9)$$
$$15{,}200\times 6 = (3800\times 7.5) + (F_0\times 9)$$
$$F_0 = (91{,}200 - 28{,}500)/9 = 6980 \text{ lb}$$

HYDRODYNAMICS

Hydrodynamics is the study of fluids in motion. Because there is flow over and through bodies, there are viscous shear stresses between the bodies and the fluid.

Example 3.3

Assuming isothermal and steady flow, a pump takes suction from a large storage tank containing sulfuric acid (sp gr = 1.84) through a 3-in. line and discharges through a 2-in. line to a point 75 ft above the level in the storage tank (Exhibit 2). The tank is under slight positive pressure with a dry gas to prevent absorption of air moisture from the atmosphere. However, this pressure may be neglected. Friction losses may be taken as 26 ft of fluid flowing. What differential pressure must the pump develop and what horsepower motor would be necessary if the pump efficiency were 60 percent and the velocity in the suction line were 3 fps?

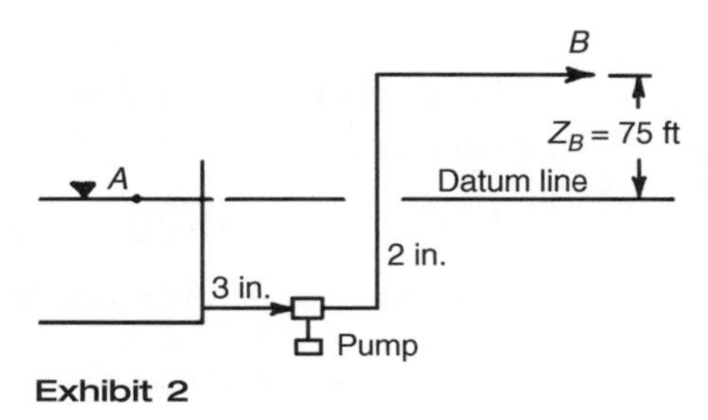

Exhibit 2

Solution

By observation Z_A is zero and v_A is zero because tank diameter is great compared to suction-pipe diameter. It may also be assumed that the water level is dropping so slowly that its velocity is negligible compared to the pipe velocity. Now, because both tanks are open to the atmosphere, $P_A = P_B$. Single fluid is flowing in the system and $w_A = w_B$, where w is density in pounds per cubic foot.

The general energy equation below,

$$h_p + \frac{P_A}{\gamma} + \frac{V_A^2}{2g} + Z_A = h_f + \frac{P_B}{\gamma} + \frac{V_B^2}{2g} + Z_B$$

$$\text{Reduces to: } h_p = Z_B + \frac{V_B^2}{2g} + 26$$

Now let us find the velocity in the 2-in. discharge line from that in the suction line.

$$v_B = 3 \times \left(\frac{3}{2}\right)^2 = 6.75 \text{ fps}$$

Rearranging the previous form and solving for W_p, it is found that

$$h_p = 75 + (6.75^2/64.4) + 26 = 101.7 \text{ ft.}$$

This is the total head required for the pump to operate properly. This is a differential head; it is neither the suction head nor the discharge head. It is the difference between the discharge and suction heads. We must convert this total head to pressure as follows:

$$(101.7\ \text{ft})\left(\frac{14.7\ \text{psi}}{34\ \text{ft}\ H_2O}\right)(1.84) = 80.9\ \text{psi}$$

For the second part of the problem we make use of the basic relation for horsepower: lb per min × total dynamic head/33,000. What the pump is really doing is, in fact, lifting so many pounds of sulfuric acid a minute through 101.7 ft.

$$\text{Flow rate} = AV = \left(\frac{\pi}{4} \times \frac{(3)^2}{144}\right)\text{ft}^2(3\ \text{ft/s})(60\ \text{s/min}) = 8.836\ \text{ft}^3/\text{min}$$

$$(8.836\ \text{ft}^3/\text{min})(62.4\ \text{lb/ft}^3)(1.84) = 1014\ \text{lb/min}$$

where gpm is derived from the equation in simple form as

$$\text{gpm} \times 0.408/d^2 = \text{fps}.$$

Here d is the nominal pipe size as given in the problem. There is no need to be concerned with actual inside diameter of the pipe.

$$\text{Hydraulic horsepower} = 1014 \times 101.7/33{,}000 = 3.12$$

We must adjust hydraulic horsepower to take into account pump efficiency.

$$\text{Hydraulic horsepower/pump efficiency} = \text{brake horsepower}$$

$$3.12/0.6 = 5.2\ \text{brake horsepower}$$

We need a 5½ horsepower motor.

Example 3.4

A pump takes 500 gpm from a sump and discharges the water from an 8-in. pipe at a point 20 ft higher. Assuming an overall efficiency of 80 percent, what horsepower is required?

Solution

Since discharge head includes head loss due to friction and actual lift we may write the equation for power required as:

$$\text{bhp} = \frac{(500\ \text{gal/min})(8.33\ \text{lb/gal})(20\ \text{ft})}{33{,}000\ \text{ft-lb/min}\ (0.80)}$$

(0.80) ← overall efficiency

$$\text{bhp} = 31.5\ \text{hp}$$

Example 3.5

Compute the discharge through an orifice whose area is 1 sq in. Coefficient of discharge C may be taken as 0.61. The water reaches the orifice through a pipe whose area is 4 sq in. A pressure gauge reads 40 psi at a point 3 ft above the orifice. Neglect the velocity of approach.

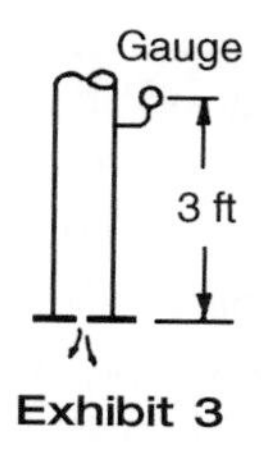

Exhibit 3

Solution

Using orifice equation

$$(\text{Flow rate})Q = (\text{Discharge coef})C\ (\text{Orifice area})A\sqrt{2gh}$$

$$g = \text{gravity}$$

$$h = \text{total head} = 3\text{ ft} + 40\left(\frac{34}{14.7}\right) = 95.5\text{ ft}$$

$$Q = (0.61)\left(\frac{1\text{ in}^2}{144\text{ in}^2/\text{ft}^2}\right)\sqrt{2(32.2)\text{ ft/s}^2(95.5)\text{ ft}}$$

$$Q = 0.332\text{ ft}^3/\text{s}$$

Natural Discharge from Tanks and Reservoirs

An orifice may be used to measure velocity of flow out of tanks/reservoirs when the fluid depth remains constant. Simply by using the equation:

$$v = C_v\sqrt{2gh}$$

to find flow rate

$$Q = C_d A_o\sqrt{2gh}$$

Q = Flow rate ft^3/s
C_d = Discharge coefficient
A_o = Area of orifice
C_v = Velocity coefficient

Example 3.6

An orifice having an area of 2 ft^2 allows water to flow from a reservoir. The head it operates under is 100 ft. The $C_v = .90$ and $C_d = .85$. Determine:
a) Velocity of flow ft/s
b) Flow rate ft^3/min

Solution

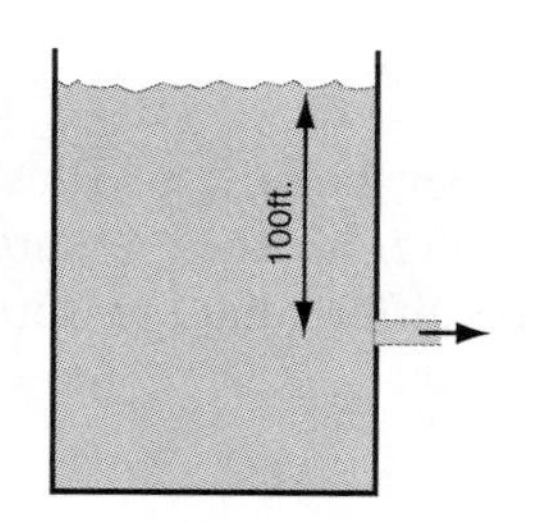

Exhibit 4

a)

$$v = C_v\sqrt{2gh}$$

$$v = 0.90\sqrt{(2)(32.2)\text{ft/s}^2(100\text{ ft})}$$

$$v = 72.2\text{ ft/s}$$

b)

$$Q = C_d A_o\sqrt{2gh}$$

$$Q\text{ ft}^3/\text{s} = (0.85)2\text{ ft}^2\sqrt{(2)(32.2)(100)}$$

$$Q\text{ ft}^3/\text{s} = 136.4\text{ ft}^3/\text{s}$$

Example 3.7

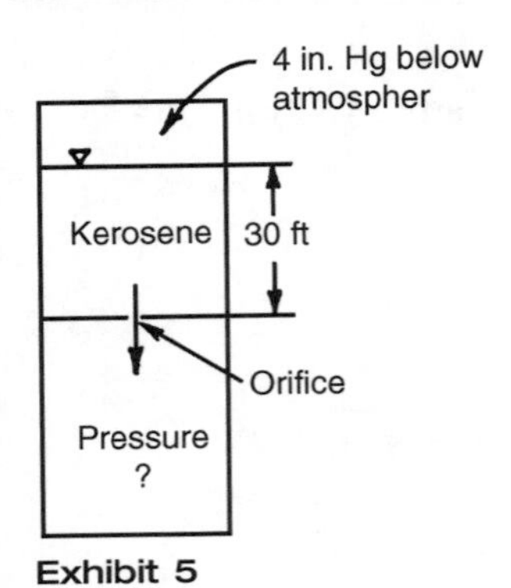

Exhibit 5

Two closed tanks are connected by a 2-in. diameter circular orifice that has a coefficient of discharge of 0.63 for kerosene (Exhibit 5). One tank contains kerosene, specific gravity 0.78, to a depth of 30 ft above the center of the orifice and the other contains air at 14.7 psia. The tank with kerosene has a pressure of 4 in. of mercury below atmosphere in the space above the kerosene. What is the pressure in psi in the empty tank when the orifice is discharging 157.2 gpm?

Solution

Convert 157.2 gpm to cubic feet per second.

$$\left(\frac{157.2\ \text{gal/min}}{60\ \text{g/min}}\right)\left(\frac{1}{7.48\ \text{gal/ft}^3}\right) = 0.350\ \text{ft}^3/\text{s}$$

The head causing the flow of 157.2 gpm may be found with the use of $Q = CA_0\sqrt{2gh}$. By rearrangement and solving for h,

$$h = \left(\frac{Q}{CA_o}\right)^2 \frac{1}{2g}$$

$$h = \left[\frac{0.350\ \text{ft}^3/\text{s}}{(0.63)\left(\frac{\pi}{4}\right)\left(\frac{2}{12}\right)^2\ \text{ft}^2}\right]^2 \left(\frac{1}{2(32.2)\ \text{ft/s}^2}\right) = 10\ \text{ft}$$

Convert the 30 ft of kerosene to its equivalent pressure in psi.

$$\frac{(30\ \text{ft})(62.4\ \text{lb/ft}^3)(0.78)}{144\ \text{in}^2/\text{ft}^2} = 10.1\ \text{psi}$$

The 10-ft head of kerosene is equivalent to $10.1/3 = 3.36$ psi. The 4 in. of mercury below atmosphere is equivalent to

$$(4\ \text{in Hg})\left(\frac{14.7\ \text{psi}}{29.92\ \text{in Hg}}\right) = 1.97\ \text{psi}$$

The vacuum effect of the 4 in. of mercury reduces the head over the orifice. Thus, $10.1 - 1.97 = 8.13$ psi, the pressure upstream of the orifice. Therefore, the pressure downstream of the orifice is the pressure upstream minus the differential pressure across the orifice.

$$8.13 - 3.36 = 4.77\ \text{psi}$$

Example 3.8

A 3-in. thin plate or sharp-edged orifice is installed in a 16-in. pipe carrying a mineral oil having a specific gravity of 0.90. A vertical manometer with mercury and oil shows 4 in. difference in mercury levels. Calculate the flow of oil as barrels per hour and the power lost due to the orifice. Assume that 95 percent of differential is lost. One barrel is equal to 42 gal.

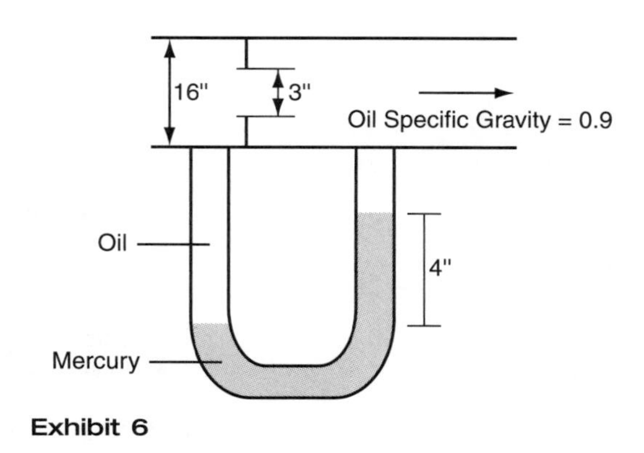

Exhibit 6

Solution

Assume coefficient of discharge is 0.61.

$$v_o \text{ ft/s} = C\sqrt{2g\Delta h}$$

$$v_o = 0.61\sqrt{2(32.2 \text{ ft/s}^2)\left(\frac{4}{12}\text{ ft}\right)\left(\frac{13.6 - 0.9}{0.9}\right)} = 10.6 \text{ ft/s}$$

Specific gravity of mercury is 13.6. The portion of the equation (13.6 – 0.9) corrects for the 4" oil in the upstream part of the nanometer. The .9 in the denominator corrects for the oil in the pipe.

The flow rate in barrels/hr is:

$$\text{Flow rate} = VA_o = \frac{(10.6 \text{ ft/s})(3600 \text{ s/hr})\left(\frac{\pi}{4}\right)\times\left(\frac{3}{12}\right)^2 \text{ft}^2(7.48 \text{ gal/ft}^3)}{42 \text{ gal/barrel}} = 334 \text{ barrel/hr}$$

To find power loss due to orifice assuring pressure differential is not recovered.

$$\text{Power loss} = \frac{\left(\frac{334}{3600}\right)\left(\frac{\text{barrel}}{\text{s}}\right)\left(42\frac{\text{gal}}{\text{barrel}}\right)\left(8.33\frac{\text{lb}}{\text{gal}}\right)H_2O(0.9\text{spgr})\left(\frac{\text{Oil}}{H_2O}\right)\left(\frac{4}{12}\right)\left(\frac{13.6-0.9}{0.9}\right)}{550 \text{ ft-lb/hp-s}}$$

$$\text{Power loss} = 0.249 \text{ hp}$$

If there is a 5% recovery, then power loss is: 0.249 hp × 0.95 = 0.236 hp

Note: if a light gas were flowing through the orifice and the nanometer fluid were mercury or water, the correction of the upstream leg of the nanometer would not be required.

Example 3.9

Water is flowing through an orifice meter in a pipeline 8-in. in diameter. The thin plate of the orifice has an opening 6 in. in diameter. Pipe taps just upstream and downstream from the orifice plate lead to a water-carbon tetrachloride (sp gr 1.6) gauge, and this shows a difference of 3 ft. How much water is flowing through the meter?

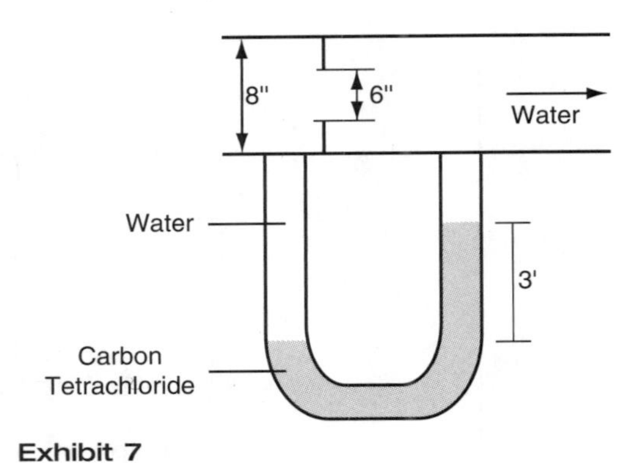

Exhibit 7

Solution

For lack of information on the orifice coefficient, 0.61 may be used. The effect of the water head over the carbon tetrachloride upstream leg of the manometer is to be corrected. Flow due to velocity of approach must also be corrected. Another useful formula to determine flow taking into account the velocity of approach is as follows:

$$Q = 0.61\left(\frac{\pi}{4}d^2\right)\ \text{ft}^2\sqrt{\frac{2g\Delta h\frac{(\text{sp gr} - 1)}{1}}{1 - \left(\frac{6}{8}\right)^4}}$$

$$Q = 0.61\left(\frac{\pi}{4}\right)\left(\frac{6}{12}\right)^2\sqrt{\frac{2(32.2)(3)\left(\frac{1.6 - 1}{1}\right)}{1 - \left(\frac{6}{8}\right)^4}}$$

$$Q = 1.56\ \text{cfs}$$

Example 3.10

Water flows through a venturi meter with throat diameter d equal to 3 in. and is installed in a 6-in. pipeline. Mercury in the manometer stands at 15 in. differential. The connecting tube is filled with water. Find the rate of discharge in gallons per minute. Do not correct for approach.

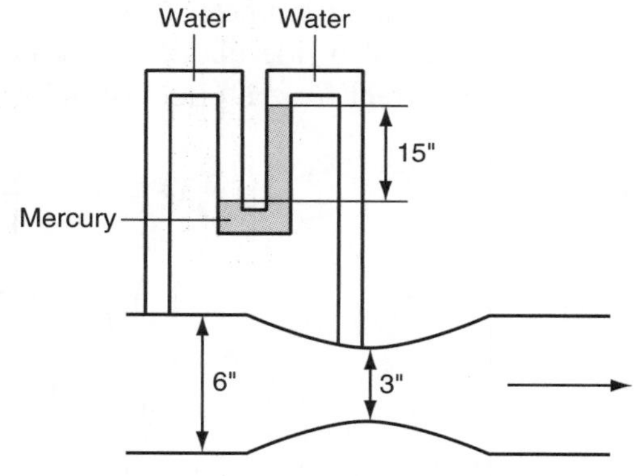

Exhibit 8

Solution

Coefficient of discharge may be taken as 0.98.

$$Q = CA_t \times \sqrt{2g\Delta h}$$

$$Q = (0.98)\left(\frac{\pi}{4}\left(\frac{3}{12}\right)^2\right)\text{ft}^2\sqrt{2(32.2)\text{ft/s}^2\left(\frac{15}{12}\text{ft}\right)\left(\frac{13.6-1}{1}\right)}$$

$$Q = 1.53 \text{ cfs}$$

The flow in gpm is:

$$(1.53 \text{ ft}^3/\text{s})(7.48 \text{ gal/ft}^3)(60 \text{ s/min}) = 686 \text{ gal/min}$$

Example **3.11**

At a water supply plant the raw water inflow is measured by means of a venturi meter located in a pipe 20 in. in diameter. The discharge coefficient is 0.95. If the difference in pressure between the upstream end and the throat is 5 psi, at what rate in gpm will water flow through the meter, if throat diameter is 8 in.?.

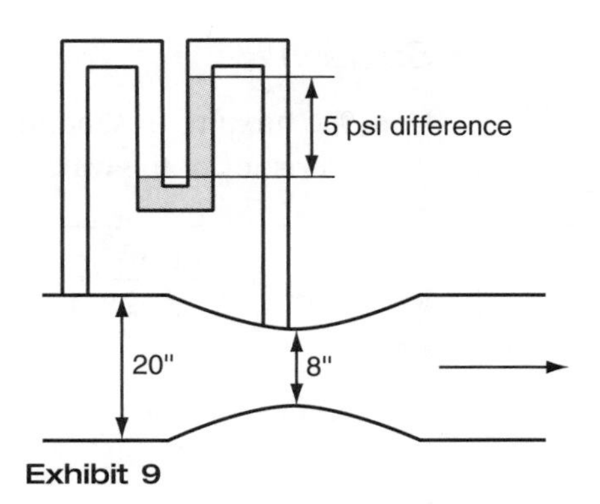

Exhibit 9

Solution

Note that in all the hydraulics problems that mention pipe diameters, the nominal size given in the wording of the problem may be taken as the inside diameter or working diameter. There is no need in most cases to concern oneself with the actual inside diameter, taking into account pipe-wall thickness. In the solution, correct for the effect of velocity of approach for accurate results.

$$Q = CA\sqrt{\frac{2g\Delta h}{1-(d/o)^4}} = (0.95)\left(\frac{\pi}{4}\right)\left(\frac{8}{12}\right)^2 \text{ft}^2 \times$$

$$\left(\sqrt{\frac{2(32.2)\text{ft/s}^2(5\text{ psi})}{1-\left(\frac{8}{20}\right)^4}\left(\frac{34\text{ ft}}{14.7\text{ psi}}\right)}\right) = 9.19\text{ ft}^3/\text{s}$$

Flow in gpm is found in the usual manner.

$$\text{gpm} = (9.19\text{ ft}^3/\text{s})(60\text{ s/min})(7.48\text{ gal/ft}^3) = 4124\text{ gal/min}$$

If the effect of velocity of approach were neglected, the flow would be 4050 gpm, for an error of 1.03 percent. This would be appreciable in practice under most conditions.

Example 3.12

In a filling operation a dredge is pumping sand, water, and mud through a steel pipe 12 in. in diameter (Exhibit 10). The discharge pipe is supported horizontally 4 ft above the ground. If the center of the jet from the pipe strikes the ground at a distance 4 ft away from the center end of the pipe horizontally, how many cubic yards of mud, water, and sand would the pipe deliver in two hours?

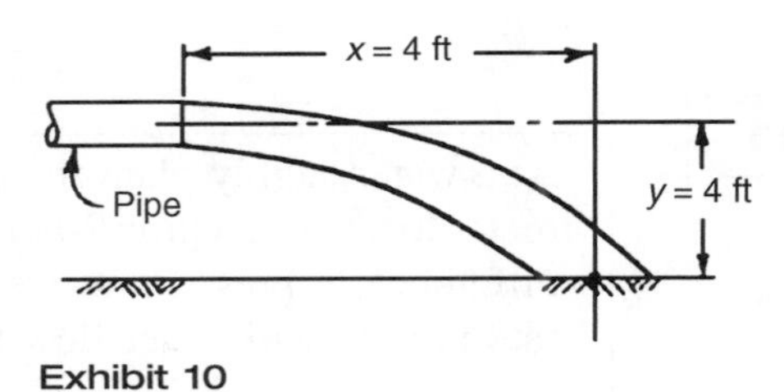

Exhibit 10

Solution

Here the mixture of sand, mud, and water may be considered as having the general characteristics of the water, which is predominant. Now, as in the previous problem,

$$v_{EL} = \sqrt{\frac{(x^2)\text{ft}^2 g\text{ ft/s}^2}{2y\text{ ft}}} = \sqrt{\frac{4^2 \times 32.2}{2 \times 4}} = 8.02\text{ fps at the outlet.}$$

$$Q = Av = \left(\frac{\pi}{4}d^2\right)\text{ft}^2(8.02)\text{ ft/s} = \left(\frac{\pi}{4}\right)\left(\frac{12^2}{144}\right)(8.02) = 6.3\text{ ft}^3/\text{s}$$

Then find cubic yards of flow in two hours:

$$\text{Cubic yards} = \frac{(6.3\text{ ft}^3/\text{s})(3600\text{ s/hr})(2\text{ hr})}{27\text{ ft}^3/\text{yd}^3} = 1680\text{ cu yards}$$

Problems of this type follow the reasoning outlined and should be readily recognizable in the examination. In the actual physical piping arrangement, the jet must project horizontally and the exit pipe must be level. Another important consideration is to provide a straight run of pipe after any obstruction or elbow in the system upstream of the aperture.

Example 3.13

A jet issuing under a head of 3 ft from a sharp-edged orifice of circular cross-section 0.1 ft in diameter in a vertical plane goes 5.9 ft horizontally to a point 3 ft below the center of the orifice. In a test 2460 lb of water is collected from the jet in 10 min. Determine the diameter of the vena contracta of the jet.

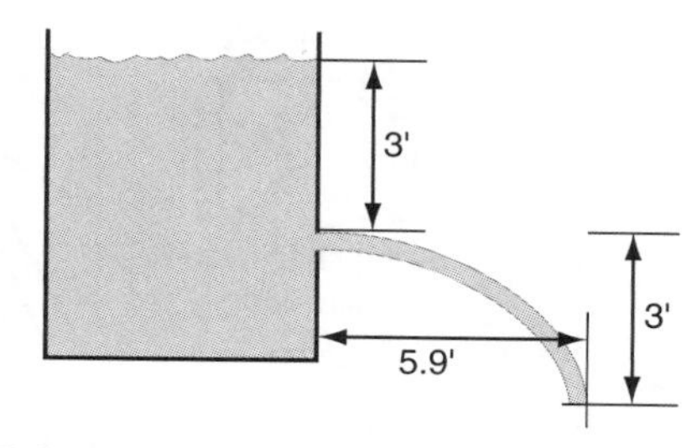

Exhibit 11

Solution

As before,

$$v = \sqrt{\frac{x^2 \text{ ft}^2 g \text{ ft/s}^2}{2y \text{ ft}}}$$

$$v = \sqrt{\frac{5.9^2 \times 32.2}{2 \times 3}} = 13.7 \text{ fps}$$

From the information given above

$$Q = \left(\frac{2460 \text{ lb}}{10 \text{ min}}\right)\left(\frac{1}{60 \text{ g/min}}\right)\left(\frac{1}{62.4 \text{ lb/ft}^3}\right) = 0.0657 \text{ ft}^3\text{/s}$$

from which d is found to be 0.0781 ft. It has been assumed that v is velocity at the vena contracta. Actually, velocity at the vena contracta is greater than at the orifice.

Also $$Q = A_o V = 0.0657 \text{ ft}^3\text{/s} = \frac{\pi}{4} d^2 (13.6) \text{ft/s}$$

$$d = 0.078 \text{ ft}$$

Example 3.14

A run consists of 37 ft of straight pipe 4 in. in diameter, three short radius elbows, two wide-open gate valves, and one wide-open globe valve. What is the total run of pipe expressed in feet that will be used to calculate head loss in the Darcy (or Fanning) formula for friction drop?

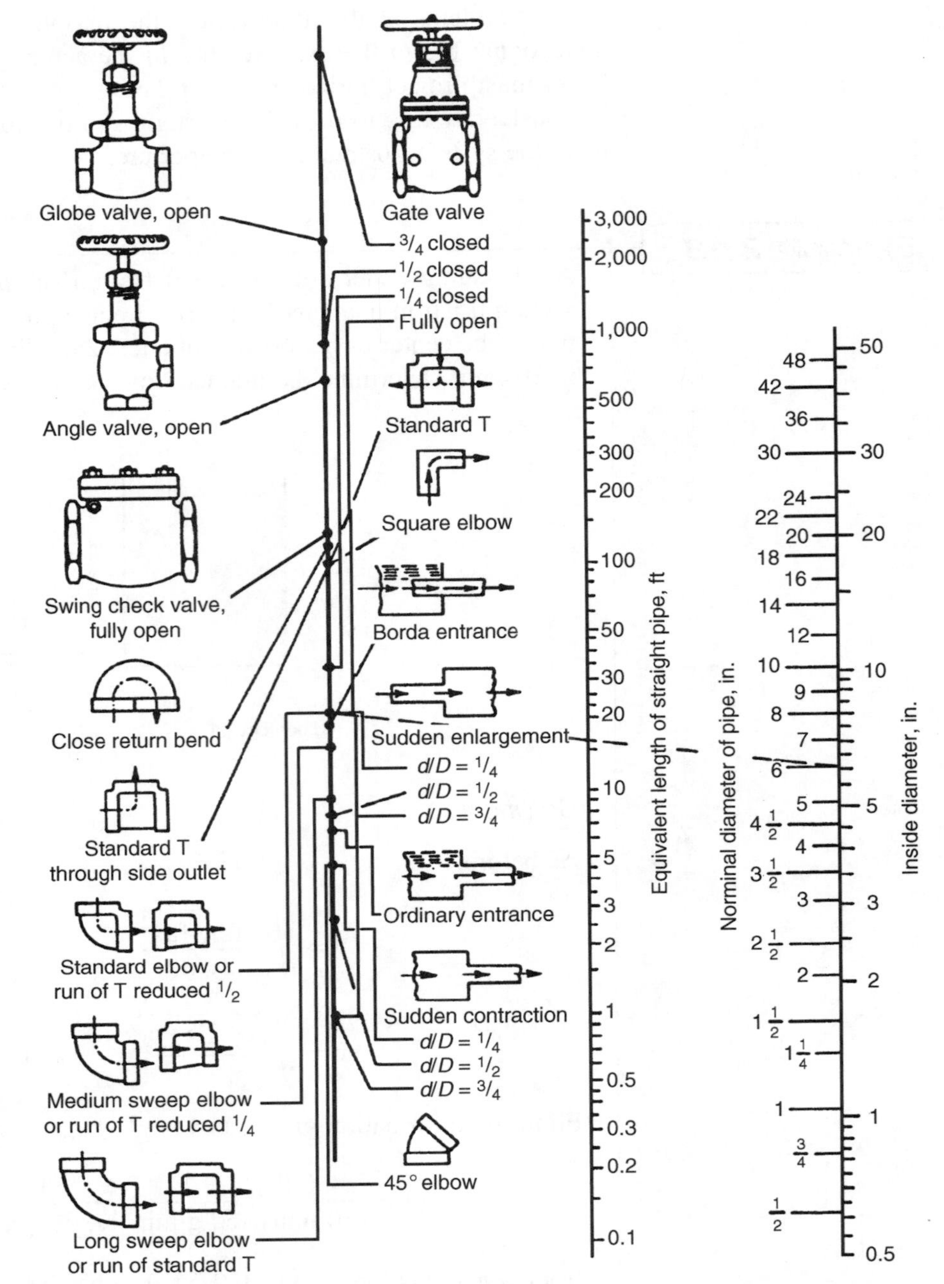

Exhibit 12 Resistance of valves and fittings to flow fluids. A simple way to account for the resistance offered to flow by valves and fittings is to add to the length of pipe in the line a length that will give a pressure drop equal to that which occurs in the valves and fittings in the line. Example: The dotted line shows that the resistance of a 6-in. standard elbow is equivalent to approximately 16 ft of 6-in. standard steel pipe. Note: For sudden enlargements or sudden contractions, use the smaller diameter on the nominal pipe-size scale. (*Reprinted with permission from "Flow of Fluids through Valves, Fittings, and Pipe," Technical Paper 410, 1988, Crane Co. All rights reserved.*)

Solution

Using the Crane Chart (Exhibit 12),

Pipe @ 37 ft	37
3 elbows @ 10 ft	30
2 gate valves @ 2.5 ft	5
1 globe valve @ 120 ft	120
Total (*L*)	192 ft

Note. When the straight pipe length is greater than 1000 times nominal pipe diameter, the effect of restrictions such as valves and fittings may be neglected.

Example 3.15

Determine in feet of oil the loss in static head when oil is pumped through a wrought-iron pipe 1 in. in diameter inside and 1000 ft long. Average velocity of oil is 1 fps at 60°F. Absolute viscosity is 4 centipoise and specific gravity of oil is 0.85 at the flowing temperature.

Solution

First let us determine the Reynolds number. For friction-loss calculations other than water it is best to use Fanning's formula and friction factor.

$$1 \text{ Centipoise} = 2.42 \text{ lb/hr ft}$$

$$R_e = \frac{\rho D V}{\mu}$$

$$R_e = \frac{(62.4 \text{ lb/ft}^3)(0.85)(\frac{1}{12}\text{ft})(1 \text{ ft/s})(3600 \text{ s/hr})}{(4 \text{ cent})(2.42 \text{ lb/hr ft-cent})} = 1643$$

since this is laminar flow, $f = \dfrac{64}{R_e} = \dfrac{64}{1643} = 0.0389$

$$\text{lost head} = f\left(\frac{L}{d}\right)\frac{v^2}{2g}$$

$$\text{lost head} = 0.0389\left(\frac{1000 \text{ ft}}{\frac{1}{12}\text{ ft}}\right)\frac{(1 \text{ ft/s})^2}{2(32.2 \text{ ft/s}^2)} = 7.25 \text{ ft oil}$$

Example 3.16

Crude oil is flowing through a pipeline 12 in. in diameter and 20 miles long at an average velocity of 2 fps. The viscosity of the oil may be taken as 2×10^{-3} lb-sec/sq ft, with a specific gravity of 0.92 with reference to water at 60°F. How much pressure would be required at a pumping station at one end of the line if the other end were 100 ft higher?

Solution

Because the length of pipeline is greater than 1000 times pipe diameter, the effect of fittings, etc., may be ignored. Pump-discharge pressure would be made up of friction head plus the static discharge heads. First determine Reynolds number.

$$R_e = \frac{\rho DV}{\mu}$$

$$R_e = \frac{(62.4\ \text{lb/ft}^3)(\frac{12}{12}\ \text{ft})(2\ \text{ft/s})(0.92)}{(2 \times 10^{-3}\ \text{lb-s/ft}^2)(32.2\ \text{ft/s}^2)} = 1783 \text{ laminar flow}$$

$$f = \frac{64}{R_e} = \frac{64}{1783} = 0.03589$$

$$\text{Head lost} = f\left(\frac{L}{d}\right)\frac{v^2}{2g} = (0.03589)\left(\frac{20 \times 5280}{12/12}\right)\left(\frac{(2\ \text{ft/s})^2}{2(32.2\ \text{ft/s}^2)}\right) = 234\ \text{ft}$$

The total dynamic head (W_p in Bernoulli's equation) is

$$234 + 100 = 334\ \text{ft}$$

The pressure required to support a head of 0.92 specific gravity oil 217 ft high is

$$(334\ \text{ft})\frac{14.7\ \text{psi}}{34\ \text{ft}} = 144\ \text{psi}$$

It is assumed that the pump is operating "flooded" with no suction pressure.

Example **3.17**

The flow in a channel is 35 cu ft per sec (cfs). It is desired to discharge 5 cfs over a 90° triangular weir in one channel and the remainder over a standard (no end contractions) rectangular weir. The crests of the weirs are to be set at the same elevation. Calculate the length of the rectangular weir and the head on both weirs, neglecting effect of velocity of approach.

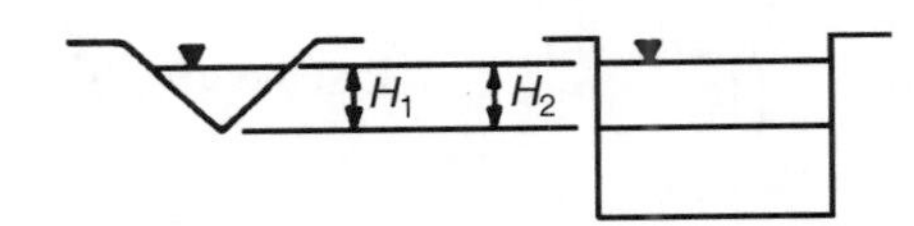

Exhibit 13

Solution

Note V-notch weirs are not appreciably affected by velocity of approach. Now refer to Exhibit 8. Let H_1 and H_2 be the heads on the triangular weir (V-notch) and rectangular weir, respectively. Then $H_1 = H_2$. Because the heads are equal, we calculate value of H_1 by use of

Triangular Weir Equation

$$H_1 = \left(\frac{Q_1}{2.53}\right)^{1/2.5} = \left(\frac{5}{2.53}\right)^{1/2.5} = 1.319\ \text{ft} = H_2$$

Because $H_1 = H_2$,

Rectangular Weir Equation

$$b = \frac{Q_2}{3.33 \times H^{3/2}} = \frac{30}{3.33 \times 1.319^{3/2}} = 5.94 \text{ ft.}$$

Example 3.18

In a certain processing plant it is desired to determine the flow rate of a gas flowing through an irregularly shaped duct in which measuring devices cannot be used very well. Gas analysis shows that the flowing gas contains 0.24 percent CO_2 by volume. It is decided to determine the flow rate by bleeding CO_2 into the gas stream from a small weighed cylinder or bottle. A constant rate is obtained by means of a flowmeter, and after thorough mixing of the gases by passing them through donut sections and bends, the average analysis of the mixture shows that it contains 1.41 percent CO_2. The loss in weight of the cylinder is 7.94 lb in 5 min, measured by a stopwatch. If the temperature of the gas stream and the mix is 120°F, what is the volume flowing in cubic feet per minute?

Solution

The ratio of CO_2 to CO_2-free gas in the original is 0.24/99.76 = 0.0024, and the same ratio after mixing is 1.41/98.59 = 0.0143. The increase in this ratio, 0.0143 – 0.0024 = 0.0119, represents moles CO_2 bled into the stream per mole CO_2-free gas. Using this figure and the data given in the problem, the flow rate is

$$\text{Flow rate} = \left(\frac{7.95 \text{ lb}}{5 \text{ min}}\right)(379 \text{ ft}^3/\text{lb mol})\left(\frac{1 \text{lb mass}}{44 \text{ lb mol}}\right)\left(\frac{1}{0.0119 \text{ mol}}\right)\left(\frac{100}{99.76}\right)\left(\frac{120+460}{60+460}\right)$$

This calculates out to 1380 cfm.

Flow rate = 1380 ft^3/min

CHAPTER 4

Heat Transfer

OUTLINE

Heat transfer is that branch of thermodynamics concerned with the transfer of energy from one point to another by virtue of temperature difference. There are three modes of heat transfer: conduction, convection, and radiation. Conduction and radiation are true forms of heat transfer because they depend only on temperature differential and the characteristics of the materials involved in the temperature exchange. Convection, however, depends not only on the materials and temperature differentials, but also on mass transport.

CONDUCTION

Conduction is the molecule-to-molecule transfer of energy. The one-directional equation of energy flow for a steady state condition is given by Fourier's Law of Conduction,

$$Q = (kA/L)(T_1 - T_2), \quad \textbf{(4.1)}$$

where

Q = Rate of heat flow in Btu per hr
k = thermal conductivity, Btu/(sq ft)(×F)(ft) thickness
A = area perpendicular to heat flow, sq ft
L = Thickness in direction of heat flow, ft
T = $(T_1 - T_2)$ difference in temperature within the material in the direction of heat flow, may be given in terms of ×F or ×R.

The coefficient of conductivity k gives the number of Btu that will flow in 1 hr through a conductor 1 ft long and 1 sq ft in cross section, when there is a temperature difference of 1°F from one end of the conductor to the other. This coefficient is a characteristic of the material of which the conductor is made and varies with temperature directly. Some texts give thermal conductivity in terms per inch of thickness. Then the length of path is in inches.

Table 4.1 gives the values of thermal conductivity for various common materials. More extensive tables may be found in the literature and in engineering handbooks.

Table 4.1 Thermal conductivities

Material	Experimental Temperature Range, °F	k per ft Thickness
Aluminum	32–212	118
Brass, yellow	32–212	63
Copper, pure	At 32	226
Wrought iron	32–527	35
Cast iron, 3.5% carbon	At 212	28
Lead	At 59	20
Nickel	32–212	34
Platinum	64–212	41
Silver	At 64	243
Steel, mild	32–212	35
Boiler scale	At 212	0.5–1.5
Asbestos	100–1000	0.04–0.12
Brick, carborundum	At 1800	5.6
Brick, building	At 70	0.4
Sil-O-Cel	At 1800	0.03
Cork	122–392	0.03
Glass, flint	50–59	0.03–0.06
Infusorial earth, 12.5 lb per cu ft	At 122	0.05
Magnesia insulation, 85%	68–310	0.04
Rubber	At 220	0.01
Fire brick	At 2300×F	1.0
Rock wool	At 212	0.023
Wood, pine	At 70	0.1–0.2
Plaster on wood	At 212	0.208
Glass wool	At 212	0.023
Concrete work	At 212	1.0
Lubricating oil	At 86	0.08
Water	At 167	0.372
Air	At 32	0.0137
	At 212	0.0174
Steam	At 32	0.0095
	At 212	0.0129
Carbon monoxide	At 32	0.131
Carbon dioxide	At 32	0.00804
Oxygen	At 32	0.0138
Hydrogen	At 32	0.092
Methane	At 32	0.0174

Example **4.1**

What is the heat flow per hour through a brick and mortar wall 9 in. thick if coefficient of thermal conductivity has been determined as 0.4 and the wall is 10 ft high by 6 ft wide, the temperature on one side of the surface being 330×F and on the other, 130×F?

Solution

Using Equation 4.1, we obtain the following:

$$Q = 0.4[(10 \times 6) \times (330 - 130)]/(9/12) = 6400 \text{ Btu per hr.}$$

Note that the solution has not taken into account the fact that there is an air film on either side of the wall.

Conductors of Nonuniform Cross Section

Where Equation 4.1 holds for constant cross-section bodies, the heat loss from an insulated pipe takes place radially from the metal surfaces through a constantly increasing area until it reaches the maximum area of the surface of the insulation. On the other hand, heat flow from a room into cold brine flowing inside a pipe covered with cork or hair-felt insulation starts with a large outside surface and passes radially inward through a constantly decreasing area until it reaches the pipe. In either case, the A term in Equation 4.1 must be expressed as an average of the section. This is the arithmetic mean average and is expressed as

$$A_a = \frac{A_1 + A_2}{2} \tag{4.2}$$

Logarithmic Mean Average of Area

In the relatively few cases in which the maximum area is very large in comparison with the minimum (that is, over twice as great), a considerable error is introduced in using the arithmetic mean as the average area. It then becomes necessary to use the logarithmic mean (log mean). Note for both Equations 4.2 and 4.3 to follow, the maximum area is A_1 and the minimum area is A_2.

$$A_m = \frac{A_1 - A_2}{2.3 \log \times A_1/A_2} \tag{4.3}$$

Conductors in Series

When heat flows through several conductors in series, it is convenient to add the resistances of all conductors and then to use the resistance type of formula. For any conductor,

$$Q = \frac{\text{driving force}}{\text{resistance}} = \frac{\Delta t}{R}. \tag{4.4}$$

Also

$$R = \frac{L}{k \times A}. \tag{4.5}$$

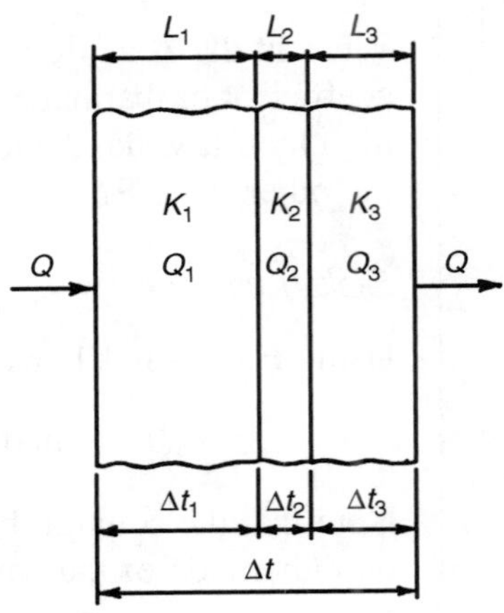

Figure 4.1

Using the units already given for these symbols, the resistances for three conductors in series are

$$R_1 = \frac{L_1}{k_1 \times A_1} \qquad R_2 = \frac{L_2}{k_2 \times A_2} \qquad R_3 = \frac{L_3}{k_3 \times A_3}.$$

Then the total resistance is the sum of all three $R = R_1 + R_2 + R_3$. And the heat transferred through the series group is

$$Q = \frac{\Delta t}{R_1 + R_2 + R_3}. \tag{4.6}$$

Note that the t in Equation 4.6 is the temperature difference across the entire series and not the individual components. The temperature drop between each individual conductor is $t = Q_1 L_1 / k_1 A_1$ and so on for each. But we know that for constant rate of transfer $Q = Q_1 = Q_2 = Q_3$ and that $t = t_1 + t_2 + t_3$. Figure 4.1 shows three conductors in series.

Example 4.2

A pipe with an outside diameter of $2\frac{1}{2}$ in. is insulated with a 2-in. layer of asbestos ($k = 0.12$), followed by a layer of cork $1\frac{1}{2}$ in. thick ($k = 0.03$). If the temperature of the outer surface of the pipe is 290×F and that of the outer surface of the cork is 90×F, calculate (a) heat loss per hour per 100 ft of insulated pipe, (b) temperature at the interface of asbestos and cork, and (c) percentage of total resistance due to asbestos and to cork.

Solution

(a) Because this problem involves an overall temperature difference and a series of two resistances, it may be solved with the use of Equation 4.6.

Inside surface of asbestos $= \pi \times 2.5/12 \times 100 = 65.4$ sq ft

Outside surface of asbestos $= 65.4 \times 6.5/2.5 = 171$ sq ft

This 171 sq ft is also the inside surface of the cork covering.

Outside surface of cork $= 65.4 \times 9.5/2.5 = 249$ sq ft

For the asbestos, the average area must be calculated as the log mean because the larger area is greater than twice the smaller.

$$A_m = \frac{171 - 65.4}{2.3\log(171/65.4)} = 111 \text{ sq ft}$$

For the cork the arithmetic average may be used.

$$A_a = \frac{171 + 249}{2} = 210 \text{ sq ft}$$

The distance through the asbestos is L_1 equal to 2/12, or 0.167 ft. Then

$$R_1 = 0.167/(0.12 \times 111) = 0.0125 \text{ for asbestos.}$$

For the cork,

$$R_2 = 0.125/(0.03 \times 210) = 0.0198.$$

Then, since $\Delta t = 290 - 90 = 200°\text{F}$,

$$Q = 200/(0.0125 + 0.0198) = 6190 \text{ Btu/hr.}$$

(b) $\Delta t_1 = QR_1 = 6{,}190 \times 0.0125 = 77.4°\text{F}$

and the temperature at interface is 290 – 77.4 = 212.6×F.

(c) $R_1/(R_1 + R_2) = 0.0125/(0.0125 + 0.0198) \times 100 = 38.8$ percent, or 38.8 percent of the resistance due to asbestos and 100 – 38.8, or 61.2 percent due to cork.

Example **4.3**

A furnace wall consists of 9-in. fire brick, $4\frac{1}{2}$-in. Sil-O-Cel brick, 4-in. red brick, and $\frac{1}{4}$-in. transit board. The k values are 0.82 at 1800×F for fire brick; 0.125 at 1800×F for Sil-O-Cel; 0.52 at 500×F for transit. Inside wall of furnace is at 1800×F and outside wall is 200×F. Calculate the heat lost per hour through a wall surface of 10 sq ft.

Solution

Let us take as the basis 1 sq ft of wall surface. The surface is the same throughout and is left out of the computations. Using Equations 4.4 and 4.5,

$$Q = \frac{1800 - 200}{[(1/0.82)(9/12)] + [(1/0.125)(4.5/12)] + [(1/0.52)(4/12)] + [(1/0.23)(0.25/12)]}$$

$$= 1600/4.646 = 344 \text{ Btu per hr per sq ft, or 3440 for 10 sq ft.}$$

Although not required, the temperature within the wall may be computed from the following:

$$\frac{\Delta t}{\Delta t_1} = \frac{R}{R_1}. \quad \textbf{(4.7)}$$

For example, between fire brick and Sil-O-Cel

$$\frac{1600}{4.646} = \frac{\Delta t_1}{0.915} \text{ from which } \Delta t_1 = 315.$$

Interface temperature is equal to 1800 – 315 = 1485×F.

Thermal resistances in series form a combination very often met in practice. In many instances the areas of the different resistances are nearly equal. Then Equation 4.6 will take on the form

$$Q = A\Delta t \frac{1}{L_1/k_1 + L_2/k_2 + L_3/k_3}. \quad \textbf{(4.8)}$$

For most engineering problems in heat transfer Equation 4.8 will give answers greatly in excess of the actual realized. What causes this discrepancy? How can this be evaluated?

FILM CONCEPT AND CONVECTION

In convection, not only are solids involved but also fluid films in series with these solids. Actually, it is of controlling importance in most cases of transfer between solids and fluids (liquids or gases).

Although there are a great many factors that influence the rate at which heat flows by convection to or from a solid in contact with a hotter or colder fluid, the most important of these is the observation that there is always a very thin film of practically stationary fluid on the surface of the solid. Through this film, heat must flow by conduction. As you can see from Table 4.1, the value of k for liquids is quite low, and for gases is exceedingly low. In the main body of the fluid, heated particles move freely and rapidly, but because the particles of the film are substantially stationary, *the film always offers the major resistance* between the solid and the main body of the fluid.

Consequently, efforts to improve that heat transfer by convection are always directed toward reducing the thickness of the highly insulating film, that is, by increasing the velocity of the moving fluid. This will take the flow out of the viscous range and into the turbulent range, where greatly improved rates of transfer can be realized.

However, once flow is turbulent, further increase in flow velocity will not reflect itself in much greater transfer rates.

Because the thickness of the fluid film on a solid is unknown and variable, and the conductivity of most liquids is likewise unknown, it is not practical to use the heat transfer equation in the form given in Equation 4.1. Instead, for calculating heat transferred by convection, the expression used is

$$Q = UA\Delta t \quad \textbf{(4.9)}$$

where U is known as the overall heat-transfer coefficient expressed as Btu/(hr)(×F)(sq ft) of surface. It has also been shown that

$$U = \frac{1}{\text{summation of resistances to heat flow}}. \quad \textbf{(4.10)}$$

Equation 4.9 may be shown to have been derived from the equation

$$Q = A\Delta t \frac{1}{1/h_1 + L_2/k_2 + 1/h_3}, \quad \textbf{(4.11)}$$

where we have a solid wall resistance (L_2/k_2) and film resistances of $1/h_1$ and $1/h_3$. The ratio within the brackets is the overall coefficient U. It is well to note that Equation 4.11 holds for thin-solid walls.

Example 4.4

Consider a tube of 16 gauge copper (thickness 0.065 in.), with a difference of 1×F between its inner and outer surfaces. On one side there is a stagnant film of water 0.01 in. thick and on the other a stagnant film of condensed steam. The tube is so thin-walled that a length sufficient to give 1 sq ft of surface on the inside will give practically 1 sq ft of surface on the outside. Calculate the heat transferred by use of Equations 4.1 and 4.11 and compare. Assume 212°F.

Solution

$$Q = (220 \times 1 \times 1)/(0.065/12) = 40{,}600 \text{ Btu per hr}$$

$$Q = \frac{A\Delta T}{\frac{L_1}{K_1} + \frac{L_2}{K_2} + \frac{L_3}{K_3}}$$

$$Q = 1 \times 1 \times \frac{1}{\frac{0.01/12}{0.417} + \frac{0.065/12}{220} + \frac{0.01/12}{0.417}}$$

$$Q = 1 \times 1 \times \frac{1}{0.00199 + 0.00002 + 0.00199} = 248.7 \text{ Btu per hr}$$

The first answer is extremely out of line compared with the second, more reasonable, rate, which might be found in practice under the conditions stated. It is also well to note that the resistance of the copper wall is so small as compared to the film resistances that it may as well be left out with little apparent effect.

Example 4.5

Assuming the same conditions as in Example 4.4, except that air is flowing on the inside of the tube (conductivity of air = 0.0174). What will be the rate of heat transfer?

Solution

$$Q = \frac{A\Delta T}{\frac{L_1}{K_1} + \frac{L_2}{K_2} + \frac{L_3}{K_3}}$$

$$Q = \frac{1}{0.00199 + 0.00002 + 0.0479} = 20 \text{ Btu per hr}$$

In practice, the thickness of the film cannot be measured directly, but can be predicated by an indirect method. As we shall see later, values of the film coefficient h can be computed and Equation 4.11 may be applied.

Thick-walled Tubes

Our entire discussion so far has been based on the assumption that the film resistances apply to surfaces that are essentially constant. This is true for flat surfaces, surfaces with only slight curvature, or for thin-walled tubes. Where the confining walls are substantially thicker, the following formula ensues:

$$Q = \frac{\Delta t}{1/h_1A_1 + L_2/k_2A_2 + 1/h_3A_3} \tag{4.12}$$

The mean area of the tube A_2 must be defined by first determining the log mean radius and then A_2 by its use in the standard cylinder area equation. The true mean radius for use in the heat-flow calculations is determined from

$$r_m = \frac{r_1 - r_2}{\ln(r_1/r_2)} \tag{4.13}$$

where

r_m = mean radius
r_1 = outside radius
r_2 = inside radius.

For rigorous accuracy, A_2 should be the area calculated using r_m for thick-walled tubes.

The log mean r_m is not too convenient in engineering calculations. If the arithmetic mean of r_1 and r_2 is used instead of r_m, the results will be in error by 10 percent or more if r_1/r_2 is 3.4 or greater. If r_1/r_2 is 1.5 or less, the results will be in error by less than 1 percent. Considering that thick-walled tubes are seldom used in heat-transfer apparatus and an accuracy of 5 percent is most likely as much as can be expected from any heat-transfer calculation, the arithmetic mean is under ordinary conditions satisfactory.

In actual heat-transfer calculations, for best results take A_1, A_2, and A_3 as the inside, mean, and outside areas per running foot of the tube, respectively. Use Equation 4.12 to obtain the heat transfer per foot of tube. Then obtain the total number of feet of tube from the total loading.

We previously mentioned that the film resistances were in most cases greater than that through the tube wall, so that the middle tem of the denominator is too small to be of practical significance. Then the results in many cases depend on the film resistances. If h_1 and h_3 in Equation 4.12 are of the same order of magnitude, then use the entire equation; but if one of the film coefficients h_1 or h_3 is very much smaller than the other, then the over-all coefficient U will approach the value of the smaller film coefficient. Thus, the smaller film coefficient will control the coinciding area, *i.e.*, if h_1 is small compared to h_3, then the term involving h_1 is the term that largely determines the value U, and, therefore, A_1 is the significant area. Simply, if the film coefficient on the inside of the tube is low, the inside surface should be used in the calculation.

Example 4.6

A certain heater has $1\frac{1}{4}$-in. standard iron pipe (Schedule 40) for its heating surface. It is operating with a liquid-film coefficient of 250 and a steam-film coefficient of 2000. What will be the relative effect of (a) increasing the liquid velocity so that the liquid film coefficient becomes 300, (b) venting the steam space more thoroughly so that the steam-film coefficient becomes 3000, or (c) substituting copper tubes 1.25 in. ID, 0.067 in. wall?

Solution

The existing overall coefficient of heat transfer is

$$U = \frac{1}{1/h + L/K + 1/h}$$

$$U = \frac{1}{1/250 + (0.140/12 \times 35) + 1/2000}$$

$$= \frac{1}{0.004 + 0.00033 + 0.0005} = 207.$$

(a) Increasing the liquid-film coefficient to 300 gives

$$U = \frac{1}{0.00333 + 0.00033 + 0.0005} = 240.$$

(b) Increasing the steam-film coefficient to 3000 gives

$$U = \frac{1}{0.004 + 0.00033 + 0.00033} = 214.$$

(c) Changing to copper tubes gives

$$U = \frac{1}{0.004 + (0.067)/(12 \times 220) + 0.0005}$$

$$= \frac{1}{0.004 + 0.000025 + 0.0005} = 221.$$

We see that increasing the steam-film coefficient by 50 percent or changing to copper tubes increases the overall coefficient only by 12 in 200, while increasing the liquid-film coefficient by 20 percent increases the overall coefficient 17 percent.

The nature and thickness of the metal wall is not always negligible. For instance, if in the preceding problem the liquid-film coefficient were 1500, changing from iron pipe to copper tubes would increase the overall coefficient from 685 to 835, a decided improvement. In general, whenever *both* the film coefficients are high, a change from a heavy metal wall of poor conductivity to a thin wall of high conductivity will be advantageous; but if *either* film coefficient is low, a change in metal would be of slight significance.

One exception may be made to this statement. A film of rust or oxide may result in a greatly thickened stagnant film. The difference between a polished steel tube and a rusty one may change the liquid-film coefficient as much as 1000 percent. Thus, in some cases the substitution of copper tubes for iron pipe may greatly improve the rate of heat transfer. This is so not because of the increased conductivity of metal (copper in this case), but because the copper stays bright and gives a thin stagnant film on the liquid side, while the iron rusts and the rust results in a thicker film, resulting in an increased resistance on the liquid side.

PIPE INSULATION

In the case of pipe insulation, the area through which heat is transferred is not constant. If the thickness of the insulation is small, compared with the diameter, the arithmetic average of the larger area and the smaller area may be used. The arithmetic mean area may be used for all cylindrical vessels and for pipe sizes down to about 2 in., if standard insulation is used and if an error of 4 percent is permitted. If greater accuracy is required, and the thickness of insulation is great compared with the diameter, the logarithmic mean area must be used.

$$\text{Logarithmic mean area} = \frac{A_2(\text{large}) - A_1(\text{small})}{2.3\log(A_2/A_1)} \qquad \mathbf{(4.14)}$$

When the ratio of A_2/A_1 is 2 or less, use the arithmetic mean area $(A_2 + A_1)/2$. But it is much simpler to use the radii as previously indicated in the section "Thick-walled Tubes" (see also Exhibit 1 in Example 4.7). Using Equation 4.13,

$$r_m = \frac{r_2 - r_1}{2.3\log(r_2/r_1)} = \frac{r_2 - r_1}{\ln(r_2/r_1)}.$$

The arithmetic mean is

$$r_a = \frac{r_2 + r_1}{2}. \qquad \mathbf{(4.15)}$$

Example 4.7

A tube 2.5 in. in diameter is lagged with a 2-in. layer of canvas-covered asbestos (conductivity 0.12) that is followed by a 1.5-in. layer of cork (conductivity 0.03). If the skin temperature of the pipe is 290×F and the canvas temperature is 90×F, calculate the loss in Btu/(hr)(ft) of lagged pipe. The various conductivities are expressed per foot units.

Solution

Refer to Exhibit 1. The layers are too thick to use the arithmetic mean so we must use the log mean. The loss in Btu per hr is determined as

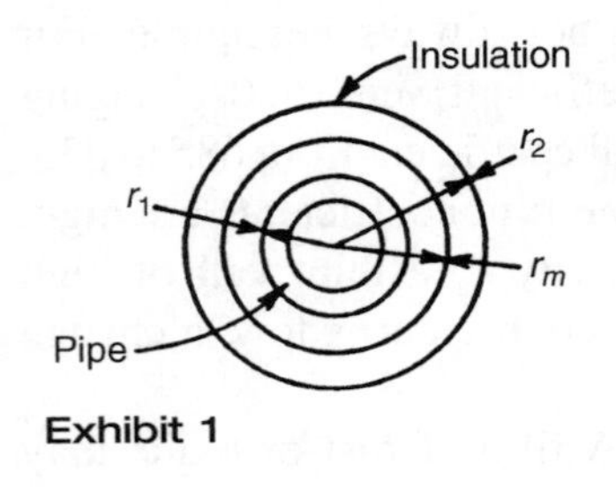

Exhibit 1

$$Q = \frac{\Delta t}{\Sigma R} = \frac{\Delta t}{R_1 + R_2} = \frac{290 - 90}{R_1 + R_2} = \text{loss in Btu per hr}$$

$$R_1 = \frac{L_1}{k_1 A_1} \text{ for asbestos layer}$$

$$R_2 = \frac{L_2}{k_2 A_2} \text{ for cork layer.}$$

The resistance of each layer is determined by using the mean radius in each case. For the asbestos,

$$r_m = \frac{3.25 - 1.25}{2.3\log(3.25/1.25)} = 2.09 \text{ in.}$$

For the cork,

$$r_m = \frac{4.75 - 3.25}{2.3\log(4.75/3.25)} = 3.95 \text{ in.}$$

$$R_1 = \frac{2/12}{0.12 \times 2\pi(2.09/12) \times 1} = 1.270$$

$$R_2 = \frac{1.5/12}{0.03 \times 2\pi \times (3.95/12) \times 1} = 2.015$$

$$Q = \frac{290 - 90}{1.27 + 2.015} = 60.9 \text{ Btu/(hr)(ft) of lagged pipe}$$

Note that the air film outside the pipe was not considered. Now if the air film resistance is equal to 1/6.0, then the heat loss would be 58, a negligible difference.

Example 4.8

Heat is being transferred from a gas through the walls of a standard 2-in. steel pipe and into water flowing on the inside of the pipe. Calculate the clean over-all coefficient of heat transfer. Let $h_0 = 6$, $h_i = 500$,

Solution

$$U = \frac{1}{1/h_1 + L_2/K_2 + 1/h_3}$$

$$U = \frac{1}{\frac{1}{6} + (0.154)/(12 \times 25) + \frac{1}{500}}$$

$$U = \frac{1}{0.1667 + 0.00051 + 0.002} = 5.91$$

Note that here the thickness is in inches and conductivity is in Btu/(sq ft)(hr)(°F)(in.) thickness. The resistance of the metal wall and the water film could be neglected with an error of only 1.52 per cent. If the tube were copper ($k = 220$), the overall coefficient would be 5.93.

FOULING FACTORS

So far we have calculated over-all coefficients that are clean, *i.e.*, no consideration has been taken to include the effects of operation and the build-up of scale or corrosion products on the tube walls. Now we shall take into account these fouling effects.

So that heat transfer surface in heat exchangers, engine coolers, and the like will have sufficient excess tube surface to maintain satisfactory performance during normal operation and with reasonable service time between cleanings, fouling factors must be applied as shown in Table 4.2. Under special conditions of fouling higher factors may be used.

Fouling factors represent the resistance r to heat transfer caused by the layer of foreign substance deposited on the surfaces of the heat transfer surfaces. Fouling resistance varies directly with the thickness of the film and inversely with the conductivity of the film k. The general equation for the over-all coefficient taking into account all factors is

$$U = \frac{1}{1/h_0 + r_0 + r_w + r_i(A_0/A_i) + (1/h_i A_i/A_0)} \tag{4.16}$$

where

U = overall coefficient of heat transfer, Btu/(hr)(×F)(sq ft) outside surface
h_0 = film coefficient of fluid medium on outside of tubing
h_i = film coefficient of fluid medium on inside of tubing
r_0 = fouling resistance on outside of tubing
r_i = fouling resistance on inside of tubing
r_w = resistance of the tube wall (L/K)
A_0/A_i = ratio of outside tube surface to inside the surface

Table 4.2 lists minimum fouling factors for some example liquids and gases to apply in Equation 4.16.

Table 4.2* Example fouling factors

Fuel oil #6	0.005
Transformer oil	0.001
Vegetable oils	0.003
Coal flue gas	0.010
Engine exhaust gas	0.010
Steam (nonoil bearing)	0.0005
CO_2 vapor	0.001
Steam, exhaust (oil-bearing)	0.0015–0.002
Refrigerating vapors (oil-bearing compressed)	0.002
Refrigerant air	0.001
Liquids	0.001
Sodium chloride solutions	0.003

* Table taken from Standards of Tubular Exchanger Manufacturers Association (TEMA).

MEAN TEMPERATURE DIFFERENCE

In our discussion so far the assumption has been that the hot fluid remains at a constant temperature and the cold fluid does likewise. In actual heat transfer equipment either one or the other, or both, are changing in temperature throughout the apparatus. In Figure 4.2 the most common arrangements of heat transfer are shown. The arrow indicates the direction of flow of the fluids with respect to each other. Almost all commercial and industrial equipment is designed for *countercurrent flow*, or at least for conditions approaching this (Fig. 4.2a). Parallel flow (Fig. 4.2b) is rarely used, while Figure 4.2c and d are simulated in all refrigeration coolers and steam condensers and wherever a fluid is condensing on one side and the other fluid is being either cooled or heated on the other. Parallel flow gives poor heat-transfer results in that it would require a greater amount of surface to accomplish the same heat transfer load as compared to countercurrent flow.

The temperature difference for a fluid system is an average quantity. If the equipment is perfectly insulated and the overall coefficient is a constant quantity, specific heat and weight rate are constant unless the process is that of evaporation or condensation, parallel path flow is taking place. Then the logarithm mean temperature is used in the transfer equation

$$Q = UAt_m \tag{4.17}$$

and the logarithm mean temperature is determined from the following equation:

$$\text{LMTD} = (\text{Large } T - \text{Small } T)/\ln(\text{large } T - \text{small } T). \tag{4.18}$$

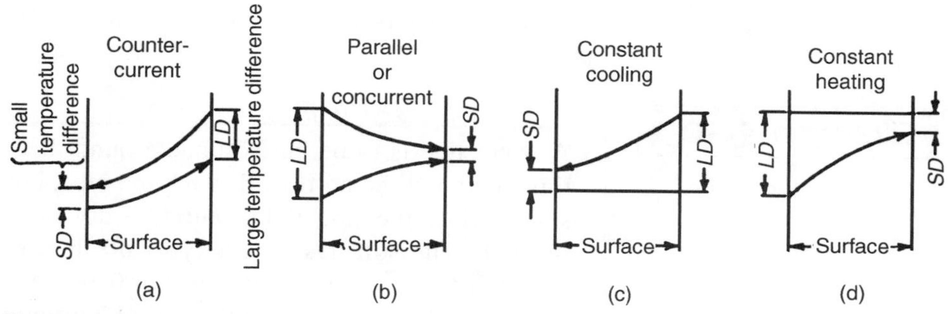

Figure 4.2

The results of using Figure 4.2 are correct when heat-transfer equipment is of the true countercurrent type. However, when the heat exchanger is of the multipass type and the flow is not strictly countercurrent, the proper t_m is less than that obtained from Figure 4.2. Factors that can be used to correct for this may be obtained directly from charts appearing in the Standards of Tubular Exchanger Manufacturers Association (TEMA).

Example **4.9**

A very long steel cylinder 3 ft in diameter, initially at 1000°F, is suddenly immersed in a fluid, and the surface temperature is instantaneously changed to 500×F. Estimate the core temperature at the end of 3 hr, using the values of physical properties tabulated below:

	Fluid	Steel
Thermal conductivity, Btuh-ft-×F	0.1	23.0
Specific heat, Btu/(lb)(×F)	0.8	0.13
Density, lb per cu ft	60	485

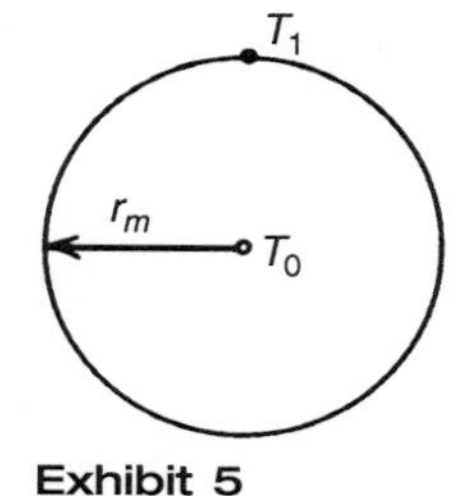

Exhibit 5

Solution

The statement "very long steel cylinder" suggests infinite extent in the axial direction. We will assume an infinite cylinder. Surface temperature is instantaneously changed from 1000×F to 500×F. Properties of fluid are therefore not needed because film resistance may be neglected. The problem will be solved using the cylinders series. Now refer to Exhibit 5. Thermal diffusivity

$$\alpha = \frac{k}{C\rho} = \frac{23.0}{(0.13)(485.0)} = 0.365 \text{ sq ft/hr}$$

Fourier modulus

$$\theta = \frac{\alpha t}{(r_m)^2} = \frac{0.365(3)}{(3/2)^2} = 0.486$$

$$C\theta = \frac{T_0 - T_1}{T_i - T_1} = \frac{T_0 - 500}{1000 - 500} = 0.0632$$

Core temperature $T_0 = 0.0632(500) + 500 = 31.6 + 500 = 531.6°\text{F}$

Example 4.10

A water line is to be buried underground. It is possible that for long periods of time there will be no flow through the pipe, but the pipe will not be drained. The soil in which the pipe will be buried is dry and has an assumed initial temperature of 40°F. The soil has a density of 40 lb per cu ft, a thermal conductivity of 0.2 Btuh-ft-×F, and a specific heat of 0.44 Btu/(lb)(×F). It is desired to design for a 36-hr period at the beginning of which the soil surface temperature suddenly drops to 0×F. Determine the minimum earth cover needed above the water pipe to prevent the possibility of freezing during the 36-hr cold spell. It is assumed that no flow occurs through the pipe during this period.

Solution

This problem considers a solid of infinite extent in the direction of the heat flow. We must find the thickness of earth cover that will keep the pipe at a temperature no lower than 32×F. Refer to Exhibit 6.

$$\text{erf}\left(\frac{x}{2\sqrt{\alpha t}}\right) = \frac{T - T_s}{T_0 - T_s} = \frac{32 - 0}{40 - 0} = 0.8$$

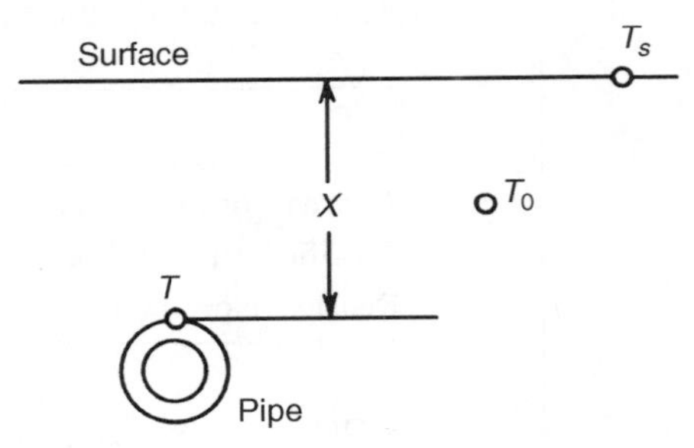

Exhibit 6

The error factor (erf) may be obtained from the reference *Heat Transfer*, by Incropera and Dewitt (Wiley, 2002). Then thermal diffusivity from given properties is given by

$$\alpha = \frac{k}{C\rho} = \frac{0.2}{0.44 \times 40} = 0.01136 \text{ sq ft per hr.}$$

If
$$\text{erf}\left(\frac{x}{2\sqrt{\alpha t}}\right) = 0.8$$

then from Incropera and Dewitt,

$$\left(\frac{x}{2\sqrt{\alpha t}}\right) = 0.906.$$

Earth cover for time t = 36 hr:

$$\begin{aligned} x &= 0.906\left(2\sqrt{\alpha t}\right) \\ &= 0.906\left[2\sqrt{0.01136 \times 36}\right] \\ &= 0.906\left[2\sqrt{0.409}\right] \\ &= 0.906\,[2(0.64)] \\ &= 1.16 \text{ ft} \end{aligned}$$

RADIATION

Radiation is the transfer of energy without the necessity of an intermediate medium. In fact, the heat transfer is more efficient if no medium is present.

The heat transfer from one body to another may be represented by the following equation:

$$Q = \varepsilon \sigma F_{1-2}\left(T_1^4 - T_2^4\right)$$

where
- Q = energy transferred in Btu/hr
- ε = emissivity of radiating body
- σ = Stefan Boltzmann constant
- F_{1-2} = shape or view factor—fraction of radiation leaving body 1 and striking body 2
- A = area ft^2
- T = absolute temperature ×R.

Example 4.11

A 6" diameter hole is bored into a large steel slab at 1040×F. The effective emissivity out of the hole is $\varepsilon_{\text{hole}} = 0.6$. If the slab is in a large room at 60×F, what is the rate of heat transfer out of the hole?

Solution

Exhibit 7 shows the schematic diagram. The next step is to draw the electric analog, as shown in Exhibit 8.

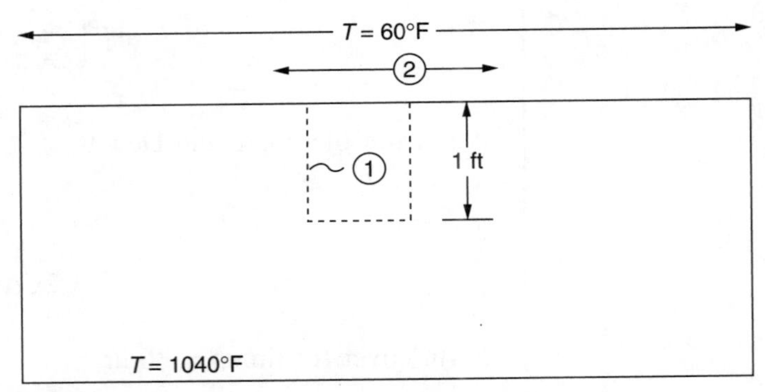

Exhibit 7

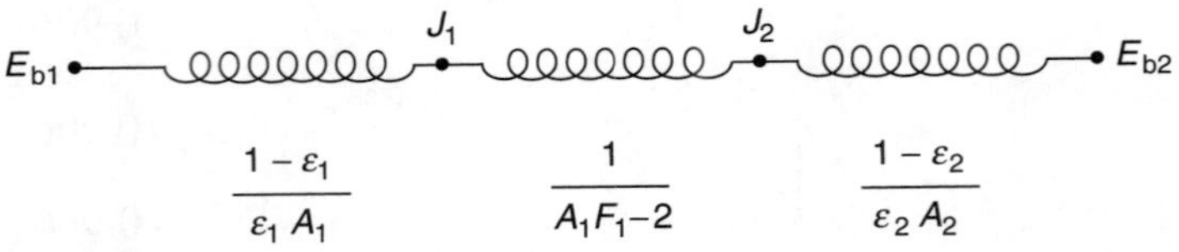

E_{b1} — $\dfrac{1-\varepsilon_1}{\varepsilon_1 A_1}$ — J_1 — $\dfrac{1}{A_1 F_{1-2}}$ — J_2 — $\dfrac{1-\varepsilon_2}{\varepsilon_2 A_2}$ — E_{b2}

Exhibit 8

Because A_2 is the large room, A_2 ×.

Therefore,

$$\frac{1-\varepsilon_2}{\varepsilon_2 A_2} \approx \infty$$

The electric analog reduces to that shown in Exhibit 9.

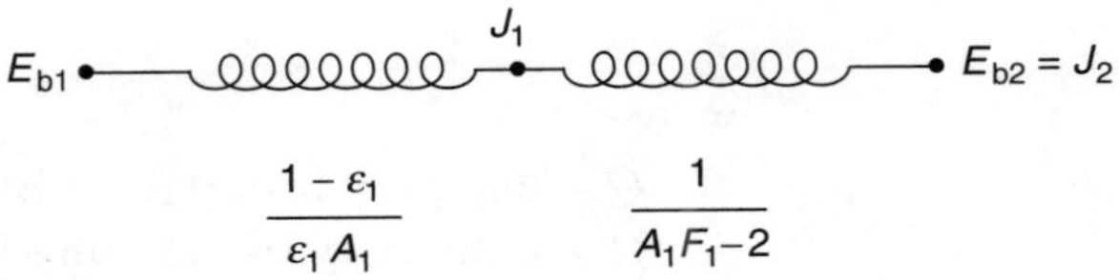

E_{b1} — $\dfrac{1-\varepsilon_1}{\varepsilon_1 A_1}$ — J_1 — $\dfrac{1}{A_1 F_{1-2}}$ — $E_{b2} = J_2$

Exhibit 9

Determining the values,

$$Q = \frac{Eb_1 - Eb_2}{\frac{1-E_1}{E_1 A_1} + \frac{1}{A_1 F_1 - 2}}$$

$$Eb_1 = (0.1713 \times 10^{-8})(1500)^4 = 8672$$

$$Eb_2 = (0.1713 \times 10^{-8})(520)^4 = 125$$

$$\frac{1-E_1}{E_1 A_1} = \frac{1-0.6}{0.6\left(\pi D^2 + \frac{\pi D^2}{4}\right)} = 0.377$$

$$A_1 F_1 - 2 = A_2 F_2 - 1 - 1$$

$$F_2' - 1 = 1$$

$$A_2' = \frac{\pi D^2}{4}$$

$$A_1 F_{1-2} = \frac{\pi}{4}\left(\frac{1}{2}\right)^2 = 0.196.$$

Therefore:

$$Q = \frac{8672 - 125}{0.377 + \frac{1}{0.196}}$$

$$Q = 1559.9 \text{ Btu/s}$$

CHAPTER 5

Machine Design

OUTLINE

Machine design combines all the disciplines in the study of mechanical engineering. A typical design problem first requires the development of a mechanism to perform the desired function. Next it must be determined that the device or component is able to achieve the intended function without failure and without injury to the user or to the public. Attention must be paid to the materials used, cost of manufacturing and, yes, even the ability to recycle the machine or device at the conclusion of its intended life.

Example 5.1

A 2-in. medium steel shaft is subjected to a turning moment of 40,000 in.-lb. Ultimate stress is 50,000. The desired factor of safety is at least 2.5. Is the shaft safe?

Solution

$$T = S_s Z_p = 40{,}000. \text{ Solve for } S_s.$$

$$Z_p = \pi \frac{d_0^3}{16} = \pi \frac{2^3}{16} = 0.5\pi \text{ in.}^3$$

$$S_s = \frac{T}{Z_p} = \frac{40{,}000}{0.5\pi} = 25{,}478 \text{ psi}$$

$$F = \frac{U_s}{S_s} = \frac{50{,}000}{25{,}478} = 1.96. \text{ Too low. Unsafe}$$

Example 5.2

A hydraulic turbine in a water-power plant is rated at 12,000 hp. The steel vertical shaft connecting the turbine and generator is 24 in. in diameter and rotates at 60 rpm. Calculate the maximum shearing stress developed in the shaft at full load.

Solution

Because the shaft is vertical, there will be no stresses caused by bending, and the maximum shearing stress will really be the maximum shaft stress. Now, because horsepower is

$$\text{hp} = \frac{2\pi NT}{33{,}000} = \frac{2\pi N\,\text{rpm}T_{\text{lb ft}}}{33{,}000\frac{\text{ft lb}}{\text{hp min}}}$$

and torque is found by rearrangement to be

$$T = \frac{12{,}000 \times 33{,}000}{6.28 \times 60} = 1.05 \times 10^6 \text{ lb-ft in shaft,}$$

the polar moment of inertia I_p is

$$\frac{\pi d^4}{64} = 3.14 \times \frac{24^4}{64} = 16{,}278 \text{ in.}^4$$

For strength, $T = S_s I_p / r$. By rearrangement and substitution in this equation

$$S_s = \frac{Tr}{I} = (1.05 \times 10^6)\text{lb ft}\left(\frac{24}{2}\right)\text{in}\left(\frac{1}{16278}\right)\frac{1}{\text{in}^4}(12 \text{ in/ft}) = 9289 \text{ psi}$$

Example 5.3

An automobile weighing 3000 lb is slowed uniformly from 40 to 10 mph in a distance of 100 ft by a brake on the drive-shaft. Rear wheels are 28 in. in diameter. Drive-shaft diameter is 1 in. Ratio differential is 40:11. Calculate extreme fiber stress in drive shaft, neglecting tire, bearing, and gear losses.

Solution

The following relation for deceleration holds:

$$-a = \frac{v^2 - v_0^2}{2s} = \frac{14.7^2 - 58.7^2}{2 \times 100} = 16.1 \text{ ft per sec}^2$$

where $(-a)$ is the deceleration, v is final velocity, and v_0 is initial velocity expressed in fps. Now, the braking force required is

$$F = \frac{W}{g}a = \frac{3000}{32.2}(-16.1) = -1500 \text{ lb.}$$

Total braking torque on driving wheels is now determined as

$$T = 1500 \times \left(\frac{28}{2}\right) = 21{,}000 \text{ in.-lb.}$$

Torque on drive shaft is

$$T = 21{,}000 \times \left(\frac{11}{40}\right) = 5775 \text{ in.-lb.}$$

Shear stress in shaft is

$$S_s = \frac{16}{\pi d^3} T = \frac{16}{\pi 1^3}(5775) = 29{,}427 \text{ psi.}$$

Example 5.4

A circular bar of solid cast iron 60 in. long carries a solid circular head 60 in. in diameter (Exhibit 1). The bar is subjected to a torsional moment of 60,000 in.-lb, which is supplied at one end. It is desired to keep the torsional deflection of the circular head below $\frac{1}{32}$ in. when the bar is transmitting power over its entire length in order to prevent the chattering of the piece. What would be the diameter of the bar if the working stress is taken as 3000 psi and the transverse modulus of elasticity is 6 million psi?

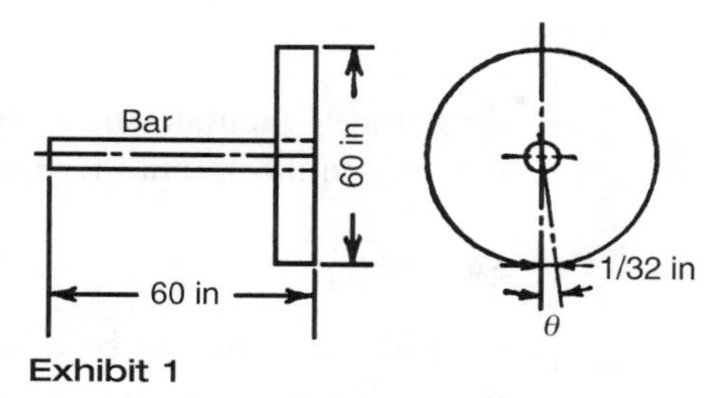

Exhibit 1

Solution

$$\text{Shearing stress } \tau = \frac{Tr}{J} = \frac{T\frac{d_o}{2}}{\frac{\pi d_o^{\,4}}{32}}$$

$$\tau = \frac{T16}{\pi d_o^{\,3}} \quad \text{so, } T = \frac{\tau \pi d_o^{\,3}}{16}$$

Solving for d_0,

$$d_o^{\,3} = \frac{(60000 \text{ in-lb})(16)}{(3000 \text{ psi})(\pi)} \qquad d_o = 4.66 \text{ in}$$

For torsional stiffness, $\theta = (\frac{1}{32})/r$. This is equal to $\frac{1}{960}$ radian. Arc length along head is θr, where r is radius of head in inches, therefore, $\frac{1}{32} = \theta r = (\theta)(30)$. Note that S equals 60 in.

$$d_0^4 = \frac{32 \times T \times S}{\pi \times E_t \times \theta}$$

Now, by direct substitution in this formula $d_0^4 = 5870$ and $d_0 = 8.8$ in. Because 8.8 in. is greater than 4.6 in., the shaft must be designed for torsional stiffness.

Example 5.5

In Exhibit 2, $1\frac{13}{16}$-in.-diameter steel shaft is supported on bearings 6 ft apart. A 24-in.-diameter pulley weighing 50 lb is attached to the center of the span. The pulley runs 400 rpm and delivers 15 hp to the shaft. The shaft weight is 8.77 lb per ft. A belt pulls on the pulley with a force of 250 lb in a vertically downward

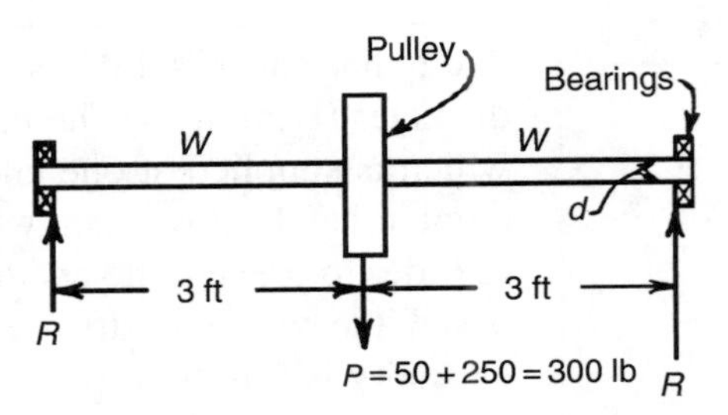

Exhibit 2

direction. Calculate the maximum stress in the shaft due to the combination of bending and torsional (twisting) stresses.

Solution

Consider the shaft to be a beam with fixed ends. The maximum bending moment due to loads occurring at bearings is

$$\frac{wl^2}{8} + \frac{Pl}{4} = \frac{8.77 \times 6^2 \times 12}{8} + \frac{300 \times 6 \times 12}{4}$$
$$= 474 + 5400 = 5874 \text{ in.-lb.}$$

The torque delivered by power is

$$T = \frac{33{,}000 \times 15 \times 12}{2\pi 400} = 2360 \text{ in.-lb.}$$

Maximum shearing stress due to combined loads is

$$S_s = \frac{16}{\pi\left(1\frac{13}{16}\right)^3} \times \sqrt{5874^2 + 2360^2} = 5415 \text{ psi.}$$

Maximum normal stress due to combined loads is

$$S_n = \frac{16}{\pi\left(1\frac{13}{16}\right)^3} \times \left[5874 + \sqrt{5874^2 + 2360^2}\right] = 10{,}439 \text{ psi.}$$

Example 5.6

A solid circular shaft is used to transmit 200 hp at 1000 rpm. (a) What diameter shaft is required if the allowable maximum shearing stress is 20,000 psi? (b) If a hollow shaft is used having an inside diameter equal to the diameter of the solid shaft determined in part (a), what must be the outside diameter of this shaft if the angular twist of the two shafts is to be equal?

Solution

The torque required to transmit 200 hp at 1000 rpm is

$$T = (33{,}000 \times 12 \times 200)/2\pi 1000) = 12{,}600 \text{ in.-lb.}$$

(a) The torsional shearing stress is first determined by use of the formula

$$S_s = \frac{T \times d}{2 \times I_p}$$

where I_p is the polar moment of inertia. Now, because $I_p = \pi d^4/32$,

$$S_s = \frac{16T}{\pi d^3}$$

$$d^3 = (16 \times 12{,}600)/(\pi \times 20{,}000) = 3.2 \quad \text{and} \quad d = 1.475 \text{ in.}$$

(b) Now, if a hollow shaft is used with an inside diameter of 1.475 in. and an outside diameter of D is to be determined such that the angular twist θ_a of the solid shaft and that of the hollow shaft θ_b are equal,

$$\theta_a = T_a \frac{L_a}{G_s \times I_{pa}}$$

$$\theta_b = T_b \frac{L_b}{G_s \times I_{pb}}$$

where G_s is the modules of elasticity in shear for the material. Now, because torque and length are the same for both shafts and both are made of the same material, by equating angular twists to each other, $I_{pa} = I_{pb}$, or

$$\frac{\pi d^4}{32} = \frac{\pi}{32}(D^4 - d^4).$$

Therefore

$$D^4 = 2d^4$$

$$D^2 = d^2\sqrt{2}$$

We find that D (outside diameter of the hollow shaft) is 1.755 in. and d (inside diameter of the hollow shaft) is equal to 1.475 in.

Example 5.7

A cylinder head of a steam engine is held by 14 bolts. The diameter of the cylinder is 14 in. and the steam pressure is 125 psi. What size of bolts is required if tensile stress is 3000 psi?

Solution

The force due to the steam pressure is pressure × area.

$$F = 125 \times \frac{\pi \times 14^2}{4} = 19{,}250 \text{ lb}$$

Because the pressure is tending to push the cylinder head away from the block, the load per bolt is simply 19,250/14 = 1375 lb. Then the root diameter of each bolt is found by

$$F = A \times S_t = \frac{\pi d^2}{4} S_t$$

$$1375 = \frac{\pi d^2}{4} \times 3000.$$

Rearranging and solving for d,

$$d = \sqrt{\frac{1375 \times 4}{3{,}000 \times \pi}} = 0.76 \text{ in.}$$

The standard root diameter that is just above the 0.76 in. is 0.8376 in. Therefore, the outside diameter is 1 in.

Example 5.8

The load on a helical spring is 1600 lb and the corresponding deflection is to be 4 in. Rigidity modulus is 11 million psi and the maximum intensity of safe torsional stress is 60,000 psi. Design the spring for the total number of turns if the wire is circular in cross section with a diameter of $\frac{5}{8}$ in. and a center-line radius of $1\frac{1}{2}$ in.

Solution

In order to find the total deflection, given the number of active coils, the following may be used:

$$y = \frac{N \times 64 \times P \times r^3}{Gd^4}$$

where

y = total deflection, in.
r = radius as axis to center line of wires, in.
N = number of active coils
G = rigidity modulus of steel in shear.

$$N = \frac{y \times G \times d^4}{64 \times P \times r^3} = \frac{4 \times 11 \times 10^6 \times 0.625^4}{64 \times 1600 \times 1.5^3} = 19.4 \text{ active coils}$$

Now let us check for safe limit of torsional stress using

$$S = \frac{16 \times 1600 \times 1.5}{\pi \times 0.625^3} = 50{,}066 \text{ psi.}$$

We see that 50,066 is less than the set limit of 60,000. Therefore, the spring is acceptable. The total number of turns for two inactive coils (one for each end) is $N + 2 = 19.4 + 2 = 21.4$, or 22 coils.

Example 5.9

A coiled spring with $1\frac{3}{4}$-in outside diameter (OD) is required to work under a load of 140 lb. Wire diameter used is 0.192 in., spring is to have seven active coils, and the ends are to be closed and ground. Determine unit deflection, total number of coils, and length of spring when under load. Assume G equal to 12 million and mean radius to be 0.779 in.

Solution

Safe shearing stress is found from the formula below to be

$$S = \frac{8 \times 140 \times 0.779 \times 2}{\pi \times 0.192^3} = 78{,}475 \text{ psi.}$$

Deflection may be found in another way than previously shown by

$$y = \frac{4\pi N r^2 S}{Gd}$$

$$y = \frac{4\pi 7 \times 0.779^2 \times 78{,}475}{12{,}000{,}000 \times 0.192} = 1.817, \text{ or about } 1\tfrac{13}{16} \text{ in.}$$

For coiled springs with circular wire, the formula for the relation between fiber stress and load is

$$S = \frac{8PD}{\pi d^3}$$

where

S = fiber stress in shear, psi
P = axial load, lb
D = mean diameter of spring (OD minus wire diameter)
d = diameter of wire, in.

For ends to be closed and ground smooth, $\frac{1}{2}$ coils should be taken as inactive. For compression springs the number of active coils depends on the style of ends as follows:

Open ends, not ground—all coils active

Open ends, ground—$\frac{1}{2}$ coil inactive

Closed ends, not ground—1 coil inactive

Closed ends, ground—$1\frac{1}{2}$ coils inactive

Squared ends, ground—2 coils inactive

$$\text{Total number of coils} = 7 + 1\frac{1}{2} = 8\frac{1}{2}, \text{ say } 9$$

The solid height of the spring, when it is entirely compressed, is

$$9 \text{ coils} \times 0.192 = 1.728 \text{ in., or about } 1\frac{3}{4} \text{ in.}$$

We determined previously that the total deflection under load of 140 lb was $1\frac{13}{16}$ in. And assuming the total free space between coils to be 1 in., the free length of the spring would be

$$1\frac{3}{4} + 1\frac{13}{16} + 1 = 4\frac{9}{16} \text{ in.}$$

Length of spring, when under load of 140 lb, is

$$4\frac{9}{16} - 1\frac{13}{16} = 2\frac{3}{4} \text{ in.}$$

Example 5.10

A cylindrical helical spring of circular cross-section wire, like ASTM 229, is to be designed to safely carry an axial compressive load of 1200 lb at a maximum stress of 110,000 psi. The spring is to have a deflection scale of about 150 lb per in. Proportions are to be as follows:

$$\frac{\text{Mean diameter of coil}}{\text{Diameter of wire}} = 6 \text{ to } 8$$

$$\frac{\text{Length closed}}{\text{Mean diameter of coil}} = 1.7 \text{ to } 2.3$$

Determine (a) mean diameter of coil, (b) diameter of wire, (c) length of coil when closed, and (d) length of coil before application of load.

Solution

Assume G to be 11,500,000. Now find trial wire size. (a) In this trial design we can assume the ratio of D/d to be 6:8, as given in the problem. Then $d = D/7$, and we can substitute the value of d in a rearrangement of the equation $S = \frac{8PD}{\pi d^3}$

$$1200 = \frac{\pi \times (D/7)^3 \times 110{,}000}{8 \times D}$$

When we solve for D, it is found to be 3.09 in. (b) Then wire diameter d is 3.09/7, or 0.441in. The nearest commercial wire size to this is 0.437 in. Then

$D/d = 3.09/0.437 = 7.07$. This lies within the limits set down in the problem. The spring scale is given by

$$\frac{P}{y} = \frac{Gd^4}{N \times 64 \times r^3}.$$

In this problem P/y is equal to 150. Now let us substitute values of d, D, G, and P/y in the preceding equation.

$$150 = \frac{11.5 \times 10^6 \times 0.437^4}{N \times 64 \times 1.545^3}$$

Solving for N, the number of active coils, we find it to be equal to 11.9. Total turns, assuming closed ends and ground, are equal to 11.9 + 1.5 = 13.4. We must now check the second limiting ratio giving in the problem: 1.7 to 2.3. Use the following relation:

$$\frac{\text{Total turns} \times d}{D} = \frac{13.4 \times 0.437}{3.09} = 1.9 \text{ (within limits)}$$

(c) Length of the closed coil is $13.4 \times 0.437 = 5.86$ plus.

In order to obtain length of coil before application of load, we add closed length to total deflection. Now, total deflection is given by

$$y = \frac{11.9 \times 64 \times 1200 \times 1.545^3}{11.5 \times 10^6 \times 0.437^4} = 8.04 \text{ in.}$$

(d) Then, length of coil before application of load is 5.86 + 8.04 = 13.9, say 14.

Example **5.11**

Determine the width and thickness of the leaves of a six-leaf steel cantilever spring 13 in. long to carry a load of 375 lb with a deflection of $1\frac{1}{4}$ in. The maximum stress in this spring is limited to 50,000 psi.

Solution

We can set up an equation relating deflection to other factors involved.

$$\delta = \frac{S \times l^2}{Et} = 1.25 = \frac{50{,}000 \times 13^2}{30 \times 10^6 \times t}$$

Rearranging and solving for thickness, t, this is found to be equal to 0.225 in. Now to solve for b, the width of each leaf. The following likewise holds:

$$W = \frac{S \times N \times bt^2}{6l} = 375 = \frac{50{,}000 \times 6 \times b \times 0.225^2}{6 \times 13}.$$

Rearranging and solving for b, this is found to be equal to 1.93 in.

Example 5.12

A solid machine shaft with a safe shearing stress of 7000 psi transmits a torque of 10,500 in.-lb. (a) Find the shaft diameter. (b) A square key is used whose width is equal to one-fourth the shaft diameter and whose length is equal to $1\frac{1}{2}$ times the shaft diameter. Find key dimensions and check the key for its induced shearing and compressive stresses. (c) Obtain the factors of safety of the key in shear and in crushing, allowing an ultimate shearing stress of 50,000 psi and a stress for compression of 60,000 psi.

Solution

(a) $d^3 = \dfrac{T \times 16}{\pi \times S} = \dfrac{10{,}500 \times 16}{\pi \times 7000}$

Solving for d, the shaft diameter, this is found to be 1.96, say, 2 in.

(b) Width of key is $\frac{1}{4}$ the shaft diameter, or $\frac{2}{4}$, *i.e.*, 1/2 in. Length of key is L equal to 1.5 × shaft diameter, or 1.5 × 2 = 3 in. Now check key for its induced shearing and compressive stresses. The tangential force set up at the outside of the shaft is P_t lb. Thus

$$P_t = \frac{\text{torque, in.-lb}}{\text{radius of shaft, in.}}$$

$$= \frac{10{,}500}{1} = 10{,}500 \text{ lb.}$$

The shearing stress of the key is given by the following relation:

$$S_s = \frac{P_t}{bL} = \frac{10{,}500}{0.5 \times 3} = 7000 \text{ psi.}$$

The crushing stress is found by

$$S_c = \frac{2P_t}{bL} = 2 \times \frac{10{,}500}{(0.5 \times 3)} = 14{,}000 \text{ psi.}$$

(c) Factor of safety $F_s = 50{,}000/7000 = 7.1$ for shear

Factor of safety $F_c = 60{,}000/14{,}000 = 4.3$ for crushing

Example 5.13

What is the minimum length of a key $\frac{7}{8}$ in. wide you would use with a gear driving shaft $3\frac{7}{16}$ in. in diameter, designed to operate at a torsional working stress of 11,350 psi?

Solution

The relation between shaft diameter d, torque T, and working stress S_s is given by

$$d^3 = 5.1\frac{T}{S_s}.$$

After rearrangement and substitution in the preceding formula,

$$T = \frac{d^3 S_s}{5.1} = \frac{3.4375^3 \times 11{,}350}{5.1} = 90{,}397 \text{ in.-lb.}$$

The tangential force on the shaft is $P = T/r$, where r is radius of shaft.

$$P = 90{,}397/1.71875 = 52{,}595$$

Length of key is then determined by use of $L = P/(W \times S_p)$.

$$L = 52{,}595/(0.875 \times 11{,}350) = 5.3 \text{ in., say 5 in.}$$

Check.

$$L = 1.5D = 1.5 \times 3.4375 = 5.15 \text{ in.}$$

Example **5.14**

A double-ply leather belt transmits 10 hp from a motor with a pulley 8 in. in diameter, running at 1700 rpm, to a 24-in.-diameter pulley. The difference in tension may be taken as 20 lb per inch of belt width. Belt thickness is 0.2 in. Calculate the width of the belt.

The power-transmitting capacity of a belt should be less than the maximum possible capacity to ensure a reasonable length of belt life and to avoid unnecessary expenses and repairs. The turning or tangential force on the rim of a pulley driven by a flat belt is equal to $T_1 - T_2$, where T_1 and T_2 are, respectively, the tension in the driving or tight side and the driven or slack side of the belt. Then, the horsepower transmitted is given by

$$\text{hp} = \frac{(T_1 - T_2)v}{33{,}000}$$

where v is belt speed in feet per minute.

Solution

Refer to the preceding equation. Then let F equal effective pull in pounds, w equal belt width in inches, and N equal to motor speed in rpm. Then, $F = T_1 - T_2 = 20w$. Substituting in the equation above, we get

$$10 = 20w\pi\left(\frac{8}{12}\right)1700/33{,}000$$

Rearranging and solving for w,

$$w = \frac{10 \times 33{,}000}{20\pi\left(\frac{8}{12}\right)1700} = 4.63 \text{ in.}$$

Example 5.15

Two shafts x ft apart are connected by a belt as shown in Exhibit 3. The angle of wrap on the small pulley is 170°. Consider the centrifugal force to be negligible. The coefficient of friction between the belt and pulley is $\mu = 0.3$. The torque transmitted is 250 ft-lb.

$r_1 = 12"$

$r_2 = 18"$

Determine P_1 and P_2.

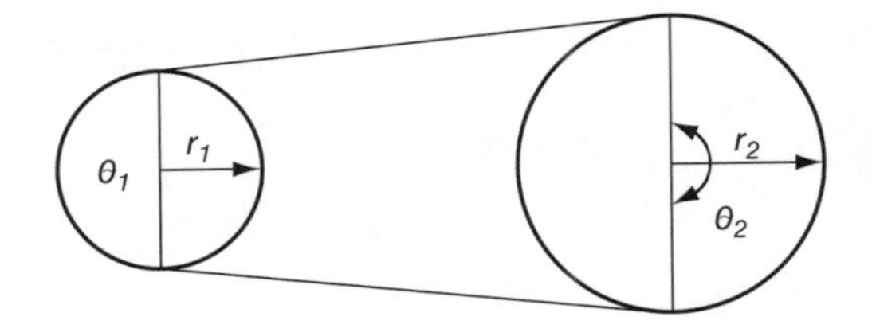

Exhibit 3

Solution

$$T = (P_1 - P_2)r \text{ ft}$$

$$250 \text{ ft-lb} = (P_1 - P_2)r \text{ ft}$$

$$(P_1 - P_2) = 250 \text{ lb}$$

$$\frac{P_1}{P_2} = e^{f\phi} = e^{(0.3)170°} = e^{(0.3)(2.967)}$$

$$\frac{P_1}{P_2} = e^{0.890} = 2.435 \text{ and } P_1 = 250 - P_2$$

$$\frac{250 - P_2}{P_2} = 2.435$$

$$\text{so } P_2 = 72.8 \text{ lb}$$

$$P_1 = 177.2 \text{ lb}$$

Example 5.16

A condensing steam engine with a bore and a stroke of 24 in. cuts off at one-third stroke and has a mean effective pressure of 50 psi. The flywheel is to be 18 ft in mean diameter and it makes 75 rpm with a variation of 1 percent. Determine the weight of the rim.

Solution

The work done on the piston is equal to the mean effective pressure times the distance traveled (twice the stroke in feet) in one revolution. Piston area may be found to be 452.4 sq in., and the distance of travel is 4 ft. Thus, work done is found to be

$$452.4 \times 50 \times 4 = 90,480 \text{ ft-lb.}$$

From handbooks, the factor of energy excess for steam engines at one-third steam cutoff is 0.163. Then the average work done by the flywheel is

$$E = 90,480 \times 0.163 = 14,748 \text{ ft-lb.}$$

The weight of the flywheel is then found to be

$$W = \frac{11,745nE}{D^2N^2} = \frac{11,745 \times 100 \times 14,748}{18^2 \times 75^2} = 9504 \text{ lb.}$$

Example 5.17

Neglecting spoke effect, calculate the energy stored in the rim of a flywheel made of cast iron 24 in. in diameter, having a rim 5 in. wide by 4 in. deep, when running at 1000 rpm. Refer to Exhibit 4.

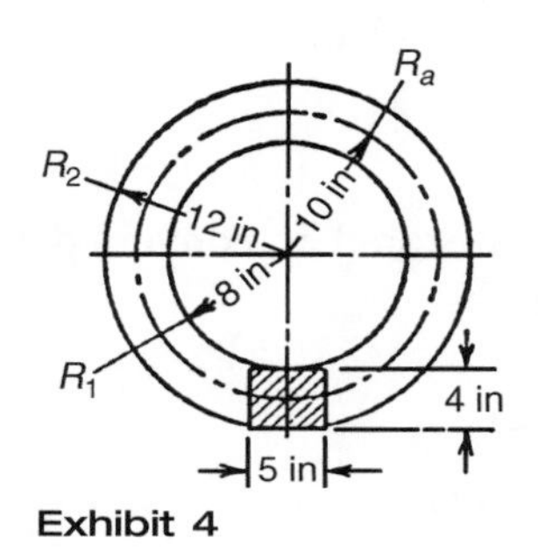

Exhibit 4

Solution

Note this is running above the safe speed listed in design books. Let W be weight of flywheel in pounds; ρ is radius of gyration in feet; R_1 is internal radius of rim in feet; R_2 is external radius of rim in feet; ω is angular velocity in radians per second; R_a is average radius in feet; density of cast iron is 450 lb per cu ft. Weight of flywheel is

$$W = 5 \times 4 \times 2 \times \pi \times 10 \times 450/1728 = 327 \text{ lb.}$$

$$\rho = \frac{1}{2}\sqrt{\left(R_1^2 + R_2^2\right)} = \frac{1}{2}\sqrt{\left[\left(\frac{8}{12}\right)^2 + \left(\frac{12}{12}\right)^2\right]} = 0.60 \text{ ft}$$

Angular velocity $\omega = \text{rpm} \times 2\pi/60 = 104.8$ radians per second as determined.

$$\omega = 1,000 \times 6.28/60 = 104.8 \text{ radians per sec}$$

Energy stored in the rim is next found in accordance with

$$\frac{W\rho^2\omega^2}{2g} = 327 \times 0.60^2 \times 104.8^2/64.4 = 20,076 \text{ ft-lb.}$$

Example 5.18

A 100-lb wheel 18 in. in diameter which is turning at 150 rpm in stationary bearings is brought to rest by pressing a brake shoe radially against a rim with a force of 20 lb. If the radius of gyration of the wheel is 7 in. and if the coefficient of friction between shoe and rim has the steady value of 0.25, how many revolutions will the wheel make in coming to rest?

Solution

Energy stored in the wheel must first be determined in the usual manner by use of the equation from the previous problem. The angular velocity may be found to be 15.7 radians per second. The energy stored in wheel is

$$E = \frac{100 \times \left(\frac{7}{12}\right)^2}{64.4} \times (5\pi)^2 = 130 \text{ ft lb.}$$

Frictional force to stop the car wheel is 20 × 0.25 = 5 lb on perimeter. Wheel perimeter is $\pi D = \pi \frac{18}{12} = 4.7$ ft. Distance force must travel is $\frac{130}{4.7} = 27.66$ linear feet. The number of turns is thus $27.66/4.7 = 5.89$ turns.

Example 5.19

A flywheel whose weight is 200 lb and whose radius of gyration is 15 in. is secured to one end of a 6-in. diameter shaft; the other end of the shaft is connected through a chain and sprocket to a motor that rotates at 1800 rpm. The motor sprocket is 6 in. in diameter and the shaft sprocket is 36 in. in diameter. The total shaft length between flywheel and sprocket is 72 in. Determine the maximum stress in the shaft resulting from instantaneous stopping of the motor drive, assuming that sprocket and chain have no ability to absorb impact loading. Assume shear modules equal to 12,000,000 psi. Neglect effect of shaft kinetic energy.

Solution

When the flywheel is stopped short, the kinetic energy stored is converted to torsional impact. The magnitude of this energy is found by

$$E = \frac{W\rho^2\omega^2}{2g} = \frac{200 \times 1.25^2 \times \left(2\pi \frac{300}{60}\right)^2}{64.4 \times \frac{1}{12}} = 57{,}471 \quad \text{in-lb}$$

The shaft offers resilience to torsional twist. Resilience U (in.-lb) is the potential energy stored up in the deformed body. The amount of resilience is equal to the work required to deform the body (here the shaft) from zero stress to stress S. The modulus of resilience U_p (in.-lb per cu in.), or unit resilience, is the elastic energy stored up in a cubic inch of material at the elastic limit. The unit resilience used in the solution of this problem (for a solid shaft) is

$$U_p = \frac{1}{4}\frac{S_t^2}{G}$$

For the full volume of the shaft, where V is $\pi/4 \times 6^2 \times 72 = 2036$ cu in.

$$U_p \text{ total} = \frac{1}{4} \times S_t^2 \times 2035 \times 1/12{,}000{,}000 = 57{,}600 \text{ in.-lb.}$$

$$\text{Solving for } S_t = \sqrt{4 \times 12{,}000{,}000 \times 57{,}600/2035} = 36{,}860 \text{ psi.}$$

Example 5.20

A journal bearing with a diameter of 2.25 in. is subjected to a load of 1000 lb while rotating at 200 rpm. If the coefficient of friction is taken as 0.02 and L/D is 3.0, find (a) projected area, (b) pressure on bearing, (c) total work of friction,

(d) work per unit area of friction, (e) total heat generated, and (f) heat generated per unit area per minute.

Solution

Because L/D is 3, then $L = 3D = 3 \times 2.25 = 6.75$ in.

(a) Projected area $= L \times D = 6.75 \times 2.25 = 15.19$ sq in.

(b) Pressure on bearing $= P/(L \times D) = 1000/(15.19) = 65.8$ psi

(c) Total work $= W = 0.02 \times 1{,}000 \times (\pi)(2.25/12)(200) = 2356$ ft-lb/min

(d) Work $= w = W/LD = 2356/15.19 = 155.1$ ft-lb per min per sq in.

(e) Total heat $= Q = W/778 = 2356/778 = 3.03$ Btu per min

(f) $q = w/778 = 155.1/778 = 0.2$ Btu/(min)(Sq in. projected area)

P = total load on bearing, lb
p = pressure or load per sq in. of the projected area, psi
L = length of bearing, in.
D = diameter of bearing, in.
N = rpm of journal
μ = coefficient of friction
W = total work of friction, ft-lb per min
w = work per sq in. of projected area, ft-lb per sq in. per min
Q = total heat generated, Btu per min
q = heat generated per sq in. of projected area, Btu per min
V = rubbing velocity, fpm

Example **5.21**

An 8-in. nominal diameter journal is designed for 140° optimum bearing when bearing length is 9 in., speed is 1800 rpm, and total load is 20,000 lb. Calculate the frictional horsepower loss when this journal operates under stated conditions with oil of optimum viscosity.

Solution

The frictional loss in horsepower is given by the equation below, where V_r is rubbing speed and other factors are known.

$$\frac{\mu P V_r}{33{,}000}$$

The rubbing speed is found to be

$$\pi D \text{ rpm} = 3.1416 \times \frac{8}{12} \times 1800 = 3770 \text{ fpm.}$$

Projected surface of rubbing area is

$$\frac{D}{2} \times \alpha \times L = \frac{8}{2} \times \frac{140°}{180°} \times \pi \times 9 = 87.97 \text{ in.}^2$$

Bearing pressure is $20,000/87.5 = 228$ psi. The coefficient of friction,* using pressure and rubbing speed as criteria, may be taken as 0.002.

$$\frac{0.002 \times 20,000 \times 3770}{33,000} = 4.56 \text{ hp}$$

Example 5.22

Calculate the torque in pound feet produced by a pressure of 200 psi acting on the piston of an automobile engine when the crank is 30° past top center. Cylinder bore is 3.25 in., stroke is 4 in., and connecting rod is 10 in. long.

Solution

Refer to Exhibit 5, and let d be cylinder bore in inches.

$$P_p = \pi/4 \times d^2 \times 200 = 1660 \text{ lb}$$

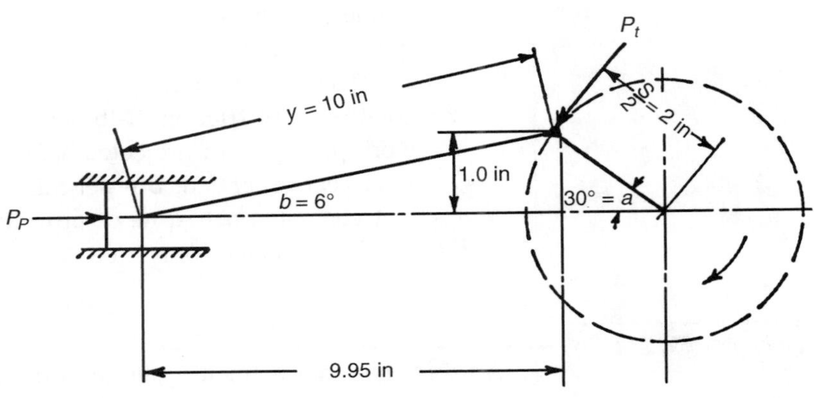

Exhibit 5

The above holds because we inserted 3.25 for d. Now determine connecting-rod pressure P_c by drawing a free-body diagram showing forces on the piston pin (see Exhibit 6). From this we observe that

$$\cos b = \frac{P_p}{P_c} = \frac{9.95}{10}$$

$$P_c = 1660/9.95 \times 10 = 1667 \text{ lb.}$$

Having determined P_c, now draw a free-body diagram of forces on the crankpin (Exhibit 7), and we see that

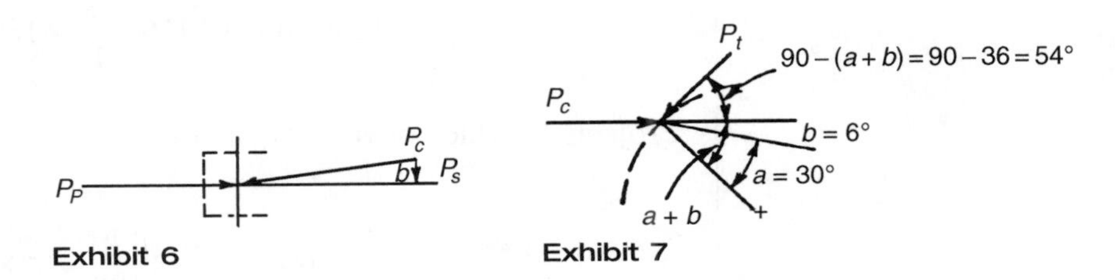

Exhibit 6 **Exhibit 7**

Machinery's Handbook, 14th ed., The Industrial Press, New York, 1960, pp. 518–20.

$$P_t = P_c \times \cos 54° = 1670 \times 0.588 = 982 \text{ lb.}$$

$$\text{Torque} = P_t \frac{s}{2} = 982 \times \frac{2}{12} = 163.7 \text{ lb-ft.}$$

Example 5.23

Calculate the side thrust against the cylinder walls of an engine of an automobile when the connecting rod is at an angle of 90° to the crank, the connecting rod being 15 in. long and the crank 3 in. The piston is 4 in. in diameter and the pressure on it is 200 psi.

Solution

Refer to Exhibit 8, and let the following considerations hold: P_p is total piston force in pounds; R is rod reaction on piston point in pounds; P_s is side thrust, in pounds.

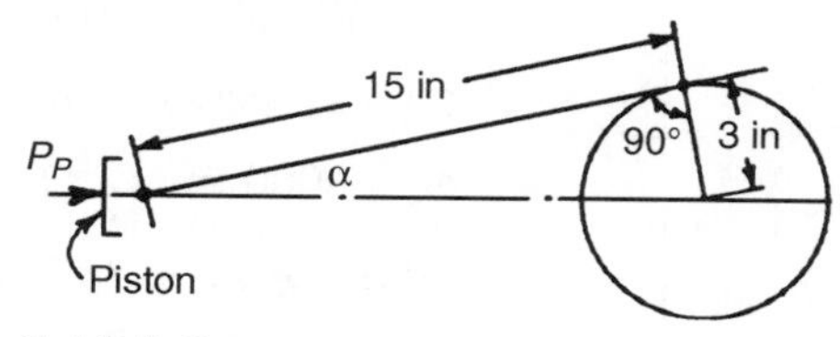

Exhibit 8

$$P_p = \pi/4 \times 4^2 \times 200 = 2513 \text{ lb}$$

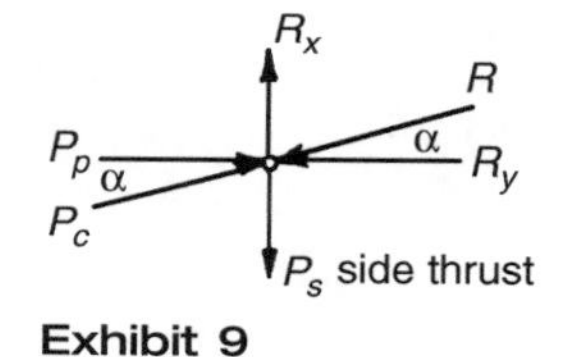

Exhibit 9

From force diagram (Exhibit 9), $P_s = R_x$, $R_y = P_p$, $R_x = R_y \tan\alpha$. Therefore,

$$P_s = P_p \tan\alpha = 2{,}513 \times \frac{3}{15} = 502 \text{ lb.}$$

Example 5.24

(a) A navy ship has a De Laval steam turbine whose rotating parts weight 10 tons, with a radius of gyration of 1 ft about its axis of rotation, and which rotates at 3000 rpm. Under the worst sea conditions the ship will pitch so that its “fore and aft” angular velocity has a maximum value of 2° per sec. Find the maximum gyroscopic couple to which the ship is subjected.

(b) In order to maintain its time schedule, an express train must run at a steady speed of 55 mph over a certain section of its journey. However, a mile of track on this section is under repair, and a distant signal will limit the train to 15 mph over it. Tests have shown that the train takes 2 mi from rest to attain a speed of 55 mph and $\frac{1}{4}$ mi to come to a stop. Determine how late the train will be as a result of this track condition, assuming uniform acceleration and deceleration.

Solution

(a) Refer to Exhibit 10 and let $n = 3000$ rpm, $W = 20{,}000$ lb, and $K = 12$ in. radius of gyration. Then the vector equation is

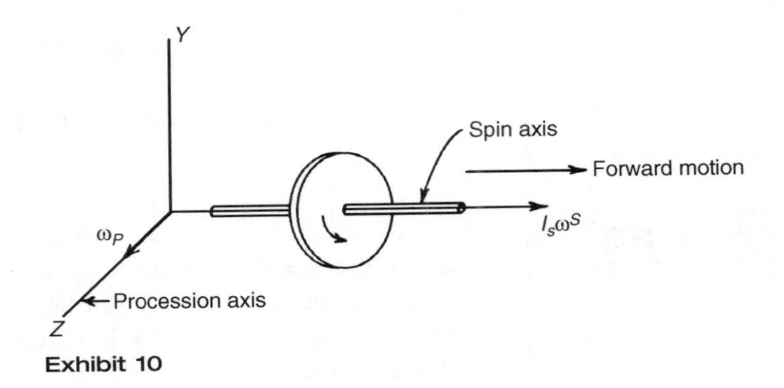

Exhibit 10

$$T = \alpha^p \times I_s \omega^s = I_s \omega^p \omega^s.$$

The scalar equation is

$$I_s = W/gK^2 = (20{,}000 \times 12^2)/(32.2 \times 12 \text{ in. per ft}) = 7453 \text{ lb-in.-sec.}$$
$$\omega^p = 2°/\text{sec}(2\pi \text{ radians}/360°) = 0.0349 \text{ radians per sec}$$
$$\omega_s = 3000 \text{ rpm}(2\pi/60) = 314 \text{ radians per sec}$$
$$T = I_s \omega^p \omega^s = 7453 \times 0.0349 \times 314 = 81{,}674 \text{ lb-in.}$$

(b) Refer to Exhibit 11

$$\text{Acceleration characteristic} = 55 \text{ mph}/2 \text{ mi} = 27.5 \text{ mph per mi}$$

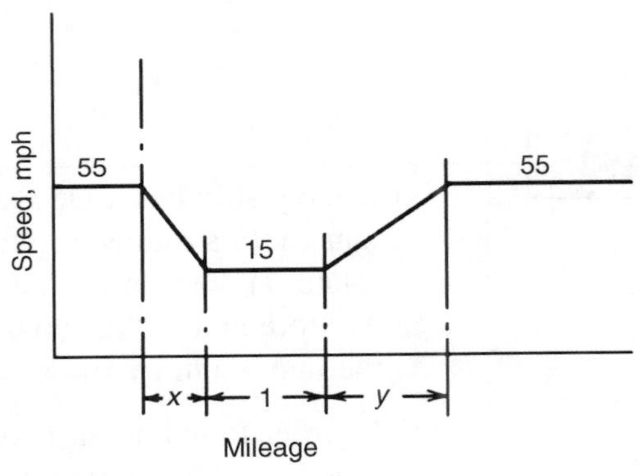

Exhibit 11

$$\text{Deceleration characteristic} = 55 \text{ mph}/\tfrac{1}{4} \text{ mi} = 220 \text{ mph per mi}$$

Therefore,

$$x = (55 - 15)/220 = 0.1818 \text{ mi}$$
$$y = (55 - 15)/27.5 = 1.454 \text{ mi}$$

Therefore, time for train to travel $(1+x+y)$ mi at reduced speed is

$$t_1 = \frac{0.1818}{(55+15)/2} + \frac{1}{15} + \frac{1.454}{(55+15)/2} = 0.113 \text{ hr.}$$

Time for train to travel $(1+x+y)$ mi at regular speed is t_0.

$$t_0 = \frac{1+x+y}{55} = \frac{2.6358}{55} = 0.048 \text{ hr}$$

Therefore, lost time $\Delta t = t_1 - t_0 = 0.113 - 0.048 = 0.065$ hr. Say 4 min.

Example 5.25

A solid steel machine shaft with a safe shearing stress of 7000 lb/in.^{2} transmits a torque of 10,500 in.-lb.

(a) Determine the diameter of the shaft.

(b) A square key is used whose width is equal to $\frac{1}{4}$ the shaft diameter and whose length is equal to $1\frac{1}{2}$ times the shaft diameter. Determine the dimensions of the key and check the key for its induced shearing and compressive stresses.

(c) Obtain the factors of safety of the key in shear and crushing, allowing an ultimate stress in shear of 50,000 lb/in^2 and an ultimate stress in crushing of 60,000 lb/in^2.

Solution

(a) Here $S_s = 7000$ lb/in.^{2}, $T = 10{,}500$ in.-lb, and $Z_p = \pi d^3/16$. Apply the following formula: $T = S_s Z_p = 10{,}500 = 7000 \times (\pi d^3/16)$, from which

$$d^3 = \frac{10{,}500 \times 16}{7000\pi} = 7.63 \quad \text{and} \quad d = 1.96, \text{ say 2 in.}$$

(b) $b = t = \frac{d}{4} = \frac{2}{4} = \frac{1}{2}$ in.

$$L = 1\frac{1}{2} \times 2 = 3 \text{ in.}$$

Key check: $P_t = T/R = 10{,}500/1 = 10{,}500$ lb

The shearing stress of the key is obtained from $S_s = P_t/bL$.

$$S_s = \frac{10{,}500}{\frac{1}{2} \times 3} = 7000 \text{ lb/in}^2.$$

The crushing stress of the key is obtained from $S_c = 2P_t/tL$.

$$S_c = \frac{2 \times 10{,}500}{\frac{1}{2} \times 3} = 14000 \text{ lb/in}^2.$$

(c) Here $U_s = 50{,}000$ lb/in.^{2}. Then $F = U_s/S_s = 50{,}000/7000 = 7.2$

With $U_c = 60{,}000$ lb/in.^{2}. Then $F = U_c/S_c = 60{,}000/14{,}000 = 4.3$
where

S_s = shearing stress, lb/in.^{2}
T = torque, in.-lb
Z_p = polar moment of inertia, in.^{3}
d = shaft diameter, in.
t = key width, in.
S_s = shearing stress, lb/in.^{2}
S_c = crushing stress, lb/in.^{2}
L = key length, in.
U_s = ultimate shearing stress, lb/in.^{2}
U_c = ultimate crushing stress, lb/in.^{2}
F = factor of safety

Example 5.26

In our plant we want to connect two 4-in. shafts by means of a cast-iron flange coupling that employs six bolts. Our stress lab people tell us the allowable shearing stress of the bolts is 6000 lb/in.^{2} while that of the shafting is 8000 lb/in.^{2}.

(a) Determine the diameter of the bolts to be used.

(b) Determine the induced crushing stress S_c if the thickness of the flange is $\frac{5}{8}$ in.

(c) Is it a safe stress?

Solution

(a) Torque $T = S_s Z_p = S_s(\pi d^3/16) = 8000(\pi 4^3/16) = 100{,}530$ in.-lb. Tangential force P_t occurring at the bolt circle

$$B = 3d = 3 \times 4 = 12 \text{ in.}$$

$$P_t = \frac{T}{\left(\frac{B}{2}\right)} = \frac{100{,}300}{\frac{12}{2}} = 16{,}755 \text{ lb}$$

To find the bolt diameter d_1, where P_t = 16,750 lb, n = 6 bolts, and S_s = 6000 lb/in.^{2}.

$$\frac{P_t}{n} = S_s \frac{\pi d_1^2}{4}$$

$$\frac{16{,}750}{6} = 6000(0.7854 \times d_1^2) \quad \text{from which } d_1^2 = 0.59$$

$d_1 = 0.77$ in, say $\frac{7}{8}$ in., which is the next higher standard diameter

(b)
$$\frac{P_t}{n} = d_t t(s_c), \qquad \frac{16{,}750}{6} = \frac{7}{8} \times \frac{5}{8} \times S_c$$

$$S_c = 5105 \text{ lb/in}^2$$

(c) This value of S_c is the induced unit crushing stress that is set up by a crushing load of (16,750/6) lb on the projected area or bearing area of each bolt. If we assume that U_c (ultimate crushing stress) for cast iron is 80,000 lb/in.^{2} and for steel is 60,000 lb/in.^{2}, it is evident there will be a greater tendency to crush the steel than to crush the cast iron. Hence we shall determine the factor of safety F for steel. $F = 60{,}000/5100 = 11.8$. This is proof that the induced stress S_c is also a safe stress.

Example 5.27

The advent of energy-conservation programs is making it more attractive to switch to synchronous belts as V-belt loss of efficiency becomes more expensive. Synchronous belts, commonly called timing belts, were developed for timing applications requiring synchronization, or "timing" between shafts. However, they have made their way into many other power-transmission applications. Timing belts are nearly 100 percent efficient, and the loss of about 0.5 percent is due to bending. V belts are usually only 95 to 96 percent efficient.

The 40-hp pump drive is to be converted to synchronous belts. The drive is belted with five C-120 belts running on two 10-ft, five-groove C-section sheaves. What is the saving for the first year of installation? Assume the following and solve accordingly:

Cost of V-belt system: sheaves and belts

Cost of synchronous system: sprockets and belts

Efficiency of V-belt system—handbooks

Efficiency of synchronous system—handbooks

Electricity costs equal 6 cents per kilowatthour

Solution

Cost of V-belt system = $250

Cost of synchronous system = $375

V-belt transmission efficiency loss = 5 percent

Synchronous system efficiency = 0.5 percent

Operating the 40-hp pump motor for 40 h per week at 6 cents per kilowatthour costs $40 \times 746/1000 \times 40 \times 0.06 = \$71.62$. With V belts, 5 percent of this cost, or $3.58 per week, is burned up in slippage. This loss represents $186.16 per year.

The synchronous system will cost $125 more, but, at 40 hp for 40 h/week, it will save $71.62 \times 0.005 = \$0.36$ burned up in 1 week in slippage and for 1 year,

$$0.36 \times 52 = \$18.72.$$

Then the savings of synchronous over V belt per year is

$$186.16 - 18.72 = \$167.44$$

Thus, the change to the synchronous system will save $167.44 in electricity in the first year, not to mention the savings in reduced maintenance. An additional advantage of timing belts is that the recommended installation belt tension is 30 to 50 percent less than that for V belts, allowing motor and sheave bearings to run under less load and pressure with considerably long life.

Savings that can be realized by converting to timing belts can be quite considerable. For example, a plant whose V belts carry 1500 hp annually is burning up about 167.44 × 1500/40 = \$6279.

GEARS

The transmission of power from one rotating shaft to another is accomplished by using gears. The ratio of the speeds of the shafts is in direct ratio to the pitch diameters of the gears.

$$N_1/N_2 = D_{p1}/D_{p2}$$

For gears to operate properly, the pitch on mating gears must be the same. As a consequence, the preceding equation may be written as the ratio of speeds of the shafts as equal to the number of teeth on each gear.

$$N_1/N_2 = t_1/t_2$$

The diametral pitch of a gear is a ratio of the number of teeth of a gear and its pitch diameter.

$$P = t_1/N_1$$

For gears to mesh, both gears must have the same pitch diameter.

Example 5.28

A 10-ton electric truck is to accelerate at the rate of 3 miles per hour per second (Mphps). Its rolling resistance is 25 lb per ton weight. If the truck employs a gear reduction between the motor and the wheels of 9 to 1, and the wheels are 36 in. in diameter, what is the power and the torque required of the motor to drive the truck at 25 mph? Assume efficiency of the reduction gear at 90 percent.

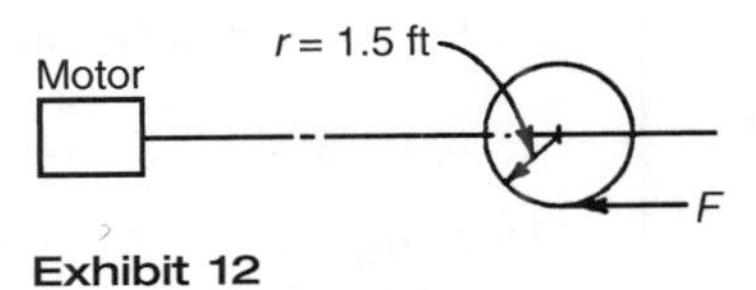

Exhibit 12

Solution

Refer to Exhibit 12, and note that the total rolling resistance is as follows:

$$F' = 25 \times 10 = 250 \text{ lb.}$$

Acceleration is found to be

$$3 \times 5280/3600 = 4.4 \text{ ft per sec}^2.$$

On a level road $F - F' = W/g \times a$, where F is the necessary propelling force. Therefore,

$$F = 2000 \times 10 \times 4.4/32.2 + 250 = 3000 \text{ lb in round numbers.}$$

Required motor horsepower is simply

$$\text{hp} = F \times V/(550 \times \text{efficiency})$$

where V is velocity in fps, and here is equivalent to $\frac{25}{60} \times 88 = 36.7$ fps.

$$\text{hp} = 3000 \times 36.7/(550 \times 0.90) = 222\,\text{hp}$$

Wheel torque = F × wheel radius in ft = 3000 × 1.5 = 4500 lb-ft. Required motor torque = wheel torque/(velocity ratio × efficiency).

$$4500/(9 \times 0.90) = 556 \text{ lb-ft}$$

Example **5.29**

Two shafts are connected by spur gears, as shown in Exhibit 13. The pitch radii of gear A and B are 4 in. and 20 in., respectively. If shaft A makes 800 rpm and is subjected to a resisting moment of 1000 in.-lb, what is (a) rpm of B, (b) torque in shaft B, (c) speed reduction factor, (d) torque multiplication factor, and (e) tooth pressure of A and B?

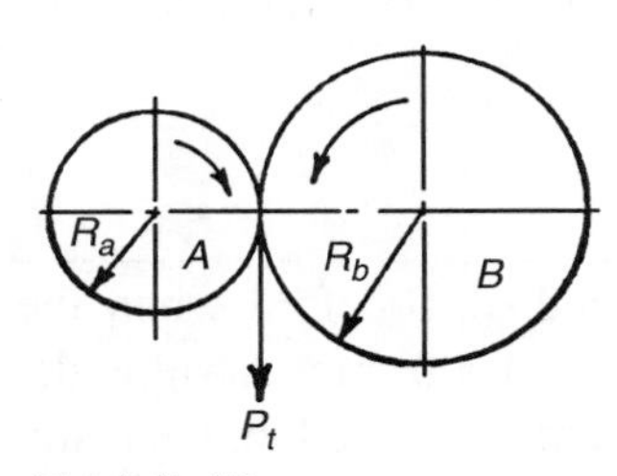

Exhibit 13

Solution

Here R_a is 4 in., R_b is 20 in., and N_a is 800 rpm.

(a) Then

$$\frac{N_b}{N_a} = \frac{R_a}{R_b}$$

$$\frac{N_b}{800} = \frac{4}{20} = \frac{1}{5},$$

from which we obtain $N_b = \frac{800}{5} = 160$ rpm. This is the speed of B.

(b) $\frac{T_b}{T_a} = \frac{R_b}{R_a} = \frac{T_b}{1000} = \frac{20}{4} = 5$

Therefore, $T_b = 5000$ in.-lb. This is the torque in shaft B.

(c) The speed reduction factor is $N_a/N_b = \frac{800}{160} = 5$.

(d) The torque multiplication factor is $T_b/T_a = 5000/1000 = 5$.

(e) Tooth pressure $P_{ta} = T_a/R_a = 1000/4 = 250$ lb.

Example 5.30

(a) A tractor is to have a drawbar pull of 5000 lb. The differential has a double-threaded worm acting on a wheel having 35 teeth. The rear wheels are 42 in. in diameter. The motor has a maximum torque of 200 lb-ft. The efficiency of the drive is assumed to be 90 percent. What gear ratio is necessary in the transmission? (b) Calculate the stress in the rear axle if the diameter is 1.75 in.

Solution

(a) For the differential (wheel teeth/worm teeth) = $\frac{35}{2}$ = 17.5. Because the total wheel reaction is 5000 lb, the total torque is

$$5{,}000 \times {}^{21}\!/_{12} = 8750 \text{ lb-ft.}$$

where the value 21 is half the rear-wheel diameter. Then from the relation

Motor torque × differential-gear ratio × transmission-gear ratio × efficiency = drive-wheel torque

we can solve for the transmission-gear ratio.

$$\text{Transmission-gear ratio} = \frac{8750}{200 \times 17.5 \times 0.90} = 2.78$$

(b) For each axle the torque is equal to 8750 × ${}^{12}\!/_{2}$ = 52,500 lb-in. Stress in axle is next determined in accordance with the following: $S = 16/(\pi \times d^3)(\text{torque in each axle}) = [16/(\pi \times 1.75^3)](52{,}500)$, where S is calculated as 49,890 psi.

Example 5.31

A worm-driven hoist raises a load of 5000 lb at a speed of 100 fpm when the driving motor is exerting a torque of 130 lb-ft at a speed of 1000 rpm. Assume the drum diameter and the type of worm, and calculate the number of teeth in the drum wheel. Determine the efficiency of the entire hoist mechanism.

Solution

Assume drum diameter to be 1.5 ft and number of worm teeth 2. Therefore, angular velocity of drum and drum wheel is

$$\frac{v}{\pi \times \text{drum diameter}} = \frac{100}{\pi \times 1.5} = 21.2$$

$$\text{Number of drum-wheel teeth} = \frac{\text{angular velocity of worm}}{\text{angular velocity of drum wheel}} \times n$$

$$\text{Number of drum-wheel teeth} = (1000/21.2)(2) = 94.3, \text{ say 94 teeth}$$

$$\text{Efficiency} = \frac{\text{output}}{\text{input}} = \frac{Wv}{2\pi ST}$$

$$\text{Efficiency} = \frac{5000 \times 100}{6.28 \times 1000 \times 130} = 0.612, \quad \text{or} \quad 61.2 \text{ percent}$$

$$\text{hp} = \frac{2\pi ST}{33{,}000} = \frac{817{,}000}{33{,}000} = 24.7, \text{ say 25 hp}$$

In this solution n is number of worm teeth, N is number of drum wheel teeth, S is motor speed in rpm, W is load, v is load speed in fpm, D is drum diameter in feet, T is motor torque in lb-ft.

Example 5.32

If one rear wheel of an automobile is jacked off the ground and the motor turns the drive shaft at 1200 rpm, calculate the speed of the floating rear wheel, of the ring gear, and of spider pinions. The numbers of teeth on the gears are drive-shaft pinion, 11; ring gear, 40; rear-axle gear, 23; spider pinion, 12.

Solution

Refer to Exhibit 14, and let

G = drive shaft
H = drive-shaft pinion, 11 teeth
K = ring gear, 40 teeth
B_1 = moving rear-axle gear, 23 teeth
CD = spider pinion, 12 teeth
B = stationary rear-axle gear, 23 teeth
N_G = drive-shaft speed in rpm
T_H = number of teeth on H; typical for other gear teeth

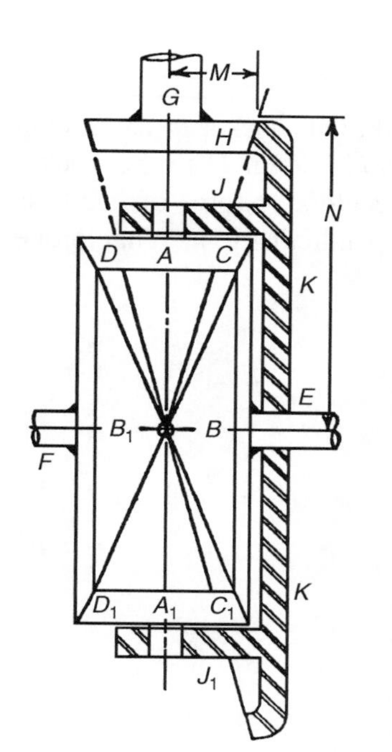

Exhibit 14

With B stationary ($N_B = 0$) and B_1 rotating

$$N_K = N_G \frac{T_H}{T_K} = \frac{1200}{\frac{40}{11}} = 330 \text{ rpm}$$

$$N_{CD} = N_K \left(\frac{T_B}{T_{CD}}\right) = (330)\left(\frac{23}{12}\right) = 632 \text{ rpm.}$$

When vehicle is moving straight ahead, $N_B = N_K$, when there is equal traction. But under conditions of this problem the moving rear wheel will have twice N_{B1} or $2 \times N_K$. So it follows that:

$$N_{B1} = N_K + \left(N_{CD}\frac{T_{CD}}{T_{B1}}\right) = 330 + \left[632\left(\frac{12}{23}\right)\right] = 660 \text{ rpm}$$

MECHANISMS

Mechanisms are components or elements of machines that perform distinct functions in the machine such as transforming one type of motion to another. Because

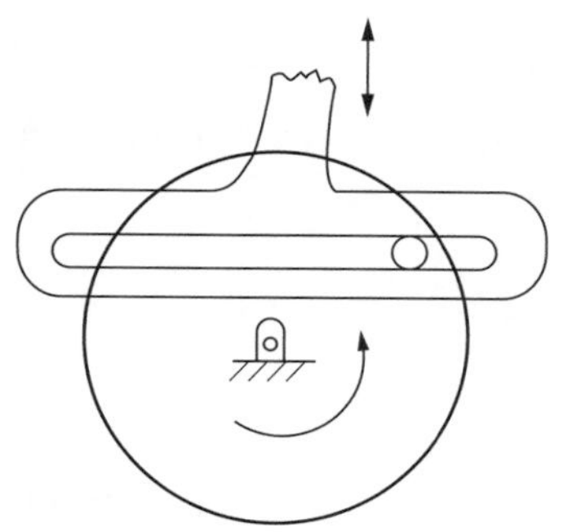

Figure 5.2

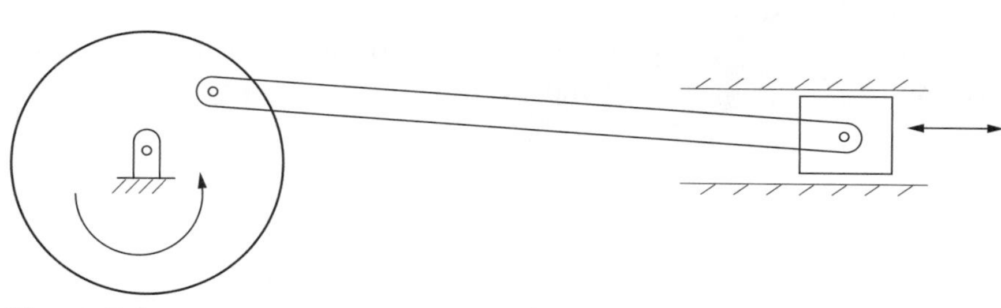

Figure 5.1

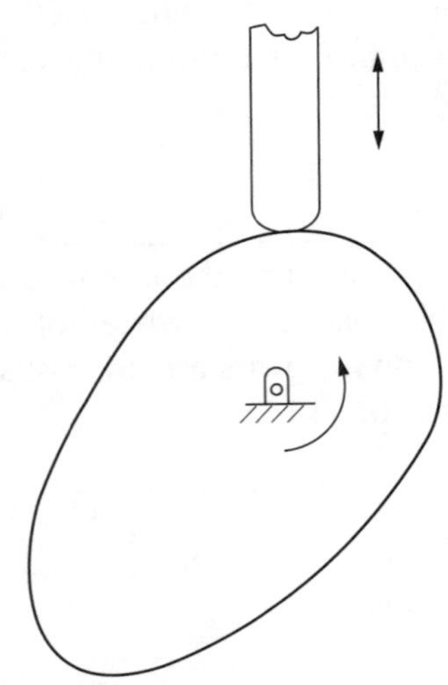

Figure 5.3

most power to machines is produced in the form of rotary motion, the mechanism is normally used to change rotational motion to some form of linear motion. Depending on the type of restrictions, such as space allowed for the mechanism, and the type of motion required, such as accelerations or velocities needed, specific mechanisms are chosen

For example, among the devices producing linear motion are

- Piston type of four bar linkage that produces simple harmonic motion (Figure 5.1)
- Scottish yoke mechanism that produces simple harmonic motion (Figure 5.2)
- Cam-follower in which the accelerations, velocities, and displacements can be varied according to what is needed (Figure 5.3).

Example 5.33

Two rockers are connected at their free ends by a link. One rocker is 12 in. long and is perpendicular to the line of centers, which is 50 in. long. The other rocker is 6 in. long and at an angle of 45° with the line of centers. A force of 100 lb is applied at the end of the 12-in. rocker and directed perpendicular to it. What torque is exerted on the shaft of the 6-in. rocker?

Solution

Refer to Exhibit 15, and let

P = force on 12-in. rocker, lb.
R = reaction of connecting link, lb.
L, x, y = distance shown, in.
T = tangential force on 6-in. rocker, lb.
a, b = angles shown on diagrams of forces, deg.

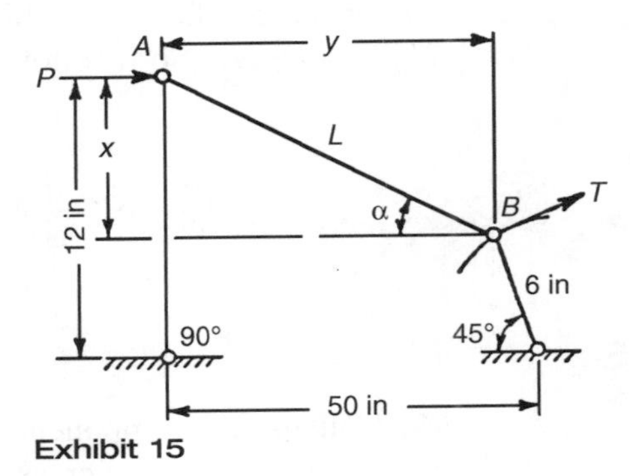

Exhibit 15

Exhibit 16

From the diagram of forces at point A (Exhibit 16) it may be seen that the horizontal component of R is

$$R_y = P = 100 \text{ lb.}$$

Also, $y = (50 - 6\ \cos 45°) = 45.76$ in.

$x = (12 - 6\ \sin 45°) = 7.76$ in.

$L = (x^2 + y^2)^{\frac{1}{2}} = (7.76^2 + 45.76^2)^{\frac{1}{2}} = 46.4$ in.

$\cos a = 45.76/46.4 = 0.987$

$R = R_y/\cos\ a = 100/0.987 = 101.3$ lb.

$a = \cos^{-1} 0.987 = 9°15'$ (see Fig. 4.4)

$b = 45° + a = 45° + 9°15' = 54°15'$

$T = R \cos b = 101.3 \times \cos\ 54°15' = 59.2$ lb.

Torque on 6-in. rocker $= T \times \frac{6}{12} = 59.2 \times 0.5 = 29.6$ lb-ft.

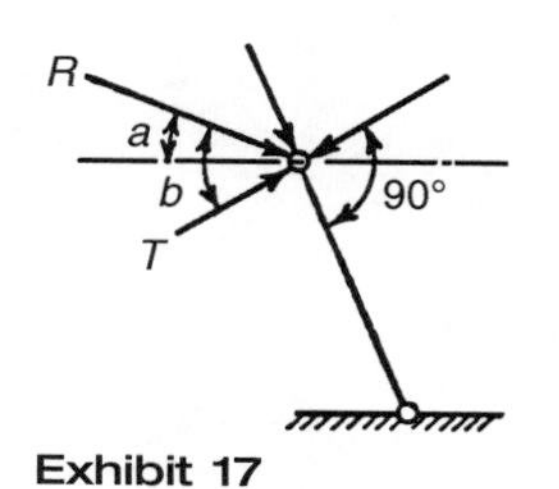

Exhibit 17

Example 5.34

A 2% C hardened steel pipe is mounted between two rigid walls eight feet apart. The pipe experiences a temperature change of 200°F.

Given Exhibit 18 and the following information:

$E = 30 \times 10^6$ psi
$\alpha_t = 6.6 \times 10^6$ in/in°F
$ID = 4.0$ in
$\varphi D = 4.2$ in

Determine a) longitudinal stress in pipe and b) load (equivalent) on pipe.

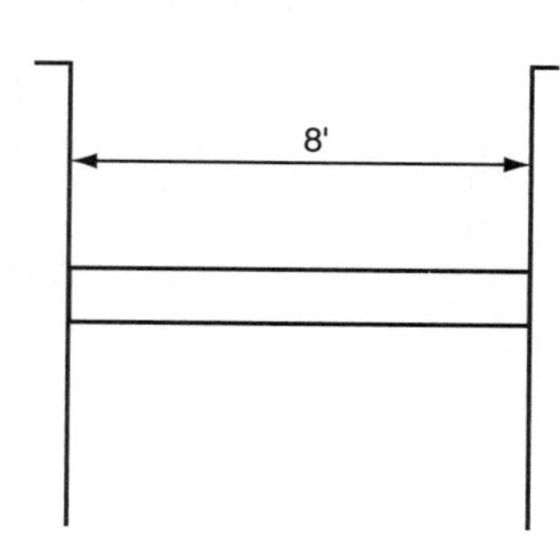

Exhibit 18

Solution

a) Elongation due to temperature is:

$$(\alpha_t)(\Delta T)(L) = \Delta L$$

$$(6.6 \times 10^{-6} \text{in/in°F})(200°\text{F})(8 \times 12)\text{in} = 0.1267 \text{ in}$$

This pipe is not allowed to expand and does not bend.

$$\sigma = \frac{P}{A} \qquad E = \frac{\Delta L}{L} = \frac{0.1267}{96} = 0.001319$$

$$E = \frac{\sigma}{E} \text{ so, } \sigma = (30 \times 10^6 \text{psi})(0.001319)$$

$$\sigma = 39{,}570 \text{ psi}$$

b) $$\sigma = \frac{P}{A} \text{ so } P = (\sigma A) = (39{,}570 \text{ lb/in}^2)\frac{\pi}{4}(d_1^2 - d_2^2)$$

$$P = (39{,}570 \text{ lb/in}^2)\left(\frac{\pi}{4}\right)(4.2^2 - 4.0^2) = 50{,}968 \text{ lb}$$

Example 5.35

A crank 4 in. long oscillates a rocker 20 in. long through a connecting link 28 in. long. The distance between crank and rocker centers is 22 in. Calculate the angle through which the rocker oscillates.

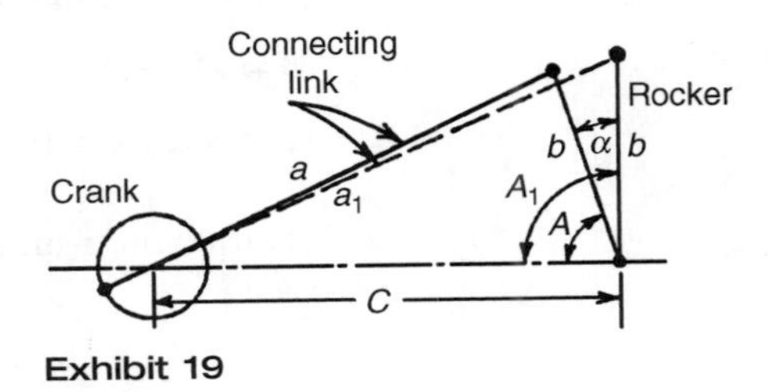

Exhibit 19

Solution

Exhibit 19 shows the limiting positions for rocker travel. From the law of cosines we have

$$\cos A = \frac{b^2 + c^2 - a^2}{2bc} = \frac{20^2 + 22^2 - (28-4)^2}{2 \times 20 \times 22} = 0.35$$

$$A = \cos^{-1} 0.35 = 69.51°$$

Similarly,

$$\cos A_1 = \frac{b^2 + c^2 - a^2}{2bc} = \frac{20^2 + 22^2 - (28+4)^2}{2 \times 20 \times 22} = -0.159$$

$$A_1 = \cos^{-1}(-0.159) = 99.14°$$

Thus, the angle moved through by rocker $A_1 - A = 99.14 - 69.51 = 29.64°$.

CHAPTER 6

Pumps and Hydraulics

OUTLINE

A pump is a device that is used to increase the pressure of a liquid. Hydraulics is the branch of physics that deals with liquids in motion. Classified by use, pumps are called low service, high service, deep well, booster, sewage, sludge, boiler feed, chemical, proportional feeders, air blowers, and so on.

Low-service pumps operate at low discharge heads to lift water from sources of supply to water-treatment works. High-service pumps operate at high discharge heads to deliver water to distribution systems. Proportional feeders are used for dosage of solutions of chemicals or liquid chemicals. Figure 6.1 shows the standard classification of pumps, with four general classes.

Because the centrifugal pump is the most common pump used in industry, we will focus our attention on this type of pump. The centrifugal pumping principle consists, essentially, of an impeller arranged to rotate within a case so that the liquid will enter at the center and be thrown out by centrifugal force to the outer periphery of the impeller and discharged into the outer case. Figure 6.2 shows the volute-type centrifugal pump. The volute converts the velocity energy of the liquid into static pressure at the discharge connection.

In the centrifugal impeller, the intake is smaller at the center than at its outer diameter. The liquid flows in at the center by suction, or from a low pressure, and is whirled by the impeller, gathering pressure from the kinetic energy by virtue

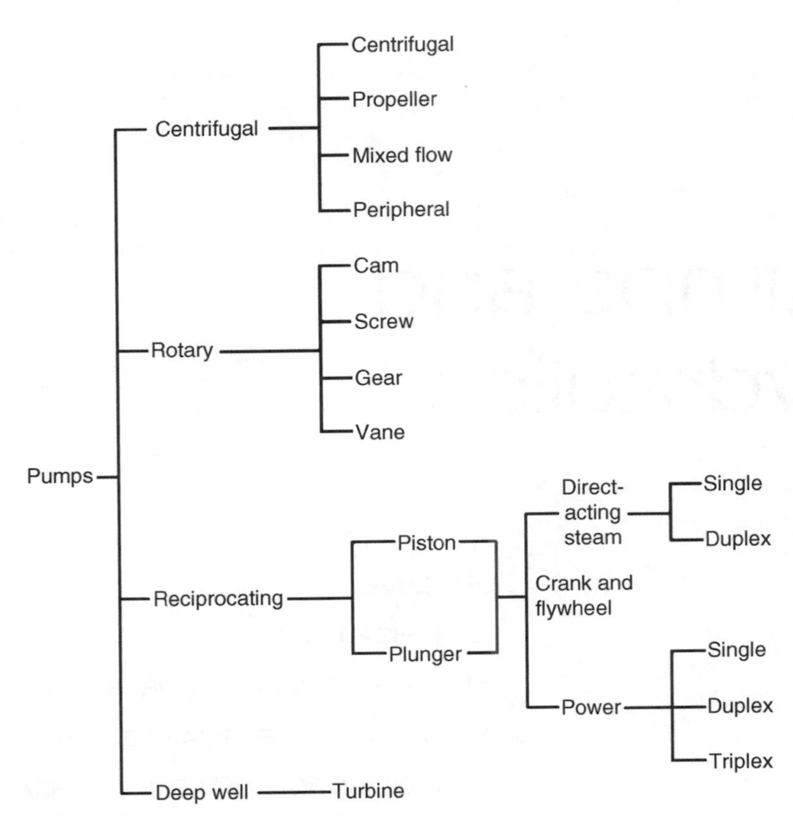

Figure 6.1

of the centrifugal force, and is discharged almost tangentially in the direction of rotation with high velocities, which velocities are decelerated and converted into pressure in the casing surrounding the impeller. The pressure head developed by the pump is entirely the result of kinetic energy in the form of velocities imparted to the water or liquid by the impeller and is not due to impact or displacement.

This pumping principle differs from others (reciprocating and the like) in that its impeller can be whirled freely in the pump even though the discharge valve has been closed. When the shutoff head has been reached (pump running and discharge valve closed) no higher pressure can be produced within the pump without increasing the speed.

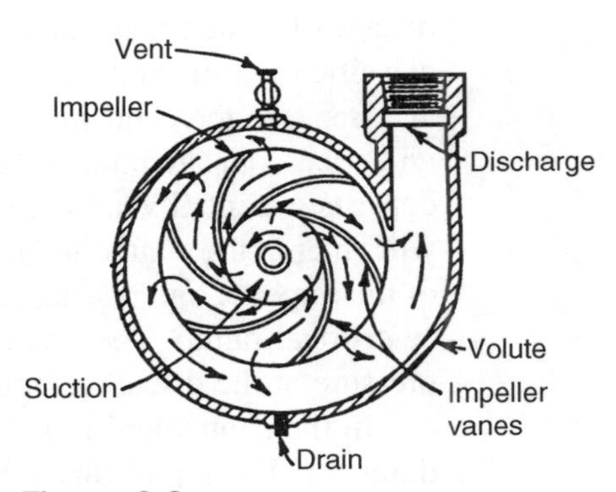

Figure 6.2

In a diffuser-type pump, stationary guide vanes surround the impeller. These gradually expanding passages change the direction of liquid flow and convert velocity head into pressure head.

In a turbine pump, the liquid is picked up by the vanes of the impeller and whirled at high velocity for nearly a complete revolution in an annular channel in which the impeller turns. Energy is added to the liquid in a number of impulses and it enters the discharge at high velocity.

Mixed-flow pumps develop their head partly by centrifugal force and partly by the lift of the vanes on the liquid being pumped.

Propeller pumps develop most of their head by the propelling or lifting action of the vanes on the liquid.

PUMP HEAD

The pressure head against which the pump will operate depends on the diameter of the impeller and the speed at which it is rotated. A centrifugal pump may actually raise the liquid or force it into a pressure vessel. It may merely give it enough head to overcome pipe friction. No matter what the service of a centrifugal pump, all forms of energy imparted to the liquid in performing this service must be accounted for in establishing the work performed. In order that all these forms of energy may be added algebraically, it is customary to express them all in terms of head in feet of liquid flowing.

Total dynamic head (W_p in Bernoulli's theorem) is made up of the total static head plus the friction head plus the pressure required at the discharge point of the system. The pressure then must be converted to feet of head in the usual manner.

A pump operating at "flooded" suction may be considered as having neither static suction lift nor static suction head. However, a slight suction head must be carried to ensure a full supply of liquid at the eye of the impeller. In most pumping applications velocity head may be neglected, except when large pumping volume rates are being handled.

PUMP PERFORMANCE CURVE

As we noted before, the pressure head against which a centrifugal pump will operate depends on the diameter of the impeller and the speed at which it is rotated. In order to determine what operating conditions a centrifugal pump is good for, a curve known as a performance curve is used. A typical performance (characteristic) curve is shown in Figure 6.3. This is a plot for an impeller of a fixed diameter rotating at a single constant speed.

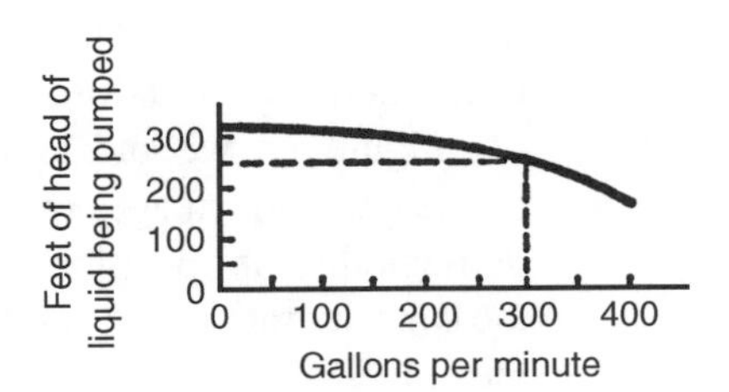

Figure 6.3

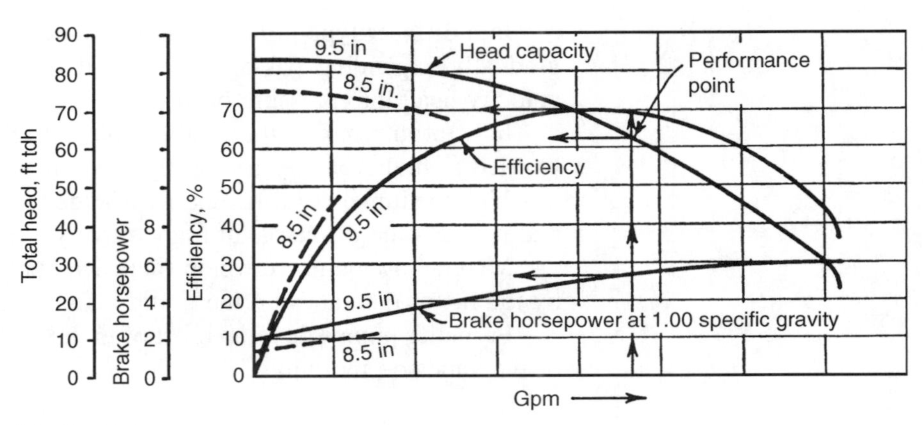

Figure 6.4

The capacity of the pump, usually in gpm, is plotted on the horizontal axis, while the pressure rise through the pump (stated in feet of head of the liquid pumped) is plotted on the vertical axis. A performance curve is determined by running the pump under actual test conditions. By throttling a valve on the discharge, various flows from zero at full shutoff to full flow at wide open can be obtained. At each flow condition, the gpm is measured with a meter and the corresponding discharge pressure is noted.

From these test results, a number of points are located and a smooth curve is drawn connecting the points. From this curve, the corresponding capacity and head for any set of operating conditions can be read. As shown in Figure 6.3, for 300-gpm flow, the pressure rise would be equal to 250 ft of the liquid being pumped. Performance curves are plotted on all centrifugal pumps.

Different curves may be obtained with the same pump by using different diameter impellers. An impeller with a larger diameter than that for which Figure 6.3 has been drawn would have a curve above that shown and parallel to it. An impeller having a smaller diameter would have a curve parallel but below. It may be noted, also, that a flat or horizontal head-capacity curve is usually most desirable from the standpoint of giving a wide range of capacities without much change in discharge pressure. This is especially so if the piping system to which the pump is connected is made up primarily of friction head and very little or no vertical lift. It must be remembered that a certain amount of slope in the curve is necessary to give proper control of flow and pressure. When a piping system is made up primarily of vertical lift and very little pipe friction, then a steep characteristic is recommended.

There is a good reason for the slope of the curve. Friction within the casing caused by the liquid (called disc friction) increases with increased flow and is reflected as actual loss in pressure or head.

A complete performance curve for a centrifugal pump is shown in Figure 6.4. In addition to the head-capacity curve we see an efficiency curve and a brake-horsepower curve. The efficiency curve rises to a peak within certain capacity limits and then falls off. Maximum efficiency lies within the design range.

Figure 6.3, we said, applied to a specified impeller diameter at constant speed. One of the great advantages of a centrifugal pump over the positive displacement pump is its flexibility of operation. This flexibility lies in the fact that with a single casing size the capacity rating of a single pump at constant rpm may be varied by simply trimming the impeller diameter to the required size. Therefore, with one set of patterns and castings for one size pump the manufacturer is able to meet any combination of head-capacity requirements within the operating range by merely machining the impeller.

If a pump is handling water and is discharging a certain flow at a certain total dynamic head requiring a definite brake horsepower, this same pump handling a liquid lighter than water, such as gasoline (sp gr 0.75), or a liquid heavier than water, such as a brine (sp gr 1.2), would discharge the same quantity of liquid at the same total dynamic head as it does for water. From the characteristic curve submitted by the manufacturer (always for clear cold water) the brake horsepower is obtained for water. The horsepower requirements for gasoline would be 75 percent of the curve reading for water; for the brine, 120 percent. This treatment holds for cases when viscosities are in the order of water. Where viscosities are much greater, special treatment of the problem is necessary because capacity and horsepower are greatly affected.

CHANGE OF PERFORMANCE

Again, we repeat, the head developed by a centrifugal pump depends on the impeller diameter and its rotational speed. According to the fundamental laws of physics, the capacity flow of the centrifugal pump will vary directly as the speed; the pressure head will vary as the square of the speed; and the power required will vary as the cube of the speed. These are known as the *laws of affinity* and apply to all velocity machines, *i.e.*, centrifugal pumps, centrifugal fans, blowers, etc.

Thus, if the speed is doubled, the flow will be doubled, the pressure head will be multiplied four times, and the power requirement will be increased eight times. Likewise, in reducing the speed one-half, the capacity will be cut in half, the pressure head will be only one-fourth, and the horsepower one-eighth.

To compute the performance of a centrifugal pump at some other speed when present values are known, for example, assume that it is desired to find what the same pump will do at the speed of 1760 rpm.

Speed	1160 rpm
Flow	300 gpm
Head	40 ft
Power input	4 hp

The flow will increase proportionately.

$$\frac{1760}{1160} \times 300 = 455 \text{ gpm}$$

The head will increase as the square.

$$\left(\frac{1760}{1160}\right)^2 \times 40 = 92 \text{ ft}$$

The power will increase as the cube.

$$\left(\frac{1760}{1160}\right)^3 \times 4 = 13.98 \text{ hp}$$

To compute in the opposite direction, from 1760 rpm to 1160 rpm, reverse the order of the divisor, place the 1160 above the line and 1760 below. Any given difference in the performance for any variation in speed can be found in the same way. In a field test, if the total head in feet, capacity in gpm, and horsepower input of the pump can be measured, its efficiency can be computed.

$$\frac{\text{gpm} \times \text{total dynamic head in ft}}{3960} = \text{water hp} \quad \textbf{(6.1)}$$

$$\frac{\text{water hp}}{\text{brake hp}} \times 100 = \text{pump efficiency} \quad \textbf{(6.2)}$$

Example 6.1

An acceptance test was conducted on a centrifugal pump having a suction pipe 10 in. in diameter and a discharge pipe 5 in. in diameter. Flow was 818 gpm of clear cold water. Pressure at suction pipe was 4.5 in. mercury vacuum and discharge pressure was 15.5 psig at a point 3 ft above that point where the suction pressure was measured. Input to pump was 15 bhp. Find pump efficiency.

Solution

We shall apply Bernoulli's theorem. Friction head h_f is considered zero because data were taken close to pump flanges. If the datum line is taken through the point of suction measurement, then Z_A is zero and Z_B is 3.

$$V_{\text{suction}} = \frac{Q}{A} = \frac{818 \text{ gpm}}{\frac{\pi}{4}\left(\frac{10}{12}\right)^2 \text{ft}^2(7.48 \text{ gal/ft}^3)(60 \text{ s/min})}$$

$$V_{\text{suction}} = 3.34 \text{ ft/s}$$

$$V_{\text{discharge}} = (V_{\text{suction}})\left(\frac{10}{5}\right)^2 = (3.34)(4) = 13.36 \text{ ft/s}$$

$$\text{Suction head} = \frac{4.5 \text{ in Hg}}{12 \text{ in/ft}}(13.6 \text{ sp gr}) = -5.1 \text{ ft water}$$

$$\text{Static discharge head} = 15.5 \text{ psig}\left(\frac{34}{14.7}\right) = 35.8 \text{ ft water}$$

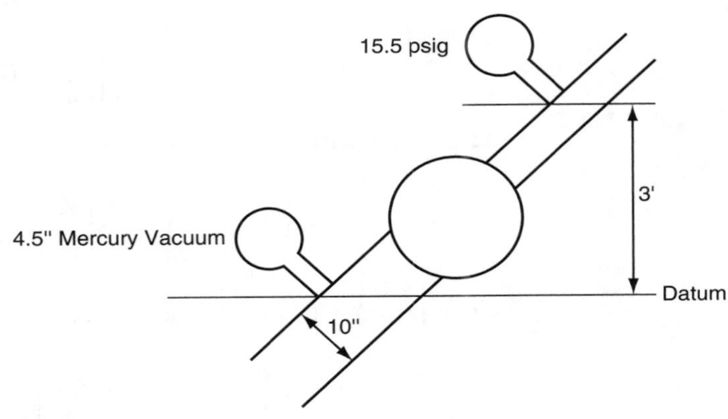

Exhibit 1

Total head pump against = Static head pump + Vacuum head + Elevation + Velocity head change

$$\text{Total head} = 35.8 \text{ ft} + 5.1 \text{ ft} + 3 \text{ ft} + \frac{13.36^2 - 3.34^2}{2g}$$

Total head = 46.5 ft

$$\text{hp} = \frac{(818\text{ gpm})(62.4\text{ lb/ft}^3)(46.5\text{ ft})}{7.48\text{ gal/ft}^3(60\text{ s/min})(550\text{ ft-lb/hp-s})} = 9.61\text{ hp}$$

$$\text{Pump efficiency} = \frac{9.61}{15} \times 100 = 64.1\%$$

Example 6.2

If the pump in Example 6.1 ran at 1750 rpm, what new gpm, head, and brake horsepower would be developed and required if the pump speed were increased to 3500 rpm? Assume constant efficiency.

Solution

In all problems involving the change in performance, and without the use of pump curves, pump efficiency is assumed to be constant for the pumping ranges involved. Now let us apply the laws of affinity.

gpm change:

$$\frac{818}{x} = \frac{1750}{3500} = 0.5 \qquad x = 1636\text{ gpm}$$

Head change:

$$\frac{46.5}{y} = \left(\frac{1750}{3500}\right)^2 = 0.25 \qquad y = 186\text{ ft}$$

bhp change:

$$\frac{15}{z} = 0.5^3 = 0.125 \qquad z = 120\text{ bhp}$$

The laws of affinity are theoretical. Not only do they apply to a change in rotative speed (constant impeller diameter), but they also apply to a change in impeller diameter for a particular pump. These laws give approximate results only, but their results are practical for everyday use in design and application. At constant rotative speed,

(a) Capacity varies directly as impeller diameter.

(b) Head varies directly as (impeller diameter)^2.

(c) Horsepower varies directly as (impeller diameter)^3.

PUMP SELECTION AND SYSTEM-HEAD CURVE

In pumping problems it is often convenient to show the relation between flow and friction head in the piping system graphically. Such a curve is shown in Figure 6.5 and is known as a system friction-head curve. It is obtained by plotting the calculated friction head in the piping, valves, and fittings of the suction and discharge lines, against the flow on which the calculation was based. In the figure, at 200 gpm the friction head is 9 ft, point *A*, and at 400 gpm the friction head is 36 ft, point *B*. Friction head varies roughly as the flow squared. This approach is acceptable whenever the flow in the pipeline is in turbulence (Reynolds number 2100 or over).

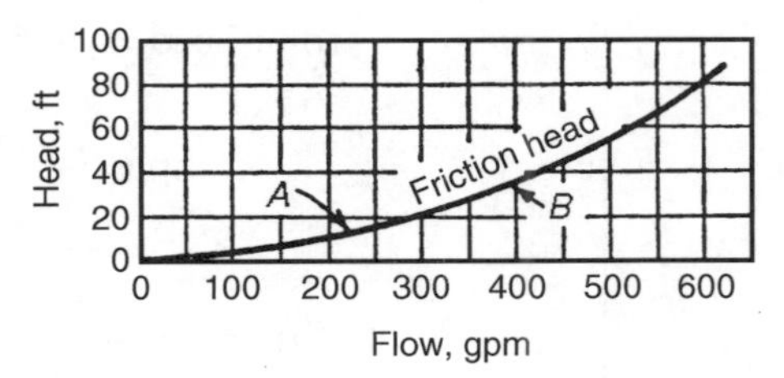

Figure 6.5

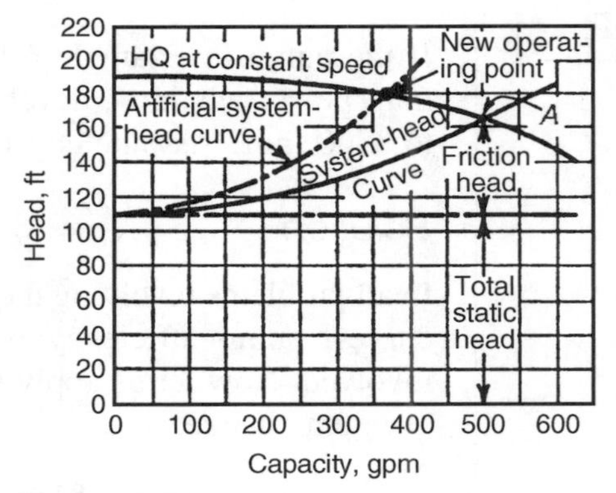

Figure 6.6

The system-head curve for a particular piping system is obtained by combining friction-head curve with static head and any difference of pressures in a pumping system. In Figure 6.6 the friction-head curve (Fig. 6.5) is combined with a total static head of 110 ft to obtain the system-head curve. Here the static head is assumed to be constant, as it is where the suction source and discharge levels are constant. Friction head increases with the flow. Superimposing the pump's head-capacity curve on the system-head curve, as in Figure 6.6, shows the head and capacity at which the pump will operate, such as point *A* where the two curves cross. In this system the pump discharges 500 gpm at a 165-ft head.

Where static or pressure varies, system-head curves may be plotted for minimum and maximum static heads or pressures. Superimposing a pump's head-capacity curve on such system-head curves permit predicting the capacity the pump delivers at different static heads.

Reduced pump capacity can be obtained with constant-speed operation by throttling its discharge to increase the friction head in the system. The capacity at which a centrifugal pump operates is determined by the intersection of its head-capacity characteristic curve with the system-head curve. Throttling the discharge of the pump increases the friction head in the system. Consequently, a new system-head curve results, such as that indicated by the artificial system-head curve in Figure 6.6. Point of intersection of the pump's head-capacity curve with the new system-head curve determines the pump's new operating capacity, as indicated.

Throttling a centrifugal's pump discharge will not build up excessive pressure. It cannot do so. For example, in Figure 6.6 if a valve is slowly closed, the head developed by the pump increases until at shutoff it reaches 190 ft, or 16.5 percent over that at design capacity. Head at shutoff varies with the pump type, or specific speed, but under no conditions will a centrifugal pump develop an excessive head at shutoff. However, a centrifugal pump should not be operated at shutoff for any great length of time, for the pump will overheat.

As we have already indicated, a system-head curve is a graphical representation of the relationship between flow and hydraulic losses in a given piping system.

Because hydraulic losses are functions of rate of flow, size and length of pipe, and size, number, and type of fittings, each system has its own characteristic curve and specific values.

In virtually all pump applications at least one point on the system curve is given to the pump manufacturer in order to help him select the pump properly. It is well to repeat here that the manufacturer will guarantee only the one point given to him by the customer. In many cases, however, it is highly desirable graphically to superimpose the entire system curve over the head-capacity curve of the candidate pump as in Figure 6.6.

CENTRIFUGAL PUMPS IN PARALLEL OR SERIES OPERATION

Frequently, where there is a wide range in demand, two or more pumps may be operated in parallel or series to satisfy the high demand, with just one of the pumps used for low demands. For proper specification of the pumps and evaluation of their performance under various conditions, the system curve should be used in conjunction with the composite pump performance curves.

For *pumps in parallel*, performance is obtained by adding the capacities at the *same head*. For *pumps in series*, performance is obtained by adding the heads *at the same capacity*. Figure 6.7 shows single pump in operation, two pumps in parallel, and two pumps in series. Figure 6.8 shows pump performance in parallel. Here, superimposing the system curve on the pump performance curves clearly indicates what flow rates can be expected and at what heads each of the pumps will be operating.

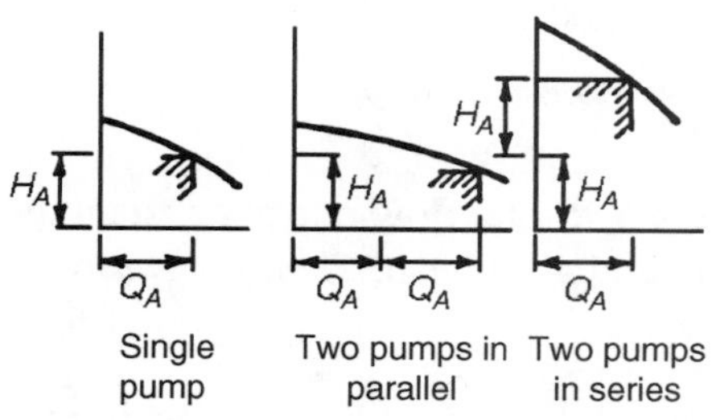

Figure 6.7

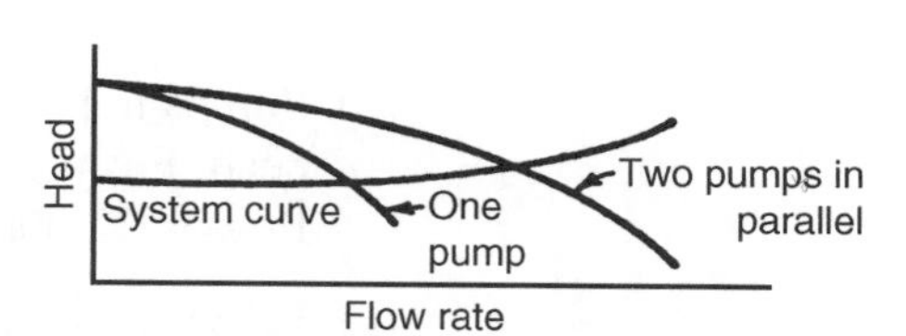

Figure 6.8

Example 6.3

Each of two variable speed centrifugal pumps to be used in a water-pumping station has the characteristic curve shown in Exhibit 2 when operating at a speed of 400 rpm. The pumps are to be arranged in series or parallel operation. Compute the following:

(a) The brake horsepower input for each pump when the pump is operating at the point of maximum efficiency at 50 percent of its rated speed.

(b) The total discharge when both pumps are operated in parallel at rated speed and when the total dynamic head is 40 ft.

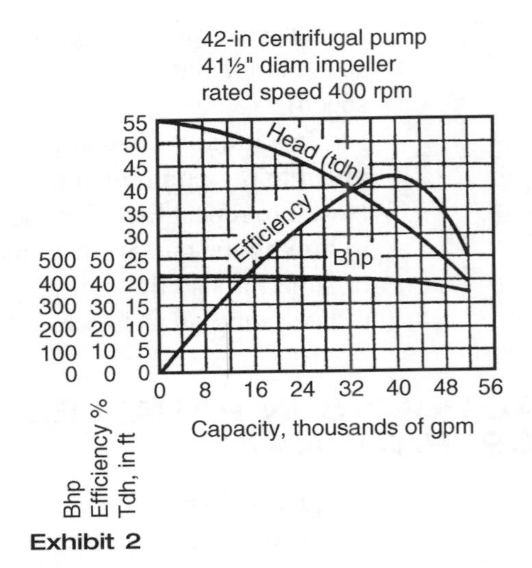

Exhibit 2

(c) The total discharge when both pumps are operated in series at rated speed, and when the total dynamic head is 50 ft.

Solution

(a) Brake horsepower is proportional to speed cubed. Apply laws of affinity and assume efficiency constant. Brake horsepower at full rated speed (from curve) is 400. Then

$$\frac{400}{z} = \left(\frac{400}{200}\right)^3 = 8$$

$$z = 400/8 = 50 \text{ bhp.}$$

(b) Pumps in parallel: each delivers half total flow rate at constant total dynamic head. From curve, 32,000 gpm and 40 ft head. Then for two pumps the total flow is 32,000 × 2 = 64,000 gpm.

(c) Pumps in series: each delivers same flow rate but at one-half of total dynamic head. From curve, at 50/2 head, flow rate is 48,000 gpm.

Example **6.4**

Two centrifugal pumps have head-capacity curves and efficiencies as shown in Exhibit 3. Pump 1 has a speed of 950 rpm and pump 2 a speed of 1150 rpm.

(a) What would be the combined discharge when both pumps work in parallel against a total dynamic head of (1) 40 ft and (2) 20 ft?

(b) Against what total dynamic head could the pumps deliver 75 gpm when working in series?

Solution

(a) Placing several pumps in parallel on the same line decreases the capacity of each in a way that varies with the pump characteristics. There is, of course,

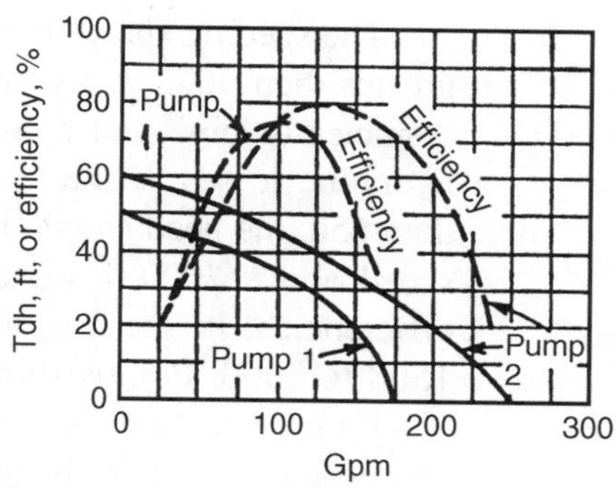

Exhibit 3

increased friction in the discharge lines when more water is put through; and, as this increases the head on the pump, there will be a corresponding decrease in discharge. To find data for parallel operation, add the discharge of each at the same pressure (head). From Exhibit 2, Pump 1 at 40 ft produces 75 gpm and Pump 2 125 gpm.

1. at 40 ft = 75 + 125 = 200 gpm.

2. at 20 ft = 150 + 200 = 350 gpm.

(b) For pumps in series, keep gpm constant and build up vertically on head. From curve read Pump 1 at 75 gpm = 40 ft, Pump 2 at 75 gpm = 50 ft. Therefore total head equals 40 ft + 50 ft = 90 ft.

SPECIFIC SPEED

For units of capacity and head (feet) used in the United States, specific speed of a centrifugal pump is the speed at which an exact model of the pump would have to run if it were designed to deliver 1 gpm against 1-ft head per stage. Thus all pump sizes can be indexed by the rotative speed of their unit capacity-head model. Specific-speed index of a pump is a guide in determining maximum head against which it can operate, modified by suction conditions. The lower the specific speed is, the higher the head per stage that can be developed. Low specific speeds range from 500 to 1000, medium specific speeds from 1500 to 4000, and high specific speeds 5000 to 20,000.

The specific speed of a pump can be calculated from the formula

$$N_s = \frac{N\sqrt{Q}}{H^{3/4}}$$

where

N = rpm
Q = gpm
H = ft head per stage.

For the correct typing of a pump design use Q and H values that give maximum efficiency.

The specific speed for efficient centrifugal pumps should never be below 650 or greater than 5000 at its rated point. For specific speed below 1000, the impeller diameter is large and narrow having excessively high disc friction and excessive hydraulic losses. If the specific speed for a set of given conditions becomes less than 650, the head should be divided between several stages. For values of specific speed above 2000, a mixed-flow impeller (Francis type) is generally used. Best efficiencies, in general, are obtained from pumps having specific speeds from 1500 to 3000. Pumps should be selected to fall within this range.

Figure 6.9 shows that specific speed is approximately related to impeller shape and efficiency. There is no sharp dividing line between various impeller designs. Ranges shown are approximate. Suction limitations of different pumps bear a relation to the specific speed. The Hydraulic Institute publishes charts giving recommended specific speed limits for various conditions.

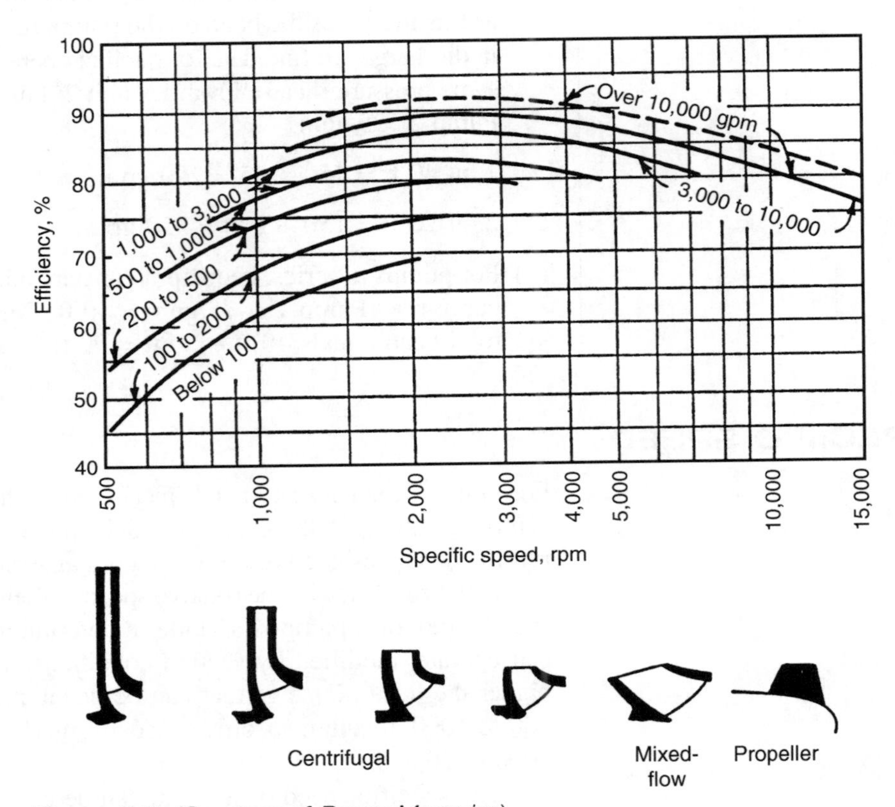

Figure 6.9 (Courtesy of *Power Magazine*)

Example 6.5

Required to pump 1000 gpm water at a total head of 500 ft. Pump will be directly connected to a 60-cycle induction motor. Assuming a speed of 1150 rpm and a single-stage pump, select the pump.

Solution

$$N_s = \frac{N\sqrt{Q}}{H^{3/4}} = \frac{1150\sqrt{1000}}{(500)^{3/4}}$$

$$N_s = 343.9 \quad \text{Speed is too slow.}$$

We must use two stages.

$$N_s = \frac{1150\sqrt{1000}}{(250)^{3/4}} = 578 \quad \text{Still too low.}$$

Try three stages,

$$N_s = \frac{1150\sqrt{1000}}{(167)^{3/4}} = 783 \quad \text{Good.}$$

Use a three stage pump.

Example 6.6

A centrifugal pump is required to develop a discharge head of 30 psi at 2000 rpm at shutoff head. Determine diameter of this impeller.

Solution

Use the following equation: Impeller diameter (in.) is equal to

$$d = C\frac{1840}{N\text{ rpm}}(\text{Head ft H}_2\text{O})^{1/2} = 1\left(\frac{1840}{2000}\right)\left[(30\text{ psi})\left(\frac{34\text{ ft H}_2\text{O}}{14.7\text{ psi}}\right)\right]^{1/2}$$

The constant C may be taken as unity but it varies between 0.95 and 1.09.

$$d = 7.66\text{ in.}$$

CAVITATION

Cavitation describes a cycle of phenomena that occurs in flowing liquid because the pressure falls below the vapor pressure of the liquid. When this occurs, liquid vapors are released in the low-pressure area and a bubble or bubbles forms. If this happens at the inlet to a centrifugal pump, the bubbles are carried into the impeller to a region of high pressure, where they suddenly collapse. Perhaps a good descriptive term for this is "implosion," the opposite of explosion.

Formation of these bubbles in a low-pressure area, and later their sudden collapse, is called *cavitation*. Erroneously the word is frequently used to designate the effects of cavitation rather than the phenomenon itself.

How does cavitation manifest itself in a centrifugal pump? Usual symptoms are noise and vibration in the pump, a drop in head and capacity with a decrease in efficiency, accompanied by pitting and corrosion of the impeller vanes. The pitting is a physical effect, which is produced by the tremendously localized compression stresses caused by the collapse of the bubbles. Corrosion follows the liberation of oxygen and other gases originally in solution in the liquid. When centrifugal pumps are being considered for handling liquids that are extremely volatile and close to their bubble points, the characteristic curve should also show the *NPSH* requirements.

There is a tendency at time to specify a small, less expensive pump at excessive speeds to compete with the higher initial cost of a larger pump operating at a lower speed. However, most manufacturers rate their pumps at safe operating speeds even in the face of low price competition.

Practical rotative speeds are inversely relative to the diameter of the impeller, therefore, a smaller pump will operate at a greater rpm than a large pump, but will also deliver less water at peak performance than a large one. When an impeller is lifting water, there is a much greater pressure on the upper or working side of the blades than on the under side, and at safe speeds this pumping action is normal and will give indefinite service. However, if the pump is operated at excessive speeds, the differential pressure becomes too great and causes a powerful pulsating vacuum on the under side of the blade tips. As each particle of water is pulled away from the blade, it takes with it a small particle of the metal and produces a peculiar pitting or grooved effect. Thus, cavitation takes its toll on the metal, and its repeated erosive action finally results in complete honey-combing and total destruction of the blade, with resultant loss in pump performance. Figure 6.10 shows the effect of cavitation on efficiency and discharge of a centrifugal pump.

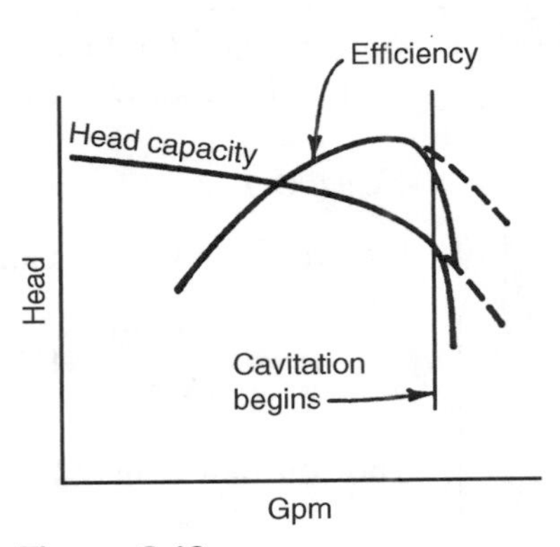

Figure 6.10

NPSH AND THE CENTRIFUGAL PUMP

A centrifugal pump is designed to handle liquids that for all practical purposes are incompressible. It will not function as a normal pump when handling compressible fluids such as gases or vapors. If such gases are present in the pump, they interfere with its normal operation and have a marked effect on its characteristics as we have previously seen. To prevent vaporization at the impeller of a centrifugal pump, it is necessary to keep the pressure at this point above the vapor pressure of the liquid at all times.

What is NPSH? The pump manufacturer knows from experience, calculations, and experiments the values of the pressure drops from the suction flange to the impeller vanes. This *internal* pressure drop can be called "suction loss in the pump," since it is a loss in available pressure to prevent vaporization at the point of low pressure. The important point to keep in mind is to maintain the pressure at the entrance to the impeller vanes above the vapor pressure of the liquid at all times. The energy available at the pump suction flange to do this and to overcome pump suction losses is called *net positive suction head over the vapor pressure.* This term is commonly abbreviated NPSH, and was originally used in conjunction with pumps handling boiling liquids, where all the available energy came from the static elevation of the liquid above the pump. It is also used to describe conditions where a volatile liquid is not at its vapor pressure at the pumping temperature, as we will show in a following example.

The suction limitations for a particular pump are often shown by the manufacturer plotted in the form of a curve giving minimum NHSH requirements for all capacities in the operating range. As long as the available energy equals or exceeds these figures there will be no undue vaporization causing limited capacity, cavitation, and accompanying troubles. This means that for all practical purposes all values stated by the pump manufacturer for minimum NPSH requirements are based on

(a) The pressure drop from the suction flange to the impeller vane. This pressure drop includes the velocity head at the pump suction flange.

(b) All values of NPSH are referenced to the pump center line. This is assumed to be 3 ft above the pump house floor.

Most pump applications resolve themselves into one of a few basic types of suction conditions. Details vary with every installation but the general method of calculation is perhaps best illustrated by a few simplified examples. In these installations the setting is at sea level and all friction losses are at the maximum flow. To simplify formulas, the following symbols are used:

S = vertical distance from liquid surface to pump house floor, ft
B = distance from pump house floor to shaft center line, ft (assumed to be 3 ft in all examples, a good average figure)
p_a = atmospheric pressure, 14.7 psia
p_{vp} = vapor pressure of liquid at pumping temperature, psia
p = pressure on surface of liquid, psig
h_f = all losses in suction line up to pump suction flange, not including drop in pressure transformed into velocity head at suction flange.

The basic formula for calculating NPSH available is

$$\text{NPSH (ft)} = \frac{(p + p_a - p_{vp})2.31}{\text{sp gr}} \pm S - B - h_f.$$

For boiling liquids the first term of this equation automatically becomes zero. The plus (+) sign is used when there is a static suction and the minus (–) sign when there is a static suction lift.

Example 6.7

A liquid is boiling in an open tank at atmospheric pressure. Refer to Exhibit 4. Liquid temperature is at 212°F boiling water whose surface is 14 ft above the floor. Suction line losses are 5 ft. Calculate the NPSH available.

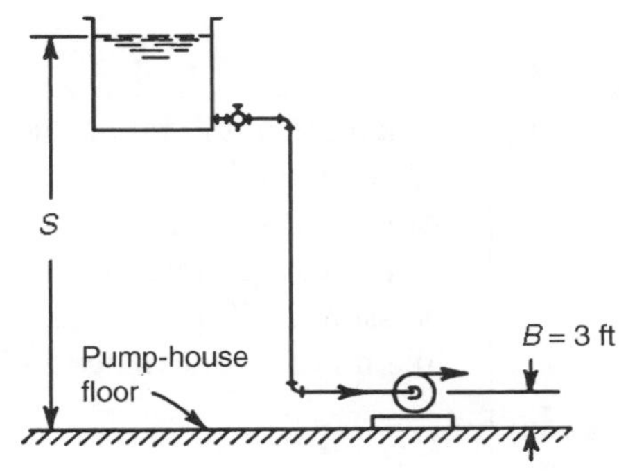

Exhibit 4

Solution

$$S = 14 \text{ ft} \quad p_a = p_{vp} \quad p = 0 \quad h_f = 5 \text{ ft}$$

Then

$$\text{NPSH} = S - B - h_f = 14 - 3 - 5 = 6 \text{ ft available.}$$

As long as a liquid is boiling at the surface the same NPSH is available regardless of whether the working pressure is 14.7 psia, or 200 psia. It is evident that if the liquid is boiling, the NPSH available is derived solely from the static head. The working pressure has no bearing on the static head, hence it cannot affect the NPSH.

Example 6.8

Assume the tank in Exhibit 5 is a closed one and contains cold liquid propane of 0.58 sp gr with the tank at 200 psia, which is the equilibrium vapor pressure of the propane at the pumping temperature. The suction line losses are 5 ft and the static head is 14 ft. Calculate the available NPSH.

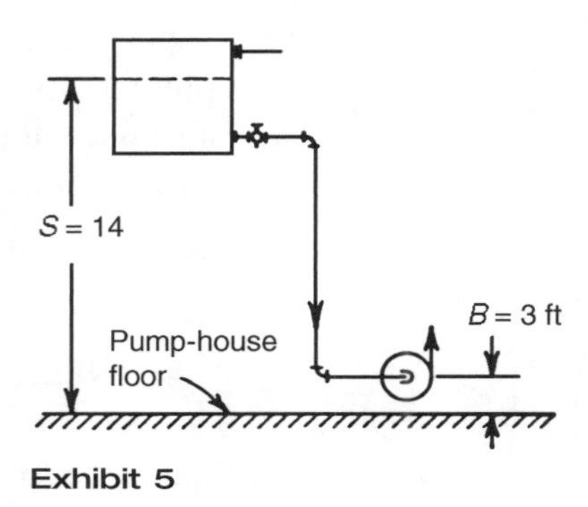

Exhibit 5

Solution

$$S = 14 \text{ ft} \quad h_f = 5 \text{ ft} \quad p + p_a = p_{vp} = 200 \text{ psia}$$

Then

$$\text{NPSH} = S - B - h_f = 14 - 3 - 5 = 6 \text{ ft available.}$$

Example 6.9

When the liquid in the suction tank is not at its boiling point, it is possible to take advantage of the energy in the difference between the working pressure and the vapor pressure. Let us assume that the tank in Exhibit 6 is a closed accumulator at a working pressure of 30 psig and contains a liquid of 0.65 sp gr whose vapor pressure is 40 psia. Again, if the static head is 14 ft and the losses in the suction line are 2 ft, calculate the NPSH available.

Solution

$$S = 14 \text{ ft} \quad h_f = 2 \text{ ft} \quad p_{vp} = 40 \text{ psia} \quad p = 30 \text{ psig}$$

Then

$$\text{NPSH} = \frac{(30+14.7-40)2.31}{0.65} + 14 - 3 - 2 = 16.8 + 9$$
$$= 25.8 \text{ ft available.}$$

Example 6.10

Tank farms, marine terminals, and bulk service stations present limiting suction conditions due to long lengths of piping. Appearances are deceiving and sometimes when there is apparently ample suction head, the piping conditions are such that they produce a suction lift condition at the pump flange. Refer to Exhibit 6. If it is full, the liquid surface is 28 ft above the pumphouse floor, and when almost empty it is 18 ft. The tank contains gasoline at 0.74 sp gr with a vapor pressure of 7.3 psia at the pumping temperature. Tank is open to the atmosphere and the line losses are 29 ft. Calculate the NPSH available.

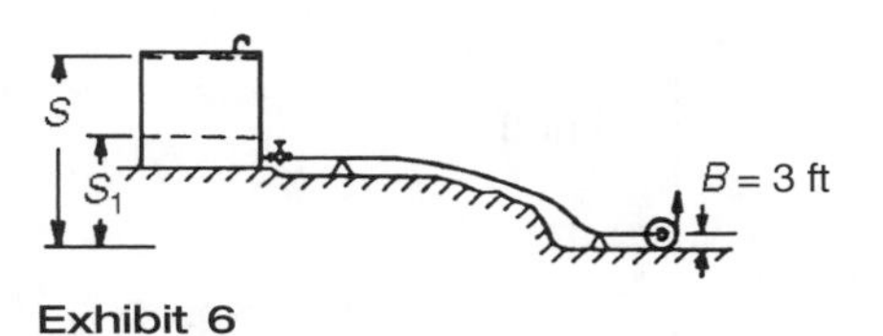

Exhibit 6

Solution

$$S = 28 \text{ ft and } 18 \text{ ft} \quad h_f = 29 \text{ ft} \quad p_{vp} = 7.3 \text{ psia}$$

Then

$$\text{NPSH} = \frac{(p_a - p_{vp})2.31}{\text{sp gr}} + S - B - h_f$$
$$= \frac{(14.7-7.3)2.31}{0.74} + 28 - 3 - 29 = 19.1 \text{ ft available with a full tank}$$
$$= 19.1 - 10 = 9.1 \text{ ft available with an empty tank.}$$

The worst condition is for the empty tank, and the pump selected should be good for these conditions. Care must be taken to prevent any pockets from developing in the suction line where vapor might collect and vapor-bind the line.

The situation often arises when pumping takes place with a centrifugal pump and there is a suction lift as from an underground tank. Such an arrangement should be avoided wherever possible. There is no static head; hence, all the energy necessary to get the liquid into the impeller eye must come from the difference between the working pressure at the surface of the liquid and its vapor pressure.

Example 6.11

The storage tank in Exhibit 7 is vented to the atmosphere and contains gasoline as in the tank in Exhibit 6. Vapor pressure is 6 psia at the pumping temperature. The surface of the liquid is 6 ft below the floor level when the tank is half full and 9 ft when at the entrance to the suction pipe. Suction line losses are 2 ft. Calculate NPSH available.

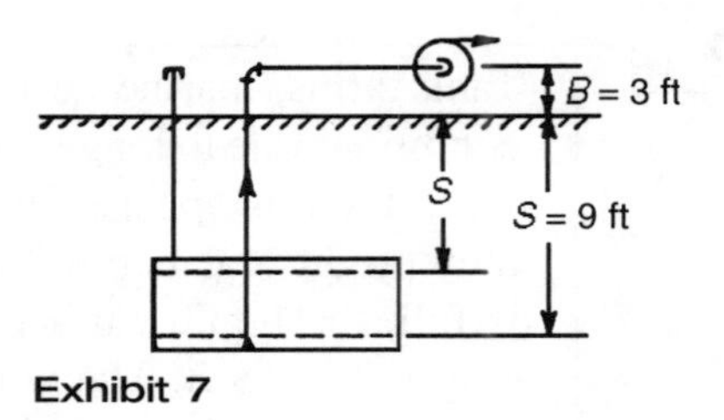

Exhibit 7

Solution

$$S = 9 \text{ ft} \quad h_f = 2 \text{ ft} \quad p_{vp} = 6 \text{ psia}$$

Then

$$\text{NPSH} = \frac{(p_a - p_{vp})2.31}{\text{sp gr}} - S - B - h_f$$

$$\text{NPSH} = \frac{(14.7 - 6)2.31}{0.74} - 9 - 3 - 2 = 13 \text{ ft available.}$$

A pump with suitable NPSH characteristics will operate under these conditions but care must be taken to make sure that the capacity does not exceed that allowed by the NPSH available. Vapor binding is sure to develop in such an installation and it is recommended that a positive displacement type of pump be installed.

Example 6.12

A motor-driven pump is required to deliver 400 gal of water per minute (gpm) against a head of 200 ft for 2000 hr each year. Bids are offered by two concerns for pumps having an expected life of 15 years with a salvage value equal to the cost of removal. The cost of interest, taxes, and insurance may be taken as 8 percent of the purchase price, and the cost of power is 2.5¢ per kwhr. Bid A guarantees an overall efficiency of 79 percent at full load, whereas bid B guarantees an overall efficiency of 75 percent. How much more is the client justified in paying for the pump and motor with the higher efficiency?

Solution

Hydraulic hp required = $(400 \times 8.33 \times 200)/33{,}000 = 20.2$ hhp
Electrical kilowatt equivalent = $20.2 \times 746 = 15.1$ kw
At 79 percent efficiency, kw for bid $A = 15.1 \times 1/0.79 = 19.1$
At 75 percent efficiency, kw for bid $B = 15.1 \times 1/0.75 = 20.1$

Annual cost saving in electrical energy by Bid A

$$= (20.1 - 19.1) \times 2000 \times 0.025 = \$50 \text{ per year saved by Bid } A.$$

If we let x = the additional capital investment made by Bid A, then $0.08x$ = additional annual fixed charges.

$$x/15 = \text{annual amount for recovery of } x.$$

Finally, the balance or "breakdown" point:

$$\text{Additional fixed annual charges} + \text{recovery} = 50$$

$$0.08x + x/15 = 50$$

Solving for x,

$$x = \$341.$$

CHAPTER 7

Fans, Blowers, and Compressors

OUTLINE

The theory of design for equipment to move gases and vapors is similar to that of pumps for liquids, with the exception that gases are compressible. Heads developed in terms of feet of fluid being pumped are almost identical for equivalent circumferential speeds as for centrifugal pump designs. Axial-flow impellers impart similar heads to both liquids and gases, and valve actions in positive displacement compressors are quite similar to those of piston water pumps.

When considering the compression of gases, the machines are usually referred to as fans (centrifugal and propeller), blowers (centrifugal and lobular), and compressors (multistage centrifugal, lobular, and piston).

FANS AND BLOWERS

We begin our review of fans and blowers with two solved examples that employ two of the basic equations of interest.

Example 7.1

Find the motor size needed to provide the forced-draft service to a boiler that burns coal at the rate of 10 tons per hr. The air requirements are 59,000 cfm, air is being provided under 6 in. water gauge (WG) by the fan, which has a mechanical efficiency of 60 percent. Assume the fan to deliver at a total pressure of 6 in. WG.

Solution

The horsepower is determined by the basic formula given as

$$\text{hp} = \frac{\text{cfm} \times \text{pressure, psf}}{(33{,}000\ \text{ft-lb/hp-min})(\text{efficiency})} \quad \textbf{(7.1)}$$

We must convert the 6 in. of water to represent a pressure in pounds per square foot.

$$\frac{6}{12} \times 62.4 = 31.2\ \text{psf}$$

Finally,

$$\text{hp} = \frac{59{,}000 \times 31.2}{33{,}000 \times 0.60} = 93\ \text{required; use 100-hp motor.}$$

Example 7.2

A blower with the inlet open to the atmosphere delivers 3000 cfm of air at a pressure of 2 in. WG through a duct 11 in. in diameter, the manometer being attached to the discharge duct at the blower. Air temperature is 70°F, and the barometer pressure is 30.2 in. Hg. Calculate the air horsepower.

Solution

$$\dot{W} = \frac{\dot{m} v \Delta P}{\eta_f} \qquad \text{Air hp} = \dot{m} v \Delta P \quad \textbf{(7.2)}$$

$$\text{bhp} = \frac{(\dot{m}\ \text{lbm/s})(v\ \text{ft}^3/\text{lbm})(\Delta P\ \text{lbf/ft}^2)}{550 \frac{\text{ft-lbf}}{\text{hp-s}} \eta_f (\text{decimal eff.})}$$

$$\begin{matrix}\text{Correct standard air density}\\ \text{to actual air density}\end{matrix} = (0.075\ \text{lbm/ft}^3)\left(\frac{30.2\ \text{actual pr}}{29.92\ \text{stand pr}}\right)$$

$$\rho = 0.0757 \quad \text{so sp. vol } v = 13.21\ \text{ft}^3/\text{lb}$$

Find total head knowing that it is the sum of static head and velocity head.

$$\text{Static head} = \left(\frac{2\ \text{in H}_2\text{O}}{12\ \text{in/ft}}\right)\left(\frac{62.4\ \text{lb/ft}^3\text{H}_2\text{O}}{0.0757\ \text{lb/ft}^3\text{air}}\right) = 137.4\ \text{ft air}$$

$$\text{Velocity} = \frac{Q}{A} = \frac{3000\ \text{ft}^3/\text{min}}{60\ \text{s/min}(\frac{\pi}{4})(\frac{11}{12})^2\text{ft}^2} = 75.76\ \text{ft/s}$$

$$\text{Velocity head} = \frac{(75.76)^2\ \text{ft}^2/\text{s}^2}{2(32.2)\ \text{ft/s}^2} = 89.1\ \text{ft air}$$

$$\text{so air hp} = \frac{\left(\frac{3000\ \text{ft}^3/\text{min}}{60\ \text{s/min}}\right)(137.4 + 89.1)\text{ft}\ (0.0757\ \text{ft}^3/\text{lb})}{(550\ \text{ft-lbf/hp-s})} = 1.56\ \text{hp}$$

Parallel Operation

What about parallel operation? It is a well-known fact that the dip in the pressure curve is a serious drawback to parallel operation when fans are operating at peak efficiency. When fans are individually motored, it is possible to have one fan carrying more than its share of the load and the other fan much less. In unitary equipment, however, where two fans are running on the same shaft, this problem is not as serious. When the air from the two fans in the unitary equipment is carried *in individual ducts for a distance before joining*, the imposed series resistance will help to stabilize operation at the design conditions.

Fan Laws

In order to determine the effect of changes in the conditions of fan operation, certain fan laws are used and apply to all types of fans. Their application is necessarily restricted not only to fans of the same shape but also to the same point of rating on the performance curve. In the majority of cases, system resistance is so small that there is no need to correct horsepower for difference in pressure. However, if there is a great difference or change in temperature through a system, the fan exhausting hot gases may require more power than a blower furnishing the same weight of air to the system. It would be helpful to review the laws of affinity on pumps.

The following constitute the several fan laws:

(a) Air or gas capacity varies directly as the fan speed.

(b) Pressure (static, velocity, and total) varies as the square of the fan speed.

(c) Power demand varies as the cube of the fan speed.

The above apply to a fan having a constant wheel diameter. When air or gas density vary, the following apply:

(d) At constant speed and capacity the pressure and power vary directly as the air or gas density.

(e) At constant pressure the speed, capacity, and power vary inversely as the square root of density.

(f) For a constant weight of air or gas, the speed, capacity, and pressure vary inversely as the density. Also, the horsepower varies inversely as the square of the density.

Example **7.3**

A certain fan delivers 12,000 cfm at a static pressure of 1 in. WG when operating at a speed of 400 rpm, and requires an input of 4 hp. If 15,000 cfm are desired in the same installation, what will be the new speed, new static pressure, and new power needs?

Solution

$$\text{New speed} = 400 \times \frac{15{,}000}{12{,}000} = 500 \text{ rpm}$$

$$\text{New static pressure} = 1 \times \left(\frac{500}{400}\right)^2 = 1.56 \text{ in.}$$

$$\text{New power} = 4 \times \left(\frac{500}{400}\right)^3 = 7.81 \text{ hp}$$

Example 7.4

A certain fan delivers 12,000 cfm at 70°F and normal barometric pressure at a static pressure of 1 in. WG when operating at 400 rpm, and requires 4 hp. If the air temperature is increased to 200°F (density 0.06018 lb per cu ft) and the speed of the fan remains the same, what will be the new static pressure and power?

Solution

$$\text{New static pressure} = 1 \times \frac{0.06081}{0.075} = 0.80 \text{ in.}$$

$$\text{New power} = 4 \times \frac{0.06081}{0.075} = 3.2 \text{ hp}$$

Example 7.5

If the speed of the fan in Example 7.4 is increased so as to produce a static pressure of 1 in. WG at 200°F, what will be the new speed, new capacity, and new power needs?

Solution

$$\text{New speed} = 400 \times \sqrt{\frac{0.075}{0.06018}} = 446 \text{ rpm}$$

$$\text{New capacity} = 12{,}000 \times \sqrt{\frac{0.075}{0.06018}} = 13{,}392 \text{ cfm (at 200°F)}$$

$$\text{New power} = 4 \times \sqrt{\frac{0.075}{0.06018}} = 4.46 \text{ hp}$$

Example 7.6

If the speed of the fan of the previous examples is increased so as to deliver the same weight of air at 200°F as at 70°F, what will be the new speed, new capacity, new static pressure, and new power?

Solution

$$\text{New speed} = 400 \times \frac{0.075}{0.06018} = 498 \text{ rpm}$$

$$\text{New capacity} = 12{,}000 \times \frac{0.075}{0.06018} = 14{,}945 \text{ cfm}$$

$$\text{New static pressure} = 1 \times \frac{0.075}{0.06018} = 1.25 \text{ in.}$$

$$\text{New power} = 4 \times \left(\frac{0.075}{0.06018}\right)^2 = 6.20 \text{ hp}$$

Duct-fan Characteristics

As in pumping, all system resistance curves pass through the origin. The curves will intersect all fan performance curves at some point. However, where the intersect takes place, operation must be stable and the efficiency must be high.

A static pressure curve can easily be drawn through a given point based on the fact that the pressure required to overcome system resistance to flow varies for all practical purposes as the square of the flow rate. When a certain fan is connected to a given system, the system characteristic may be used to learn what will happen. Suppose we want to know what such an arrangement will look like when handling 12,500 cfm against a static pressure of 0.95 in. WG (see Figure 7.1). We see that but one condition satisfies both fan and system. This is point *A*. A higher-speed fan having a definite wheel diameter when attached to the same system would show higher cfm capacity, higher static pressure, and greater horsepower requirements. In order to realize the needed capacity at the static pressure resulting from this flow, the duct system must be dampered or the fan speed reduced. Normally, in practice the fan is selected to produce the correct flow and develop the desired static pressure at the selection point where both curves cross. Because dampering is a waste of power when carried on in the duct system, this mode of operation should be avoided as much as possible. For best total results, radial dampering at fan inlet is suggested.

Fans in Series

When low-pressure fans are in series, the weight of flow is substantially that of one fan but the total pressure is the sum of the total pressures of the fans in series. The installation of two identical fans in series obviously does not double the

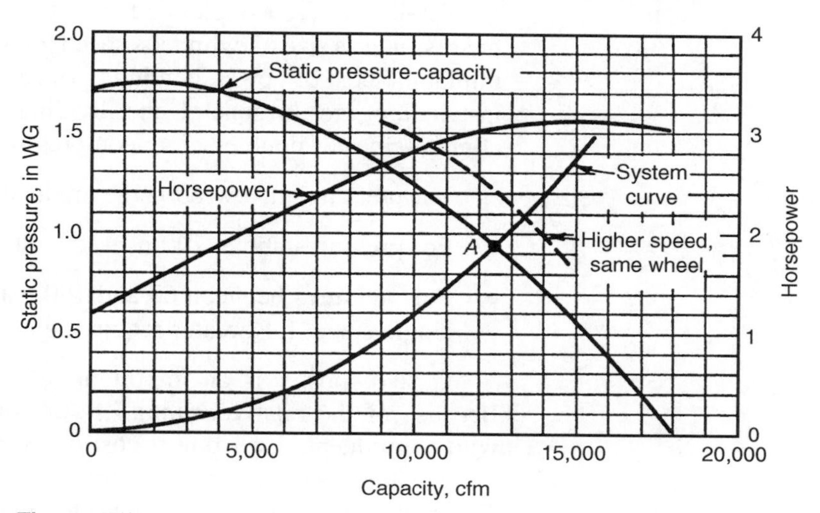

Figure 7.1

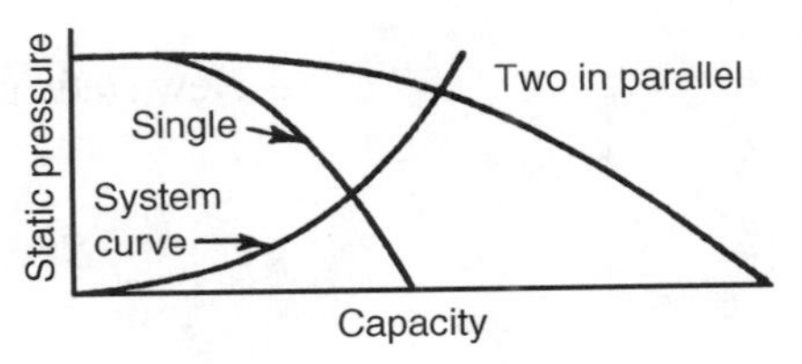

Figure 7.3

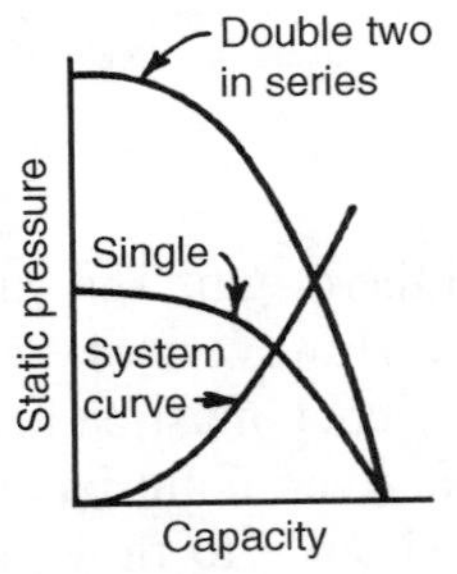

Figure 7.2

quantity of flow through a given system. Double the air quantity would require some four times the pressure and eight times the horsepower. Placing two identical fans in series a little more than doubles the horsepower and increases the capacity and static pressure as indicated in Figure 7.2. This placing of two identical low-pressure fans in series is often termed staging.

Fans in Parallel

Previously we passed over paralleling of fans. We shall now show how two identical fans can be hooked up characteristically. The characteristic of the combined set is such that the capacity is the sum of the separate capacities for a given static or total pressure (see Fig. 7.3). It is well to note that the placing of fans in parallel has been a more common practice than placing them in series.

AIR AND GAS COMPRESSORS

Standard types of stationary reciprocating compressors include (a) vertical and V-type single-acting units in sizes up to 100-hp single-stage, and two-stage, either air- or water-cooled; (b) vertical or V-type semi-radial and right-angle type double-acting machines, available in sizes of 60 hp and above, single-stage and multistage, and water-cooled only; (c) single-frame horizontal or vertical double-acting compressors, available in sizes up to 125 hp, single-stage and multistage, and water-cooled only; (d) duplex, horizontal, or vertical double-acting machines in sizes of 75 hp and over, single-stage and multistage, and water-cooled only.

Choice between single- and two-stage compression depends on many varying factors such as size of cylinders, speed, ratio of compression, discharge temperature limitation, cost of power, continuity of service, method of cooling, permanence of installation, etc. In general, the dividing line between single- and two-stage air compression for double-acting compressors may be drawn as follows:

For pressures below 60 psig, single-stage

For pressures above 100 psig, two-stage

For pressures between 60 and 100 psig and capacities below 300 cfm use single stage, for greater requirements use two stage.

Cost of power is an important factor in selecting the type of compressor. Because of the relatively long life of compressor equipment, higher efficiency with consequent lower power cost often justifies a higher initial investment.

Treatment of compressor problems will be confined to the reciprocating type. The student is referred to standard texts for the many other types.

Example **7.7**

Air-compressor capacity is the quantity of air compressed and delivered per unit of time. It is usually expressed in cfm at intake pressure and temperature. Assuming that the intake temperature is at 68°F (528°R) and 14.7 psia, how much compressed air at 250 psia and 80°F is delivered by a compressor with a rated capacity of 700 cfm?

Solution

Assuming that the perfect-gas law applies, we can proceed as follows:

$$\frac{P_1V_1}{T_1} = \frac{P_2V_2}{T_2} \quad \text{so} \quad \frac{(14.7\text{ psia})(700\text{ cfm})}{(68+460)R} = \frac{(250\text{ psia})V_2}{(460+80)}$$

Solving for V, we find it to be equal to 42 cfm.

Work of Compressor without Clearance

Refer to Figure 7.4. The work done on the air (or gas) is the area enclosed within the diagram and is expressed by the following equation for polytropic processes on ideal gases:

$$W = \frac{nwR(T_2 - T_1)}{1-n}\text{ ft-lb.} \tag{7.3}$$

A more convenient form is

$$W = \frac{nP_1V_1}{1-n}\left[\left(\frac{P_2}{P_1}\right)^{(n-1)/n} - 1\right]\text{ft-lb,} \tag{7.4}$$

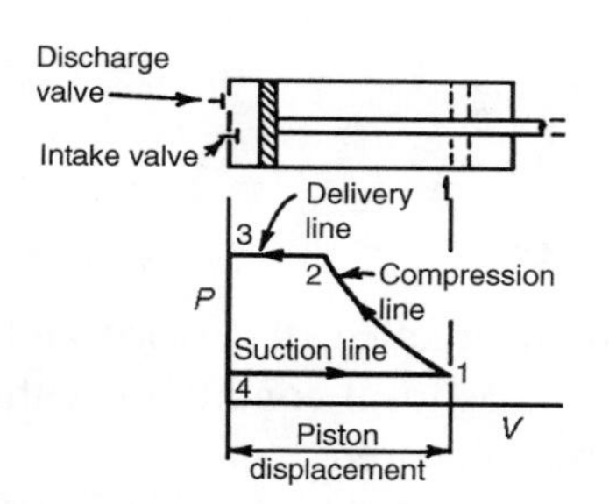

Figure 7.4

where V_1 is volume drawn into cylinder measured in cubic feet and w is the weight of this charge of gas, which passes through the compressor. For isentropic compression

$$W = \frac{kP_1V_1}{1-k}\left[\left(\frac{P_2}{P_1}\right)^{(k-1)/k} - 1\right]\text{ ft-lb.} \tag{7.5}$$

For isothermal compression

$$W = wRT_1 \ln\frac{P_1}{P_2}\text{ ft-lb.} \tag{7.6}$$

Example **7.8**

Calculate the power required to compress 5000 cu ft of "free" air per hour, initially at 14.5 psia and 70°F, to 100 psia, assuming nonclearance compressor (Fig. 7.5) and single-stage compression, (a) isothermal compression and (b) isentropic compression.

Solution

(a) We need to use the isothermal work equation so weight of air needs to be calculated using the ideal gas law:

$$PV = WRT$$

$$\dot{W} = \frac{P\dot{V}}{RT} = \frac{(14.5 \text{ psia})(144 \text{ in}^2/\text{ft}^2)(5000 \text{ ft}^3/\text{hr})}{53.3\frac{\text{ft-lb}}{\text{lb R}}(70 + 460)°\text{R}}$$

$$\dot{W} = 369.6 \text{ lb/hr}$$

Using isothermal work equation:

$$\dot{W} = \dot{w}RT_1 \ln\frac{P_2}{P_1} = (369.6 \text{ lb/hr})\left(53.3\frac{\text{ft-lb}}{\text{lb°R}}\right)(530°\text{R}) \ln\frac{100 \text{ psia}}{14.5 \text{ psia}}$$

$$\dot{W} = 20.2 \times 10^6 \text{ ft-lb/hr} = \frac{20.2 \times 10^6 \text{ ft-lb/hr}}{(3600 \text{ s/hr})(530 \text{ ft-lb/hps})}$$

$$\dot{W} = 10.2 \text{ hp}$$

(b) Now we use the equation for isentropic work.

$$\dot{W} = -\dot{w}\left(\frac{KRT_1}{K - T_1}\right)\left[\left(\frac{P_2}{P_1}\right)^{\frac{k-1}{k}} - 1\right]$$

$$\dot{W} = -369.6 \text{ lb/hr}\left[\frac{1.4(53.3\frac{\text{ft-lb}}{\text{lb°R}})(530°\text{R})}{1.4 - 1}\right]\left[\left(\frac{100 \text{ psia}}{14.7 \text{ psia}}\right)^{\frac{1.4-1}{1.4}} - 1\right]$$

$$\dot{W} = 26 \times 10^6 \frac{\text{ft-lb}}{\text{hr}} \quad \text{or} \quad \frac{26 \times 10^6 \text{ ft-lb/hr}}{(3600 \text{ s/hr})(550)\frac{\text{ft-lb}}{\text{hp-s}}} = 13.1 \text{ hp}$$

Isothermal Compression Horsepower

Although isothermal compression is never attained in practice, the work and horsepower required by the process are quite often used as ideal conditions with which the actual performance of a compressor may be compared.

As we have indicated previously, the net work represented by an ideal *PV* diagram is equal to the area enclosed in the diagram. By integration, it can be shown that the area is equal to the work expressed in foot-pounds.

$$W = wRT_1 \ln\frac{P_2}{P_1} = P_1V_1 \ln\frac{P_2}{P_1} \quad \textbf{(7.7)}$$

where

W = work of isothermal compression, ft-lb
P_1 = intake pressure, lb per sq ft abs
V_1 = volume of gas compressed measured at intake conditions, cu ft

As we know, gas volumes are ordinarily measured in standard cubic feet so that Equation 7.7 can be made into a more useful form by substituting pressure and volume at standard conditions 14.7 psia and cubic feet measured at 60°F. Then Equation 7.7 becomes

$$W = (144)\text{in}^2/\text{ft}^2(14.7)\text{lb/in}^2(V_{ST})\ \ln\frac{P_2}{P_1} \tag{7.8}$$

Combining all constants,

$$W = (2116.8\ \text{lb/ft}^2)V_s\ \text{ft}^3/\text{lb}\ \ln\frac{P_2}{P_1} \tag{7.9}$$

This equation can be used to calculate the work required to compress any volume of gas isothermally or the horsepower according to the following:

$$\text{hp} = \frac{2116.8}{33000}V_s\ \ln\frac{P_2}{P_1} \tag{7.10}$$

$$\text{hp} = 0.064V_s\ \ln\frac{P_2}{P_1} \tag{7.11}$$

Example 7.9

What is the isothermal horsepower required for compressing 1000 scf of air per minute from an intake pressure of 14.0 psia to a discharge pressure of 60 psia?

Solution

You may either use Equation 7.7 or Equation 7.11. Equation 7.11 will be used here because it is simpler.

$$\text{hp} = 0.064V_s\ \ln\frac{P_2}{P_1} = 0.064(1000)\ \ln\frac{60}{14.7}$$

hp = 93.1 power required to compress the gas.

Example 7.10

A compressor operating isothermally compresses 5000 ft³/min of air from 14.7 psia to 450 psia. Entering temperature is 60°F.

Determine hp required.

Solution

Use the isothermal compressor equation:

$$\text{hp} = wRT_1\ \ln\frac{P_1}{P_2} = P_1V_1\ \ln\frac{P_1}{P_2} \tag{7.12}$$

$$\text{hp} = \frac{(14.7\ \text{lbf/in}^2)(5000\ \text{ft}^3/\text{min})(144\ \text{in}^2/\text{ft}^2)}{33000\dfrac{\text{ft-lbs}}{\text{hp-min}}}\ \ln\frac{14.7}{450}$$

hp = –1097 input to compressor.

Polytropic and Isentropic Compression Horsepower

With a given value of n, the actual compressor cylinder horsepower required for the compression of a given quantity of gas in a given length of time can be calculated. This indicated horsepower of the compressor should not be confused with the brake hp of compression. The brake horsepower of compression is somewhat larger than the indicated horsepower and represents the brake horsepower that an engine must deliver to a compressor not only to compress the gas but also to overcome the friction in the moving parts of the compressor.

Because of the similarity of the calculations involved, the determination of polytropic and isentropic hp will be discussed jointly.

The relation for work is given by

$$W = \frac{n}{n-1} P_1 V_1 \left[\left(\frac{P_2}{P_1} \right)^{\frac{n-1}{n}} - 1 \right] \tag{7.13}$$

For isentropic work substitute k for n.

If V_1 is the volume handled per minute, the preceding equation will give foot-pounds of work done per minute.

Because a common standard pressure for gas measurement is 14.7 psia, Equation 7.13 can be changed to a more convenient form by substituting this standard pressure in the equation as follows:

$$W = \frac{n}{n-1} \times 144 \times 14.7 V_s \left[\left(\frac{P_2}{P_1} \right)^{\frac{n-1}{n}} - 1 \right] \tag{7.14}$$

$$\text{or} \quad W = 2{,}117 V_s \frac{n}{n-1} \left[\left(\frac{P_2}{P_1} \right)^{\frac{n-1}{n}} - 1 \right] \tag{7.15}$$

where V_s is the volume of gas handled measured at 14.7 psia. When V_s is in terms of cubic feet per minute, the horsepower required for compression is

$$\text{hp} = \frac{2117}{33{,}000} V_s \frac{n}{n-1} \left[\left(\frac{P_2}{P_1} \right)^{\frac{n-1}{n}} - 1 \right] \tag{7.16}$$

Gas Compressor with Clearance

Refer to Figure 7.5. The events shown there are the same as for those in the case of no clearance, except that because the piston does not force all the gas from the cylinder at pressure P_2, the remaining gas must reexpand to the intake pressure, process 3-4, before intake starts again.

Because the value of n on the expansion curve has little effect on the results, it is taken as being the *same for both compression and expansion*, although actually the values are different. Without clearance, the volume of air drawn into the cylinder is the same as the piston displacement. The work of compression required is given by

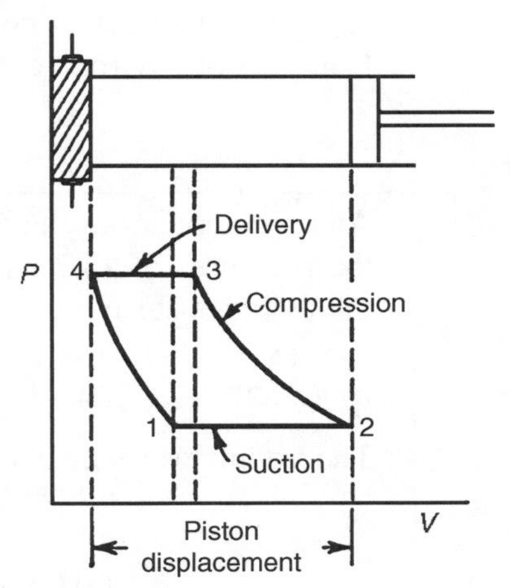

Figure 7.5

$$W = \frac{n}{n-1} w' RT_1 \left[\left(\frac{P_2}{P_1} \right)^{(n-1)/n} - 1 \right] \text{ ft-lb} \tag{7.17}$$

where w' is the weight of actual volume drawn into cylinder along 4-1. In an actual compressor with clearance, the piston displacement must be greater than volume drawn in, for a given capacity; this means a larger machine than without clearance, costing more and having greater mechanical friction.

Volumetric Efficiency

Volumetric efficiency ranges from 65 to 85 percent and is determined from the following:

$$E_v = \frac{\text{actual volume drawn into compressor, cu ft}}{\text{piston displacement, cu ft}} \tag{7.18}$$

$$\text{or} \quad E_v = 1 + c - \left[c \left(\frac{P_2}{P_1} \right)^{1/n} \right] \tag{7.19}$$

where c is clearance percent as a decimal.

Capacity of Gas Compressor

Let us assume the gas is air. The capacity is actual volume of "free" air delivered. Altitude and ambient air temperature affect capacity. The higher above sea level and the greater the ambient temperature, the lower the delivered capacity. In these cases, a booster compressor is required to bring the main compressor up to capacity. In an actual compressor installation the air must be filtered and drawn from a cool environment to realize best performance.

Example 7.11

An air compressor with 6 percent clearance handles 50 lb per min for air between the pressures of 14.7 and 64.7 psia with n equal to 1.33. What is the weight in the cylinder per minute?

Solution

The volumetric efficiency is

$$E_v = 1 + 0.06 - \left[0.06\left(\frac{64.7}{14.7}\right)^{1/1.33}\right] = 0.877, \quad \text{or} \quad 87.7 \text{ percent.}$$

Piston displacement weight = 50/0.877 = 57.1 lb. The weight corresponding to total volume including clearance volume is 57.1 × 1.06 = 60.5 lb.

Effect of Clearance on Compressor Performance

In a single-stage compressor, clearance reduces volumetric efficiency. The percent of capacity reduction is greater than the percent of cylinder clearance because the piston must travel back part of its return stroke before clearance-space air has expanded to atmospheric pressure, permitting free air to flow into the cylinder. Clearance may be so great that no air is discharged from the compressor. This characteristic is sometimes used to control the output of the compressor by increasing the clearance when a reduced output is desired. Observe from Equation 7.19 that volumetric efficiency goes down as the pressure ratio goes up.

The volume occupied by expanded clearance air is in proportion to its discharge pressure; and the loss in compressor capacity due to clearance is less for two-stage than for single-stage compression, volume and terminal pressure being equal.

We can see that it is desirable to have the clearance as small as practicable. However, it has been shown that because there is no significant variation in the actual horsepower required for small variations of the clearance, there is no need to increase the cost of manufacturing just to reduce the clearance by a small amount. Clearances vary from about 1 percent in some very large compressors to 8 percent or more in other compressors, with clearances of 4 to 8 percent being common. Neither clearance nor volumetric efficiency is a reliable indicator of the quality of a compressor. The user is most concerned about the power consumed for a given capacity. An increase in clearance requires a larger compressor to deliver the same amount of gas, hence requires more power.

Example 7.12

Air is compressed in an adiabatic compressor at a rate of 500 ft^3/s. The air enters at 14.7 psia and 60°F and exits at 380°F. Isentropic efficiency = 90%.

Determine

a) exit air pressure if the air has a variable special heat.

b) power input to compressor.

Solution

a) Obtain values from air tables

$520°R - h_1 = 124.27 \quad Pr_1 = 1.2147$

$840°R - h_{2g} = 201.56 \quad Pr_{2g} = 6.573$

$$n_c = \frac{h_{2g} - h_1}{h_{2a} - h_1}$$

$$0.90 = \frac{201.56 - 124.27}{h_{2a} - 124.27}$$

$h_{2a} = 210.14$ from air tables $Pr_{2a} = 2.614$

$$\text{so knowing: } \frac{P_1}{P_2} = \frac{Pr_1}{Pr_{2a}} = \frac{1.2147}{7.614} = \frac{14.7}{P_2}$$

$$P_2 = 92 \text{ psia}$$

b) $\text{hp} = \dot{m}(h_{2a} - h_1) \quad \dot{m} = \dfrac{P_1 V_1}{RT_1} = \dfrac{(14.7 \text{ psia})(144 \text{ in}^2/\text{ft}^2)(500 \text{ ft}^3/\text{s})}{53.3 \frac{\text{ft-lbf}}{\text{lbm°R}}(520°\text{R})}$

$\text{hp} = 38.2 \text{ lb/s } (210.14 - 124.27) \text{ Btu/lb}$

$\text{hp} = \dfrac{3280 \text{ Btu/s}}{0.70696 \text{ Btu/hp-s}} \quad \dot{m} = 38.2 \text{ lb/s}$

$\text{hp} = 4{,}640$

Actual Indicator Card for Compressor

Because of fluid friction and inertia of the valves and frictional resistance of valves to motion, the actual indicator card differs from the ideal (*PV* diagram). Figure 7.6

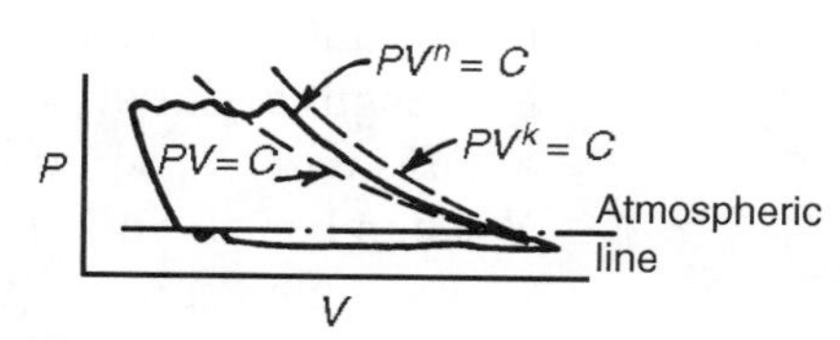

Figure 7.6

shows the effect of the fluttering of valve operation so that the area of the diagram is increased; and thus the work of the cycle.

Efficiency of Compressor

Compression efficiency is found from the air horsepowers we have been calculating and given by previous equations.

$$\text{Compression efficiency} = \frac{\text{air hp}}{\text{ihp}} \tag{7.20}$$

For mechanical efficiency:

$$E_m = \frac{\text{ihp}}{\text{bhp}} \tag{7.21}$$

For compressor efficiency:

$$E_c = \text{Compression eff.} \times \text{mech eff.} = \frac{\text{air hp}}{\text{bhp}} \tag{7.22}$$

Air horsepower may be required for an isothermal, isentropic, or polytropic compressor. There are times when the efficiency given in an examination problem may be called adiabatic efficiency. If such a problem involves a reciprocating compressor and mechanical efficiency is mentioned, the adiabatic efficiency is the compression efficiency referred to isentropic compression. If the problem involves a centrifugal compression, or other rotary compressor, as in a gas-turbine application and where mechanical friction is small and mechanical efficiency is high and indicated horsepower is not mentioned, then adiabatic efficiency is compressor efficiency referred to isentropic compression.

The indicated horsepower for a compressor of the reciprocating type may be determined from the relation very familiar to all,

$$\text{ihp} = \frac{PLAN}{33{,}000} \tag{7.23}$$

where

P = mean effective pressure, psi
L = length of stroke, ft
A = Piston area, sq in.
N = number of working strokes per minute (equal to two times the rpm for a double-acting single-cylinder (compressor).

Example 7.13

Air is compressed from an inlet state of 15 psia and 60°F to an exit condition of 175 psia. Find work of compression per unit mass.

a) Isothermal process
b) Isentropic process
c) Polytropic process, n = 1.2

Solution

a) Isothermal compression

$$\text{work} = P_1 V_1 \ln\frac{P_2}{P_1} = R_1 T_1 \ln\frac{P_2}{P_1}$$

$$\text{work} = \left(53.3\frac{\text{ft-lbs}}{\text{lbm}\,^\circ\text{R}}\right)(520^\circ\text{R}) \ln\frac{175}{15}$$

$$\text{work} = 68{,}090 \text{ ft-lbs/lbm.}$$

b) Isentropic process

$$\text{work} = \frac{KRT_1}{K-1}\left[\frac{P_2}{P_1}^{\frac{k-1}{k}} - 1\right] = \frac{(1.4)(53.3)(520)}{1.4-1}\left[\frac{175}{15}^{\frac{1.4-1}{1.4}} - 1\right]$$

$$\text{work} = 98{,}709 \text{ ft-lbf/lbm}$$

c) Polytropic process, $n = 1.2$

$$\text{work} = \frac{nRT_1}{n-1}\left[\frac{P_2}{P_1}^{\frac{n-1}{n}} - 1\right] = \frac{(1.2)(53.3)(520)}{1.2-1}\left[\frac{175}{15}^{\frac{1.2-1}{1.2}} - 1\right]$$

$$\text{work} = 84{,}141 \text{ ft-lbf/lbm}$$

Note: as heat transfer increases, work decreases.

CHAPTER 8

Fuels and Combustion Products

OUTLINE

The source of heat used to produce steam in a boiler is the fuel. The cost of fuel is by far the greatest single item of expense in the production of power. Fuels may be solid, liquid, or gaseous. The principal solid fuels are wood, peat, lignite, and coal. For liquid fuels we have fuel oil (Bunker C for industrial use and No. 2 oil for commercial use), tar, and other unrefined petroleum oils. The gaseous fuels consist of natural gas, refinery and betterment gases, coke-oven gas, and blast-furnace gas. Sawdust, bagasse (sugar-cane pulp), and garbage and sewer disposal wastes may be additional fuel options. The principal ingredients of all these fuels are carbon and hydrogen.

In atomic power plants the one natural fuel is U^{235}, making up 0.7 percent of natural uranium with the balance U^{238}. U^{235} is the only naturally occurring, readily fissionable nuclear fuel. Enriched uranium fuels may also be used.

THEORETICAL AIR FOR COMBUSTION

In order to select certain equipment in a boiler installation the amount of theoretical air required for combustion is of great significance. Let us take, for example, the combustion of carbon to carbon dioxide with the theoretical amount of air.

$$C + O_2 \rightarrow CO_2$$

One mole of each of the reactants (fuel and air) and the product of combustion are involved. This means that 12 lb carbon will react with 32 lb of oxygen to

Table 8.1 Consolidated Combustion Data: Air Required and Combustion Products

	For 1 Mole of Fuel					For 1 cu ft of Fuel					For 1 lb of Fuel					
	Air		Other Products (than N_2)			Air		Other Products (than N_2)			Air		Other Products (than N_2)			
Fuel	O_2	N_2	CO_2	H_2O	SO_2	O_2	N_2	CO_2	HO_2	SO_2	O_2	N_2	CO_2	H_2O	SO_2	
C	1.0	3.76	1.0								0.0833	0.313	0.0833			Moles
	379	1425	379								31.6	118.8	31.6			Cu ft
	32.0	105	44.0								2.67	8.78	3.67			Pounds
H_2†	0.5	1.88		1.0		0.00132	0.00496		0.00264		0.250	0.940		0.5		Moles
	189.5	712		379*		0.5	1.88		1.0*		94.8	356		189.5*		Cu ft
	16.0	52.6		18		0.0422	0.139		0.0475		8.0	26.3		9.0		Pounds
S	1.0	3.76			1.0						0.0312	0.1176			0.0312	Moles
	379	1425			379						11.84	44.6			11.84	Cu ft
	32.0	105			64						1.0	3.29			2.0	Pounds
CO	0.5	1.88	1.0			0.00132	0.00496	0.00264			0.179	0.0672	0.0357			Moles
	189.5	712	379			0.5	1.88	1.0			6.77	25.4	13.53			Cu ft
	16.0	52.6	44.0			0.0422	0.139	0.116			0.571	1.88	1.57			Pounds
CH_4	2.0	7.52	1.0	2.0		0.00528	0.0198	0.00264	0.00528		0.125	0.470	0.0625	0.125		Moles
	758	2850	379	758*		2.0	7.52	1.0	2.0*		47.4	178	23.7	47.4*		Cu ft
	64.0	210	44.0	36.0		0.169	0.556	0.116	0.0950		4.0	13.17	2.75	2.25		Pounds
C_2H_2	2.5	9.40	2.0	1.0		0.0066	0.0248	0.00528	0.00264		0.0962	0.362	0.0769	0.0385		Moles
	947	3560	758*	379*		2.5	9.40	2.0	1.0*		36.4	137	29.15	14.58*		Cu ft
	80.0	263	88.0	18.0		0.211	0.694	0.232	0.0475		3.08	10.13	3.38	0.692		Pounds
C_2H_4	3.0	11.29	2.0	2.0		0.00792	0.0298	0.00528	0.00528		0.1071	0.403	0.0714	0.0713		Moles
	1137	4280	758	758*		3.0	11.29	2.0	2.0*		40.6	153	27.1	27.1*		Cu ft
	96.0	316	88.0	36.0		0.253	0.834	0.232	0.0950		3.43	11.29	3.14	1.286		Pounds
C_2H_6	3.5	13.17	2.0	3.0		0.00923	0.0347	0.00528	0.0079		0.1167	0.439	0.0667	0.10		Moles
	1326	4990	758	1137*		3.5	13.17	2.0	3.0*		44.2	166.3	25.3	37.9*		Cu ft
	112.0	369	88.0	54.0		0.296	0.972	0.0232	0.1425		3.73	12.29	2.93	1.8		Pounds

* The volumes shown for H_2O apply only where the combustion products are at such high temperature that all the H_2O is a gas.

† Varying assumptions for molecular weight introduce a slight inconsistency in the values of air and combustion products from the burning of hydrogen. True molecular weight of hydrogen is 2.02 but the approximate value of 2 is used in figuring the air and combustion products. Courtesy of *Power Magazine*.

form 44 lb of carbon dioxide. Thus, for each pound of carbon for theoretical combustion,

$$\frac{32}{12} = 2.666 \text{ lb oxygen required}$$
$$\frac{44}{12} = 3.666 \text{ lb carbon dioxide formed.}$$

Now, because air by weight has about 23.13 percent oxygen and 76.87 percent nitrogen,

$$\frac{2.666}{0.2313} = 11.541 \text{ lb air needed}$$

$11.541 \times 0.7687 = 8.87$ lb nitrogen in the products of combustion. If hydrogen and hydrocarbons are involved in the combustion reaction, similar treatment may be accorded them with their products of combustion, carbon dioxide and water. In order to facilitate calculations see Table 8.1.

EXCESS AIR

In actual practice, an amount of air in excess of the theoretical is introduced with the fuel to influence complete combustion. While the theoretical air is readily calculated when the fuel analysis is given, the amount actually required for a given fuel and installation depends on experience and economical limitations. Table 8.1 gives required combustion air and products for common combustibles burned with theoretical air requirements. Air and products are given in moles, cu ft and lb (see right-hand column) for 1 mole, 1 cu ft, and 1 lb of fuel.

Table 8.2 gives practical excess air in percent by weight for various fuels.

Table 8.2 Excess Air for Various Fuels

Fuel	Excess Percent by Weight
Bunker C oil	20
Natural gas	15
Refinery gas	15
Coke-oven gas	20
Blast-furnace gas	20
Wood	50
Tar	30
Coal (pulverized)	25
Coal (stoker)	40

The theoretical amount of air required to burn a fuel completely to CO_2 and water vapor is very small, but in order to obtain maximum steam output, ensure complete combustion, avoid smoke, avoid overheating, minimize slagging (coal burning), and ensure adequate mixing of the combustion gases and air, it is necessary to use from 10 to 75 percent excess air. If this is not done, some of the oxygen will escape without combining with its allotted share of the combustible matter so that some of the hydrogen will escape unburned or there will be only partial combustion of the carbon into CO. Under these undesirable conditions the Orsat analysis and other methods will show CO, O_2, and N_2 levels in the exhaust.

Theoretically, maximum furnace efficiency is highest when a minimum of excess air is used. Practical difficulties in securing close regulation of air may, however, dictate a large amount of air to ensure some excess air at all times, particularly when the air flow is subject to varying conditions of draft.

The degree of efficiency of fuel combustion is determined by calculations involving the flue-gas analysis. This gas analysis is usually obtained by the fundamental Orsat principle which involves absorbing the oxygen, carbon dioxide, and carbon monoxide individually from a gas sample, thus obtaining their percent of the original sample. Other methods of analysis are based on the specific gravity or the electrical conductivity of the flue gas.

When the rate of fuel firing is held constant, the effect of decreasing excess air is as follows:

(a) Amount of flue gas will decrease, thus saving heat which would be swept up the stack.

(b) Stack temperature may either increase or decrease, depending on the number of tubes in the convection section of the boiler.

(c) Firebox temperature will increase because there is less cooling by the excess air.

(d) The capacity of the furnace will increase. The increased firebox temperature allows a higher charge rate. The decreased flue-gas volume allows a higher flue-fired rate. This does not hold if firebox temperature is already at its maximum.

(e) Furnace efficiency increases. The ideal condition is approached where there is a ratio of fuel and air perfectly mixed and burned so that there will be neither an excess of fuel nor an excess of air.

Example 8.1

Calculate the pounds of air and the products of combustion required and formed, respectively, when 1 lb of a gasoline composed of 85 percent carbon and 15 percent hydrogen is burned in the theoretical amount of air. What percentage of CO_2 exists in the products of combustion by volume?

Solution

Basis: 1 lb fuel and standard gas conditions.
Since we may change mass to volume (which is also molar) write the equation:

$$\frac{0.85}{12}\,C + \frac{0.15}{2}\,H_2 + (0.1083)3.76\;\;N_2 + 0.1083\;O_2 \rightarrow$$

$$0.0708\;CO_2 + 0.075\;H_2O + (0.4072)N_2$$

Calculate volume of products; note no water is included because it condenses at room temperature

$$\%\text{ by volume of } CO_2 = \frac{0.0708}{(0.0708 + 0.4072)}$$

$$\%\text{ by volume of } CO_2 = 14.8\%$$

Example **8.2**

An analysis of the flue gases of a boiler shows 14 percent CO_2, 2 percent CO, 5 percent O_2, and 79 percent nitrogen. Calculate the percentage of heat lost in the unburned CO up the stack.

Solution

The solution of this problem may take the following form:

$$\frac{\text{Heat generated if all C burned to } CO_2 - \text{actual heat generated}}{\text{Heat generated if all C burned to } CO_2} \times 100.$$

Combustions reactions for carbon in problem:

$$C + O_2 \rightarrow CO_2 + 14{,}544 \text{ Btu per lb C}$$
$$2C + O_2 \rightarrow 2CO + 4350 \text{ Btu per lb C}$$

From the wording of the problem, for every 100 moles of flue gas there are included 14 moles of CO_2 and 2 moles of CO. There are 14 lb atoms C in 14 moles of CO_2 and 2 lb atoms C in 2 moles of CO. Then, substituting in the heat-generation ratio

$$\frac{[(14+2) \times 12 \times 14{,}544] - [(2 \times 12 \times 4350) + (14 \times 12 \times 14{,}544)]}{(14+2) \times 12 \times 14{,}544} \times 100$$

or 9.6 percent heat lost up stack.

MOLAR ANALYSIS

Problems may also be worked using molar analysis (by volume). This method requires tables that give the energy values on a molar basis. Either mass analysis or molar analysis is acceptable, but at this time molar analysis is the more common method.

Example **8.3**

A natural gas containing 75 percent methane and 25 percent ethane by volume is burned in 200 percent theoretical air. Heat is transferred out until the products reach a temperature of 700×R. The fuel and dry air enter into the combustion chamber at 537°R. Determine the following: (a) stoichiometric *A*/*F*; (b) actual air *A*/*F*; (c) heat transfer out per lb mole of fuel.

Solution

(a) To get the stoichiometric A/F ratio we need the theoretical equation

$$0.75CH_4 + 0.25C_2H_6 + 2.375(O_2 + 3.76N_2) \quad 1.25CO_2 + 2.25H_2O + 8.93N_2.$$

Mass air fuel ratio = mass of air/mass of fuel

$$A/F = \frac{\text{(number of moles air)(mol wt air)}}{\text{(number of moles fuel)(mol wt fuel)}}$$

$$A/F = \frac{(2.375)(1+3.76)(29)}{0.75(16)+0.25(30)} = \frac{327.8}{19.5} = 16.8$$

(b) The acutal equation uses 200% theorectical air. This means two times the air for same fuel. We thus take the equation for (a) and multiply the numerator by 2; thus

$$A/F_{\text{actual}} = 33.6$$

(c) Obtain the chemical equation in actual terms, that is, 200 percent theoretical air.

$$0.75CH_4 + 0.25C_2H_6 + [2.375(O_2 + 3.76N_2)] \times 2$$
$$1.25CO_2 + 2.25H_2O + 2.375O_2 + 17.86N_2$$

To obtain heat transfer put the equation in reactants and products form.

for reactants where:
h_f = Heat of formation
$\bar{h}$ = Enthalpy of gas at entering temperature
$\bar{h}^0$ = Enthalpy of gas at datum (S.T.P.) since fuel and air both enter at the datum (S.T.P.) then $\bar{h} = \bar{h}^0$

Therefore equation reduces to

$$0.75(h_f)_{CH_4} + 0.25(h_f)_{C_2H_2} = 0.75(-32,210) + 0.25(-36,420)$$
$$\text{reactants} = -33,263 \text{ Btu/lb mol}$$

For products: Now $\bar{h}$ is at 700°R

$$1.25[h_f + \bar{h} - \overline{h^\circ}]_{CO_2} + 2.25[h_f + \bar{h} - \overline{h^\circ}]_{H_2O} + 2.375[\overset{0}{\cancel{h_f}} + \bar{h} - \overline{h^\circ}]_{O_2}$$
$$+ 17.86[\overset{0}{\cancel{h_f}} + \bar{h} - \overline{h^\circ}]_{N_2} + Q$$

$$1.25[-169,300 + 5552.0 - 4027.5] + 2.25[-47540 + 4886 - 3725.1]$$
$$+ 2.375[4879.3 - 3725.1] + 17.86[4864.9 - 3729.5] + Q$$

$$\text{Products} = -291,053 + Q$$

Equate products to reactants

$$-33,263 = -291,053 + Q$$

$$Q = 257,790 \text{ Btu/lb mol}$$

Example 8.4

Given complete analysis of coal and flue gas, determine per pound of coal (a) the theoretical air, (b) actual air, (c) dry flue gas, (d) moisture in flue gas, (e) percent excess air, and (f) percent loss from incomplete combustion.

Solution

The ultimate analysis (which is the chemical analysis by constituent) of the coal as fired, percent by weight,

Ash	10.49
Sulfur	1.20
Hydrogen	6.47
Carbon	71.98
Nitrogen	1.16
Oxygen	8.70
Total	100.00

Heating value as higher heating value of the coal is 13,800 Btu per lb coal. The flue gas analysis, percent by volume is

CO_2	10.2
CO	0.6
O_2	9.2
N_2	80.0
Total	100.0

Moles of carbon and nitrogen in flue gas is

Flue Gas	%	Moles Gas	Moles C	Moles N_2
CO_2	10.2	10.2	10.2	
CO	0.6	0.6	0.6	
O_2	9.2	9.2		
N_2	80.0	80.0		80.0
Total moles		100.0	10.8	80.0

$$\text{Weight of carbon burned} = 10.8 \times 12 = 129.6 \text{ lb}$$
$$\text{Weight of coal burned} = 129.6/0.7198 = 180.0 \text{ lb}$$
$$\text{Moles air supplied} = 80.0/0.79 = 101.3 \text{ moles}$$
$$\text{Weight of air supplied} = 101.3 \times 29 = 2938 \text{ lb}$$
$$\text{(a) } \textit{Weight of air supplied per lb coal} = 2938/180 = 16.32 \text{ lb}$$
$$\text{(d) } \textit{Moisture produced per lb coal} = 0.0647 \times 9 = 0.58 \text{ lb}$$
$$\text{(c) } \textit{Dry flue gas per lb coal} = \textit{coal} + \textit{air} - \textit{ash} - \textit{moisture}$$
$$= 1 + 16.32 - 0.1049 - 0.58 = 16.64 \text{ lb}$$

$$\text{Fraction carbon burning to CO} = 0.6/(10.2 + 0.6) = 0.0555$$
$$\text{Weight of carbon burning to CO} = 0.0555 \times 0.7198 = 0.040 \text{ lb}$$
$$\text{Heat loss from CO} = 0.040 \times 10{,}160 = 406 \text{ Btu}$$
$$\text{(f) } \textit{Percent heat loss from } \text{CO} = (406/13{,}800) \times 100 = 2.94 \text{ percent}$$

Theoretical air per lb coal is

Combustible	Lb of Combustible	Lb O_2 Required Per lb Combustible (from Table 8.1)	Lb O_2 Required
Carbon	0.7198	2.67	1.922
Hydrogen	0.0647	8.00	0.518
Sulfur	0.012	1.00	0.012
Total			2.452
Less oxygen in the coal			0.087
Net oxygen needed			2.365

(b) Air required: 2.365 × 4.32 = 10.22 lb
Excess air: 16.32 – 10.22 = 6.10 lb

(e) *Percent excess air*: 6.10/10.22, or 59.7 percent.

Example **8.5**

Given the volumetric analysis of a natural gas, determine the volumetric analysis of the flue gas if the fuel is burned with 40 percent excess air by volume.

Solution

Gas analysis by volume percent is

CO	0.60
H_2	1.62
CH_4	94.30
C_2H_4	0.15
H_2S	0.25
O_2	0.40
CO_2	0.85
N_2	1.83
Total	100.00

Oxygen, combustion products of 100 moles of fuel are
Nitrogen in flue gas:

$$N_2 \text{ required for 190.13 moles } O_2 = 190.13 \times 3.76 = 714.9 \text{ moles}$$
$$40\% \text{ excess } N_2 \text{ (by volume)} = 0.4 \times 714.9 = 286.0 \text{ moles}$$
$$N_2 \text{ in fuel} = 1.8 \text{ moles}$$
$$N_2 \text{ in flue gas by addition} = 1002.7 \text{ moles}$$

Gas	Moles	Moles O_2 Needed	Moles Produced CO_2	H_2O	SO_2
CO	0.60	0.30	0.60		
H_2	1.62	0.81		1.62	
CH_4	94.30	188.60	94.30	188.60	
C_2H_4	0.15	0.45	0.30	0.30	
H_2S	0.25	0.37		0.25	0.25
CO_2	0.85		0.85		
Subtotal		190.53	96.05	190.77	0.25
Minus O_2 in fuel		– 0.40			
Total		190.13	96.05	190.77	0.25

Unburned O_2 in flue gas: $0.40 \times 190.13 = 76.0$ moles

Volumetric gas analysis (H_2O disappears) is

Gas	Moles	Mole %	Vol %
CO_2	96.1	8.2	8.2
O_2	76.0	6.5	6.5
SO_2	0.2	0.0	0.0
N_2	1002.7	85.3	85.3
Total	1175.0		100.00

Example 8.6

An Orsat analysis of a flue gas shows that it contains 12.5 percent CO_2, 4.1 percent O_2, 83.4 percent N_2. From this information determine (a) gravimetric analysis, (b) percent excess air, (c) volume of air per pound of fuel fired, (d) volume of total flue gases per pound of fuel fire, and (e) percentage of gross heating value lost through noncondensation of water vapor in flue gases. Air leakage into furnace is negligible; fuel has no nitrogen content. Gross heating value of fuel may be taken as 18,300 Btu per lb. Flue-gas temperature may be taken as 490°F and 29.5 in. Hg. Barometer is 30.0 in. Hg.

Solution

By inspection of the heating value the fuel appears to be a fuel oil. Now let us take as a basis 100 moles of the flue gas. Then we see that there are 83.4 moles of N_2 leaving and the same entering because there is no leakage. Also N_2 undergoes no chemical reaction in the combustion process. We also know that mole percent equals volume percent. The moles of air coming in $83.4/0.79 = 105.4$ moles air needed to give 100 moles of flue gas.

Moles of oxygen in the flue gas consist of 12.5 moles in CO_2 and 4.1 moles in oxygen, or a total of 16.6 moles. This was determined on the basis of the theoretical combustion equations. Moles of O_2 entering are $105.4 \times 0.21 = 22$ moles. But we can account for only 16.6 moles from the flue-gas analysis. Therefore, there are 22 16.6 = 5.4 moles of O_2 coming in with the air which react chemically with another substance within the furnace. This must be hydrogen because it would not show as water vapor in the Orsat analysis.

Referring to the theoretical combustion of H_2 and O_2, 2 moles of H_2 plus 1 mole O_2 gives 2 moles of water vapor, the 5.4 moles of O_2 must give 10.8 moles of water vapor. Thus, the total fuel composition is made up of carbon and hydrogen, and nothing else.

Fuel analysis:

$$\text{Weight of carbon in 12.5 moles flue gas} = 12.5 \times 12 = 150 \text{ lb}$$
$$\text{Weight of hydrogen in 10.8 moles water vapor} = 10.8 \times 2 = 21.6 \text{ lb}$$
$$\text{Total weight of fuel} = 171.6 \text{ lb}$$
$$\text{Percent } carbon = 150/171.6 = 100 = 87.5 \text{ percent}$$
$$\text{Percent } hydrogen \text{ by difference} = 100 - 87.5 = 12.5 \text{ percent}$$

Percent excess air may be found from the following relation:

$$\frac{\text{Unnecessary } O_2 \text{ from flue-gas analysis}}{\text{Necessary } O_2 \text{ for combustion}}.$$

$$\frac{4.1}{12.5 + 5.4} \times 100 = 22.9 \text{ percent}$$

Volume of air required per pound of fuel fired taken at 78×F and 30 in. Hg for boiler-room air is

$$105.4 \times 1/171.6 \times 379 \times \frac{78+460}{520} \times \frac{30}{30} = 240 \text{ cu ft.}$$

Volume of total flue gases per pound of fuel fired:

$$\frac{\text{Moles total flue gases}}{\text{lb fuel}} = \frac{100+10.8}{171.6} = 0.654$$

$$0.654 \times 379 \times \frac{490+460}{520} \times \frac{30}{29.5} = 455 \text{ cu ft}$$

Percentage gross heating value of fuel lost through noncondensation of water vapor formed during combustion:

$$\text{Mole water vapor per lb fuel} = 10.8/171.6 = 0.063$$
$$\text{Weight of hydrogen so involved} = 0.063 \times 18 = 1.13 \text{ lb}$$
$$\text{Thus } 1.13 \times 1058.2/18{,}300 \times 100 = 6.55 \text{ percent}$$

Note that heat of condensation is taken as 1058.2 Btu per lb vapor.

Example **8.7**

A steam boiler is required to deliver 50,000 lb of steam per hour (evaporation rate) at 400 psia and a total steam temperature of 600×F. Assuming an efficiency of 80 percent, calculate how many tons of coal containing 5 percent moisture and having a dry calorific value of 14,800 Btu per lb must be supplied to the furnace per hour.

Solution

Some high-volatile coals are wetted down (tempered) to improve their burning characteristics. Now, the required heat output is the evaporation rate × the enthalpy of the steam at the superheater outlet.

$$50{,}000 \times (1{,}306.9 - 68) = 61.9 \times 10^6 \text{ Btu per hr}$$

The value of 68 is the enthalpy of the feed water assumed to enter the boiler at 100×F. Thus, it is perfectly acceptable to say 100 – 32 = 68 Btu per lb. The Btu per hour under actual fuel conditions is found in accordance with

$$14{,}800 \times 0.80 \times (1 - 0.05) \times 2000 = 22.5 \times 10^6 \text{ per ton coal.}$$

Finally, the firing rate is

$$61.9/22.5 = 2.76 \text{ tons tempered coal.}$$

Up to 5 percent moisture is added to reduce the rate of combustion of volatile matter in a coal so as to reduce smoking. Western coals usually show this characteristic to "smoke."

Example **8.8**

The air entering a boiler has a relative humidity of 70 percent. Atmospheric pressure is 14.42 psia. Boiler-room temperature is 90×F. The gas used as fuel is 75 percent methane and 25 percent ethane by volume. The gas enters the burners with a relative humidity of 80 percent at a pressure of 5 psig and 76×F. The amount of air actually available for combustion is 15 percent in excess of that theoretically required. Calculate the volumetric percentage of water vapor in the flue gas leaving through the boiler breeching.

Solution

It is assumed that the boiler setting is tight and that all air used in the combustion calculations enters with the fuel. Using the concept of moles or pound moles, we shall take as a basis 1 mole of the fuel. Then the amount of "bone-dry" air used per mole of fuel may be calculated.

$$\frac{(0.75 \times 2) + (0.25 \times 3.5)}{0.21} \times 1.15 = 12.95 \text{ moles "bone-dry" air}$$

This follows, because 2 moles of oxygen and 3.5 moles of oxygen are needed theoretically to combine with a mole of methane and ethane, respectively. This gives the amount of theoretical total oxygen, and dividing by the percentage by volume of oxygen in dry air gives the theoretical dry air needs. Increasing this by 15 percent gives actual dry air amount as 12.95 moles.

The flue gas from the boiler contains nitrogen, oxygen, carbon dioxide, and water vapor. The nitrogen comes through the combustion reaction chemically unchanged with the incoming air stream. The oxygen appears in the flue gas due to the fact that air in excess of theoretical requirements was used to promote combustion. If no excess air were used and intimate mixing of air and fuel took place, then no oxygen would appear in the flue gas. Carbon dioxide appears because of the complete combustion of the carbon portion of the ethane molecule and methane molecule with the oxygen entering with the air. Water vapor is the result of the combustion of the hydrogen portions of the fuel. The absence of carbon monoxide indicates complete combustion of the carbon portion of the fuel.

We shall now calculate the relative amounts of each of the constituents in the flue gas.

$$N_2 = 0.79 \times 12.95 = 10.23 \text{ moles per mole fuel}$$
$$O_2 = (0.21 \times 12.95) - [(0.75 \times 2) + (0.25 \times 3.5)]$$
$$= 0.345 \text{ mole per mole fuel}$$
$$CO_2 = 0.75 + (2 \times 0.25) = 1.25 \text{ moles per mole fuel}$$

This last factor is true because an analysis of the chemical equation after balancing shows that for each mole of methane burned 1 mole of CO_2 is formed,

$$CH_4 + 2O_2 \rightarrow CO_2 + 2H_2O$$

and that 2 moles of CO_2 are formed when 1 mole of ethane is burned.

$$2C_2H_6 + 7O_2 \rightarrow 4CO_2 + 6H_2O$$

The water content in the flue gas is derived from three sources, namely, moisture in the fuel, moisture in the entering air, and water formed by combustion. Let us first calculate the water entering with the fuel. From steam tables the saturation vapor pressure of water at 76×F is 0.444 psia. Then the water-vapor content in the fuel is by humidity ratio

$$\frac{0.444}{(5 + 14.42) - 0.444} \times 0.80 = 0.019 \text{ mole vapor per mole dry fuel.}$$

The water-vapor content of the incoming combustion air is

$$\frac{0.698}{14.42 - 0.698} \times 0.70 \times 12.95 = 0.461 \text{ mole vapor per mole dry fuel}$$

where vapor pressure of water at 90×F is 0.698 psia from steam tables.

The water formed through combustion is

$$(0.75 \times 2) + (0.25 \times 3) = 2.25 \text{ moles vapor per mole dry fuel.}$$

This may be seen by referring back to the preceding combustion formulas.

Then the total water vapor in the flue gas is

$$0.019 + 0.461 + 2.25 = 2.73 \text{ moles vapor per mole dry fuel.}$$

The volumes of the various constituents in the flue gas are
The volumetric percentage of water vapor may be now determined as follows:

$$(2.73)/(10.23 + 0.345 + 1.25 + 2.73) \times 100 = 18.8 \text{ percent}$$

N_2	10.23	moles per mole dry fuel
O_2	0.345	mole per mole dry fuel
CO_2	1.25	moles per mole dry fuel
H_2O	2.73	moles per mole dry fuel

An examination of the above problem and solution would provide the necessary information to calculate the dew point of the flue gas leaving the boiler. Assuming that the breeching pressure is 13 psia due to stack effect or because of an induced draft fan, then the partial pressure of the water vapor in the mixture is $0.18 \times 13 = 2.34$ psia.

Now refer to the steam tables; this corresponds to a saturation temperature of approximately 132×F. This is the dew point of the mixture below which water would condense out. In practice, an air heater or economizer placed in the exit flue-gas stream would be designed not to cool the exit gases below 132°F. In actual installations, the flue gas leaving such a heat reclaimer would hardly ever drop below 300×F for good operating and maintenance reasons. This lower temperature limit of 300×F is far enough above the safe limit to prevent corrosion of the cooling surfaces.

Example **8.9**

A producer gas made from coke has the following composition by weight: CO = 27.3 percent, CO_2 = 5.4 percent, O_2 = 0.6 percent, N_2 = 66.7 percent. This gas is burned with 20 percent excess air. If combustion is 98 percent complete, calculate the weight and composition of the gaseous products formed per 100 lb of producer gas burned.

Solution

The combustion equation is $CO + 1/2O_2 = CO_2$. Take the basis of calculation to be 100 lb of the original producer gas.

CO present = 27.3/28 = 0.975 lb mol
O_2 required for combustion = 0.975/2 = 0.487 lb mol
O_2 supplied with 20 percent excess air = 1.2 × 0.487 = 0.585 lb mol
O_2 already present in gas = 0.6/32 = 0.019 lb mol
O_2 to be supplied from air = 0.566 lb mol or 18.2 lb
Weight of air used = 18.2/0.232 = 78.2 lb
Weight of N_2 introduced = 78.2 – 18.2 = 60.0 lb
CO_2 formed in combustion = 0.98 × 0.975 = 0.995 lb mol or 42.1 lb
O_2 assumed in combustion = 0.487 × 0.98 = 0.477 lb mol or 15.3 lb
Total N_2 present in gases = 60 + 66.7 = 126.7 lb
Total CO_2 present in gases = 42.1 + 5.4 = 47.5 lb
Total O_2 present in gases = 18/1 + 0.6 – 15.3 = 3.4 lb
CO present in gases = 0.02 × 27.3 = 0.5 lb
Total weight of flue gases = 126.7 + 47.5 + 3.4 + 0.5 = 178.1 lb
Total weight of gas plus air used = 178.1 lb

Composition of flue gases by weight:

CO_2	47.5/178.1 or 26.6 percent
O_2	3.4/178.1 or 1.9 percent
CO	0.5/178.1 or 0.3 percent
N_2	126.7/178.1 or 71.2 percent

CHAPTER 9

Steam Power

OUTLINE

TESTING OF STEAM-GENERATING UNITS

The Power Test Codes of the American Society of Mechanical Engineers (ASME) go into quite a detailed discussion for the determination of the efficiency of steam-generating units. There are simpler test forms that may be used, but only for routine plant operating purposes. The ASME method should be used for contract testing and where more precise data are needed.

In order to determine the efficiency of a steam-generating unit, the *direct method* simply involves measurement of the energy input and output. This is given in equation form as follows:

$$\text{Efficiency} = \frac{\text{Wt of steam} \times (\text{enthalpy of steam} - \text{heat of feedwater})}{\text{wt of fuel} \times \text{gross heating value of fuel}} \times 100 \quad (9.1)$$

This determination is usually checked by means of a *heat balance*.

However, because of the inaccuracy of instrumentation (weighing scales and flowmeters), accepted practice to determine efficiency has become the *indirect method.* Thus,

$$\text{Efficiency} = \frac{\text{heating value of the fuel} - \text{losses}}{\text{heating value of the fuel}} \times 100 \quad (9.2)$$

In this method, steam is flowmetered or the fuel is weighed only to establish the capacity at which the test is made.

In the performance test report the following should be covered: (a) the apparatus under test should be described; (b) purpose of test should be stated; (c) conditions of the test should be indicated, for example, capacity, type of fuel, feed-water temperature, steam temperature, excess air; (d) duration of test should be decided on—a 24-hr test for direct method and especially if boiler is stoker fired; otherwise a 12-hr test would be acceptable for other fuels and methods of firing. For the indirect method, the duration may be from 4 to 6 hr.

For the *heat balance,* the following points must be determined:

(a) Heat absorbed by boiler unit

(b) Heat lost in dry flue gas

(c) Heat lost due to moisture in fuel

(d) Heat lost due to moisture from H_2 in fuel

(e) Heat lost due to moisture in air

(f) Heat lost due to incomplete combustion of carbon

(g) Heat lost due to unburned carbon

(h) Heat lost due to radiation

For formulas and methods for determining the above points, the license candidate or student is referred to the Power Test Code or the standards of the industry.

Steam quality may be determined with throttling calorimeter or the separating calorimeter.

Boiler output may be obtained from the following relationship:

$$\frac{\text{Evaporation lb per hr }(H_2 - H_1)}{1000} = \text{kilo Btu per hr.} \tag{9.3}$$

Also

$$\text{Boiler hp} = \frac{\text{evaporation lb per hr }(H_2 - H_1)}{33{,}480} \tag{9.4}$$

$$\text{Factor of evaporation} = \frac{H_2 - H_1}{970.3} \tag{9.5}$$

where 970.3 is the heat of vaporization of steam at atmospheric pressure, H_1 is enthalpy of feedwater, and H_2 is enthalpy of steam at outlet.

Example 9.1

A steam generator converts 40,000 lb of water from a feed-water temperature of 220°F to steam at 180 psia and a quality of 97 percent. The weight of coal fired per hour is 4500 lb, gross heating value of coal as fired is 11,800 Btu per lb. Determine the rate of heat absorption in Btu per hr and the boiler horsepower developed.

Solution

From steam tables H_1 is 188.06 Btu per lb and H_2 is 1171 Btu per lb. Then the rate of heat absorption is found from

$$40{,}000(1171 - 188.06) = 39.32 \times 10^6 \text{ Btu per hr}$$

$$\text{Boiler hp} = \frac{39.32 \times 10^6}{33{,}480} = 1174 \text{ boiler hp}$$

PERFORMANCE OF BOILERS

The overall efficiency of a boiler at any operating condition is the percentage of the heating value of the fuel that is transferred after combustion to the working substance (steam and water). Efficiency is output over input based on the gross heating value of the fuel. If we let W_f be equal to pounds of fuel fired per hour and H_f be the gross heating value of the fuel as fired in Btu per lb, then

$$\text{Over-all boiler efficiency} = \frac{\text{evaporation lb per hr }(H_2 - H_1)}{W_f H_f} \times 100 \qquad (9.6)$$

Equation 9.6 includes the effect not only of the furnace, boiler, and grate, but also of heat-transfer accessories, such as superheater, water walls, economizer, and air preheater. Heat release, pounds of steam generated per pound of fuel fired, rate of heat transfer per square foot of heating surface, temperature of stack gases, and percent CO_2 in stack gases all must be noted.

After installation and drying out of the boiler setting, an efficiency test is made on the boiler. The amount of CO_2 in the flue gases at the point of maximum efficiency is noted. This is then used as the control point and is an indication of excess air. Curves plotting all these variables against rate of evaporation or boiler horsepower may be made for future reference.

Example 9.2

Calculate the efficiency of a steam-generating unit consisting of an economizer, boiler, and superheater. One hundred thousand pounds of feed water enter the economizer per hour at a pressure of 200 psia and a temperature of 190°F. The steam leaves the superheater at 200 psia and a total steam temperature of 500°F. The coal as fired has a gross heating value of 12,500 Btu per lb and 5.8 short tons are fired per hour.

Solution

With the use of the steam tables and Mollier diagram and letting $H_1 = (190 - 32) = 158$ Btu per lb, $H_2 = 1269$ Btu per lb, $W_f = 5.8 \times 2000 = 11{,}600$ lb of coal fired per hr,

$$H_f = 12{,}500 \text{ Btu per lb}$$

And the overall efficiency is

$$\frac{100{,}000 \times (1269 - 158)}{11{,}600 \times 12{,}500} \times 100 = 76.5 \text{ percent.}$$

Example 9.3

A gas-fired steam boiler delivers 250 lb of steam per hour of 98 percent quality. Calculate the boiler efficiency at the following conditions: feed-water temperature 60°F, steam pressure 25 psig, barometer 30.2 in. Hg, fuel-gas pressure 4 in. water gauge (WG) at burner, gas temperature 80°F, gas consumption 800 cu ft per hr, gross heating value of fuel gas 540 Btu per cu ft at 60°F and 30 in Hg.

Solution

The approach to this problem is quite similar to the previous one except for the determination of the proper heating value to use in the efficiency formula. The output of the boiler may be found in accordance with

$$\text{Boiler output} = (W \text{ lb/hr})[H_f + (xH_{fg}) - H_{\text{feed water}}] \text{ Btu/lb} =$$

$$250[236 + (0.98 \times 934) - 28] = 281{,}000 \text{ Btu per hr.}$$

The input may be found by first correcting the gas volume from the "as-fired" condition to standard conditions, for the heating value given in the problem is for standard conditions. Another way to do this would be to correct the heating value to the gas at the actual as-fired condition. Our method of solution is to follow the first approach, *i.e.*, correct the gas volume to standard conditions. Now, the actual gas pressure in absolute measurements is

$$30.2 + \frac{4}{13.6} = 30.5 \text{ in. Hg.}$$

Then the gas volume as corrected is
s = standard
a = actual

$$\frac{P_s V_s}{T_s} = \frac{P_a V_a}{T_a}$$

$$V_s = V_a \left(\frac{P_a}{P_s}\right)\left(\frac{T_s}{T_a}\right)$$

$$800 \times \frac{30.5}{30.0} \times \frac{460 + 60}{460 + 80} = 782 \text{ cu ft per hr.}$$

Therefore, the input will be 782 × 540 = 422,000 Btu per hr and the efficiency

$$\frac{281{,}000}{422{,}000} \times 100 = 66.7 \text{ percent.}$$

Example 9.4

Exhaust steam from an engine goes into an open-type feed-water heater at 1 psig and a moisture content of 12 percent. How many pounds of this steam are needed to raise the temperature of 1000 lb of feedwater from 60 to 200°F?

Solution

If this were a closed feed-water heater the solution to the problem would be a rather simple heat balance, *i.e.*, heat absorbed by the feedwater being equal to the heat released by the steam. However, because this is an open feed-water heater, the steam mixes with the feed water so that both weights of steam condensed and feed water are combined in the solution. The 1000 lb of heated feed water are composed of actual initial feed water plus the weight of steam condensed. Let w equal weight of steam condensed, W equal weight of actual feed water, enthalpy of feed water equal to $200 - 32 = 168$ Btu per lb at final conditions of 200°F, enthalpy of initial feed water at 60°F equal to $60 - 32 = 28$ Btu per lb. Then the heat balance is

$$(W + w)168 = (1000 - w)28 + w(185 + 0.88 \times 967).$$

Solving for w, this is found to be equal to 166.7 lb of steam.

Example **9.5**

A single-effect horizontal tube evaporator is designed to produce 30,000 lb per hr of distillate (water) at a pressure of 30 psig using steam at a pressure of 45 psig. Assume feed water at 100°F containing 150 ppm of solids and continuous blowdown limiting the concentration to 2000 ppm. How much steam will be required?

Solution

Refer to Exhibit 1. The following is the heat balance across the system:

Heat in = heat in steam + heat in feed water
Heat out = heat in distillate + heat in continuous blowdown
+ heat in steam heating the condensate.

Thus

Heat in = heat out, neglecting radiation and other losses.

If we let W be the amount of steam used, then we strike the balance

$$\text{Heat in} = (W \times 1177) + (32{,}250 \times 68)$$
$$= (30{,}000 \times 1171.5) + (2250 \times 243) + (W \times 262) = \text{Heat out}$$

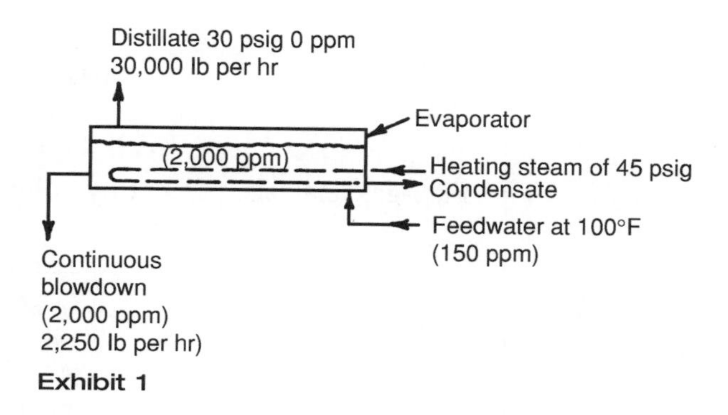

Exhibit 1

Solving for W, this is found to be 36,611 lb of steam. However, because the feed-water rate is given by

W_T of steam in = W_T of condensate out + W_T of blow down.

$$W = 30{,}000 + \left[\left(\frac{150}{2000}\right)(30{,}000)\right]$$

$$= 30{,}000 + 2250 = 32{,}250 \text{ lb per hr.}$$

The continuous blowdown must be the difference

$$32{,}250 - 30{,}000 = 2250 \text{ lb per hr.}$$

Obtain enthalpies from the steam tables and Mollier diagram.

Example 9.6

The following data were obtained during a test of a boiler using natural gas as fuel:

Volume of fuel used, standard cubic feet per hour	2895
Water rate of boiler, lb per hr	1690
Steam pressure, psia	170
Steam temperature, °F	485
Flue-gas temperature, °F	505
Feed-water temperature, °F	145
Temperature of fuel entering burners, °F	90
Gross heating value of fuel, Btu per standard cubic feet	1024
Standard conditions of gas measurement, psia at 60°F	14.7
Volume ratio flue gas/fuel	10.53

Fuel analysis, percent by volume:

CH_4	93.7 percent
C_2H_6	5.3
CO_2	1.0
Total	100.00

Fuel-gas analysis, percent by volume:

CO_2	10.0
O_2	3.4
N_2	86.6
Total	100.00

Prepare a heat balance based on the above data.

Solution

Assume as a time basis a 1-hr period. A temperature of 60°F will be selected as a datum above which to calculate heat input and output.

Heat input.

(a) Heat in the fuel: volume of fuel in standard cubic feet (scf) multiplied by Btu per scf

Heat in fuel = (volume of fuel)(Btu content)

2895 scfh × 1024 Btu/scf = 2,964,480 Btu for 1 hr

(b) Heat in feed water: pounds of water used per hour multiplied by the difference between the actual feed-water temperature and 60°F.

$$1690 \times (145 - 60) = 143{,}650 \text{ Btu}$$
$$\text{Total heat input} = 2{,}964{,}480 + 143{,}650 = 3{,}108{,}130 \text{ Btu}$$

Heat output.

(a) Heat in steam: from the steam tables, it is found that saturated steam at an absolute pressure of 170 has a temperature of 368.4F. Then the degree superheat is the difference,

$$485 - 368.4 = 116.6°\text{F}.$$

The enthalpy of saturated steam (above 32°F on which the steam tables are based) at 170 psia is 1196.3 Btu per lb. In obtaining steam-table values or with the use of the Mollier diagram there may be slight differences from values in this presentation, but this is a natural occurrence and depends on the tables or diagram used. Insofar as the examination is concerned, this does not make the slightest bit of difference in the credit given the written solution.

The heat of superheat may be either obtained from the steam tables or from the Mollier diagram. It may also be calculated, as we shall see here. The heat of superheat is calculated by multiplying the degrees of superheat by the mean specific heat of steam which may be obtained from tables. Then the heat of superheat is

$$116.6 \times 0.582 = 67.9 \text{ Btu per lb.}$$

The total heat of the steam above the 32°F datum is
Total heat = saturation enthalpy + heat of superheat

$$1196.3 + 67.9 = 1264.2 \text{ Btu per lb.}$$

Now, because the steam tables are based on a datum of 32°F and these calculations are based on a datum of 60°F, a correction must be made corresponding to the heat of the liquid at 60°F, namely, 28.1 Btu per lb. Thus, the heat in each pound of steam above the 60°F datum is 1264.2 – 28.1 = 1236.1 Btu per lb.

The total heat in the steam on the hourly basis is 1236.1 × 1690, or 2,089,000 Btu.

(b) Sensible heat loss in the dry flue gases: We remember that one mole of gas at 14.7 psia and 60°F will occupy 379 cu ft. Then, on this test, 2895/379 = 7.638 moles of gas are burned per hour. It is given that the ratio of flue gas to fuel is 10.53. So 7.638 × 10.53 = 80.43 moles of flue gas leave the boiler each hour.

By breakdown:

$$CO_2 = 80.43 \times 0.1 = 8.043 \text{ moles}$$
$$O_2 = 80.43 \times 0.034 = 2.735 \text{ moles}$$
$$N_2 = 80.43 \times 0.866 = 69.652 \text{ moles}$$

Then the sensible heat loss in the dry flue gas may be obtained by converting moles to pounds and multiplying by the specific heat at constant pressure and by the temperature difference.

$$CO_2 = 8.043 \times 44 \times 0.202 \times (505 - 60) = 31{,}811 \text{ Btu}$$
$$O_2 = 2.735 \times 32 \times 0.219 \times (505 - 60) = 8529 \text{ Btu}$$
$$N_2 = 69.652 \times 28 \times 0.248 \times (505 - 60) = 215{,}230 \text{ Btu}$$

Total sensible heat loss = 31,811 + 8529 + 215,230 = 255,570 Btu.

(c) Heat loss to water vapor formed during combustion: It is given that for each mole of flue gas 0.193 mole of water vapor was formed. Then the total water formed per hour is

$$0.193 \times 80.43 = 15.52 \text{ moles water.}$$

The sensible heat in the water formed by combustion may be determined as

$$15.52 \times 18 \times 0.30 \times (505 - 60) = 37{,}295 \text{ Btu.}$$

Because the gross heating value was used in calculating the heat input, the heat of vaporization of the water must be included as a part of the heat output. From the steam tables the heat of vaporization of water at 60°F is 1059 Btu per lb. Then this item of heat loss is 15.52 × 18 × 1059 = 295,842 Btu.

The total heat loss to water formed during combustion, then, is 37,295 + 295,842 = 333,137 Btu.

(d) Heat loss due to incomplete combustion: None, since the flue gas contains no CO.

(e) Radiation and other losses: The difference between the heat input and other items of heat output is the radiation and other losses, which in this problem are

$$3{,}108{,}130 - (2{,}089{,}000 + 257{,}200 + 333{,}137) = 428{,}793 \text{ Btu.}$$

The completed heat balance is shown in Table 9.1.

Table 9.1

		Btu per hr	Percent
Input:			
Heat in fuel		2,964,480	95.38
Heat in feed water		143,650	4.62
Total		3,108,130	100.00
Output:			
Heat in steam		2,089,000	67.21
Heat loss in flue gas			
Sensible heat in dry gas	257,200	590,337	19.05
Heat in water vapor	333,137		
Due to incomplete combustion (none)			
Radiation and unaccounted for losses		428,793	13.74
Total		3,108,130	100.00

Example **9.7**

A boiler plant can burn coal costing $65 per ton, or oil costing $1.00 per gallon, or natural gas costing $1.50 per 1000 cu ft. Assume Btu values and combustion efficiency ratings for the fuels listed and choose the most economical fuel for producing 1000 lb of steam at 250 psig and 50°F superheat, using feed water at a temperature of 180°F.

Solution

The solution process to this problem is the same even though the unit fuel costs may vary.

Steam at 250 psig and 150°F superheat1292Btu per lb
Feed water at 180°F (180−32)−148Btu per lb

1144Btu per lb

Total Btu to be absorbed = 1144 × 1000 = 1,144,000

Coal. Assume 13,500 Btu per lb and 85 percent efficiency:

$$\frac{1,144,000 \times \$65.00}{13,500 \times 0.85 \times 2000} = \$3.24$$

Oil. Assume 18,500 Btu per lb and 85 percent efficiency:

$$\frac{1,144,000 \times \$1.00}{18,500 \times 0.85 \times 8.33 \times 0.80} = \$10.92$$

Gas. Assume 1000 Btu per cu ft and 75 percent efficiency:

$$\frac{1,144,000 \times \$1.50}{1000 \times 0.75 \times 1000} = \$2.28$$

Therefore, *gas* is the most economical fuel under these conditions.

STEAM ENGINES

A steam engine is a reciprocating machine having displacement in which work is done by a piston acted on by pressure moving it. Steam engines are generally classified as to construction, operation, or type of valve gear. The construction may be horizontal, vertical, or angular. They may be single-acting, double-acting, reciprocating, or rotary. Their operation may be condensing, noncondensing, bleeder, or extraction, simple or multiple stage.

According to the type of valves employed, the classification may be divided into simple, compound, uniflow, D-shaped slide valve, rotary valves, poppet, and piston valves.

Example **9.8**

A simple double-acting steam engine of 15-in. bore, 16-in. stroke, and 2.5-in.-diameter piston rod is operating at 225 rpm. Find the indicated horsepower if the scale or range of the spring used is 80 psi and the mean effective pressure determined by the usual method is 25.3 at the head end and 25.9 at the crank end.

Solution

Note carefully that the area of the rod must be subtracted from the piston area when determining the crank-end horsepower.

$$\text{Piston area at head end} = 15^2 \times 0.785 = 176.5 \text{ sq in.}$$

$$\text{Piston-rod area} = 2.5^2 \times 0.785 = 4.9$$

$$\text{Piston area at crank end by difference} = 176.5 - 4.9 = 171.6$$

$$\text{Stroke} = \frac{16}{12} = 1.333 \text{ ft}$$

$$\text{Head end ihp} = 25.3 \times 1.333 \times 176.5 \times \frac{225}{33{,}000} = 40.6 \text{ ihp}$$

$$\text{Crank end ihp} = 25.9 \times 1.333 \times 171.6 \times \frac{225}{33{,}000} = 40.4 \text{ ihp}$$

$$\text{Total ihp} = 40.6 + 40.4 = 81.0 \text{ ihp}$$

If friction hp were 10, then the brake (shaft) hp would be by difference

$$\text{bhp} = 81 - 10 = 71 \text{ bhp.}$$

Example **9.9**

A test of a steam engine shows that it uses 1560 lb of steam per hr when developing 40 bhp. At the same time, the engine is developing an indicated horsepower of 50. The inlet steam is 95 percent quality at 250 psia. Exhaust or back pressure is 14.7 psia. Calculate (a) brake thermal efficiency and (b) indicated thermal efficiency.

Solution

(a) Steam consumption of engine is W_s = 1560/40, or 39 lb per bhp-hr. The enthalpy in 1 lb of steam at throttle conditions is given by

$$H_1 = 376.04 + 0.95(825.4) = 1160.17 \text{ Btu per lb.}$$

The enthalpy of the liquid for 1 lb of exhaust steam at 14.7 psia and operating condensing is 180.07 Btu.
Then, *brake thermal efficiency* is

$$E_t = \frac{2545}{39\ (1160.17 - 180.07)} \times 100 = 6.66 \text{ percent.}$$

(b) The steam consumption based on the indicated hp is 1560/50, or 31.2 lb per bhp-hr. The *indicated thermal efficiency* is

$$E_t = \frac{2545}{31.2(1160.17 - 180.07)} \times 100 = 8.32 \text{ percent.}$$

Example **9.10**

What is the horsepower constant of an 18-in. piston by 48-in. stroke double acting engine running at 74 rpm, and what would be the mean effective pressure for the development of 375 indicated horsepower?

Solution

$$\text{Horsepower constant} = \frac{1 \times \frac{48}{12} \times (18 \times 18 \times 0.7854) \times 74 \times 2}{33{,}000} = 4.565 \text{ hp/psi mep.}$$

Development of 375 ihp would require 375/4.565 = 82.15 psi mep.

Example 9.11

An engine uses 25 lb of steam per indicated horsepower-hour and the evaporative economy of the boiler, under the operating conditions, is 8 lb of steam per pound of coal. Assume that the heating value of the coal is 10,000 Btu per lb, what percentage of energy contained in the coal is realized by the engine?

Solution

The fuel consumption is $\frac{25}{8}$, or 3.125 lb coal per hp-hr. Now, because 1 hp-hr is equal to 33,000 × 60, or 1,980,000 ft-lb, there is 1,980,000/3.125, or 633,600 ft-lb realized per lb of coal. One Btu is equivalent to 778 ft-lb, and since each lb of coal has 10,000 Btu, the energy in a pound of coal would be 7,778,000 ft-lb. Finally, the energy realized by the engine per pound of coal used would be

$$\frac{633{,}600 \times 100}{7{,}778{,}000} = 8.15 \text{ percent.}$$

STEAM TURBINES AND CYCLES

Steam turbines have replaced steam engines in many applications in power plant services, for driving centrifugal pumps, for driving generators in electric power stations, and in many industrial plant applications. This is due primarily to the turbine's compactness, higher speed, and fewer moving parts. The steam turbine has about the same economy as the steam engine when operating with vacuums from 24 to 26 in. Hg. However, the turbine can operate successfully with 29 in. Hg vacuum. These latter conditions do not materially improve performance of the engine because of the increased specific steam volumes. The economy of non-condensing steam turbines is considerably below that of high-grade reciprocating steam engines. The steam turbine gives better speed regulation and its exhaust is not contaminated by cylinder lubricating oils. The small turbine is found useful where its speed can be utilized advantageously as in driving rotary pumps and blowers. Turbine exhaust steam may be used for process and heating purposes such as is common in oil refinery practice. Overall heat economy is helped considerably in this way. Very often steam turbines are used for standby service for electric motors and electric generators for emergency purposes.

Example 9.12

Determine the efficiency of a condensing turbine operating on a Rankine cycle as follows:

Throttle pressure	200 psia
Steam temperature	600°F
Exhaust (back) pressure	2 in. Hg
Exhaust temperature	101°F
Steam rate	10.5 lb/hp-hr

Solution

From the Mollier diagram:

Enthalpy at inlet (H_1)	1322 Btu/lb
Enthalpy at outlet (H_2)	936 Btu/lb
Available energy	386 Btu/lb

The theoretical steam rate is $\frac{2545 \text{ Btu/hp-hr}}{386 \text{ Btu/lb}} = 6.6$ lb/hp-hr. Thus, the *Rankine efficiency* of the turbine is 6.6/10.5, or 63 percent.

The *theoretical thermal efficiency* of the cycle is the ratio between the available (1322–936 = 386 Btu per lb) and the energy supplied to the condensate above 32°F, which is

$$\text{(enthalpy at inlet)} - \text{condensate enthalpy} = \text{energy supplied}$$

$$1322 \text{ Btu/lb} - 69 \text{ Btu/lb} = 1253 \text{ Btu/lb}$$

Then the *ideal Rankine cycle* is determined as 386/1253, or 30.8 percent.

The Rankine cycle may be improved by increasing the initial pressure and temperature. Then, in order to avoid excessive condensation in the lower pressure stages, this pressure increase must go hand-in-hand with an increase in temperature.

Example 9.13

In a power plant, the steam supply is at 400 psia and a temperature of 700°F. After expansion in the turbine until the pressure is reduced to 80 psia, the steam is reheated to 700°F. Upon completing the expansion in the last stages of the turbine, the steam is exhausted at 1 in. Hg absolute (Exhibit 2). (a) For the corresponding ideal vapor cycle, find the cycle efficiency. (b) If the steam were not reheated, what would be the corresponding Rankine vapor cycle efficiency? (c) What is the chief advantage due to reheating?

Solution

Refer to Exhibits 2 through 4. Make free use of Mollier diagram and steam tables. Now note the following points:

$$H_a = 1362.3 \text{ Btu/lb} \quad S_d = S_e = 1.8274$$
$$H_b = 1199.38 \text{ Btu/lb} \quad S_a = S_b = 1.6396$$
$$H_d = 1379.2 \text{ Btu/lb}$$

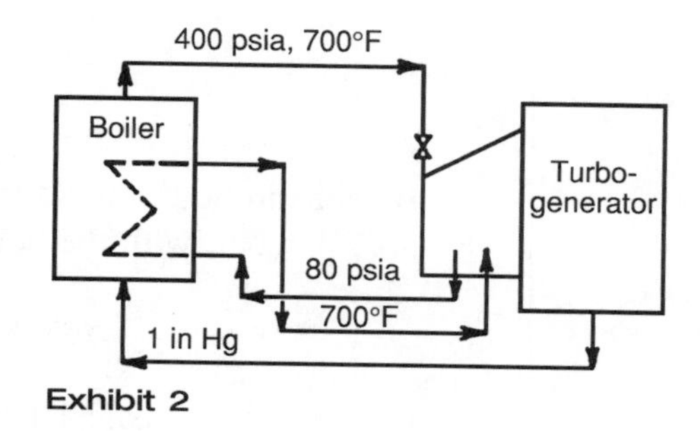

Exhibit 2

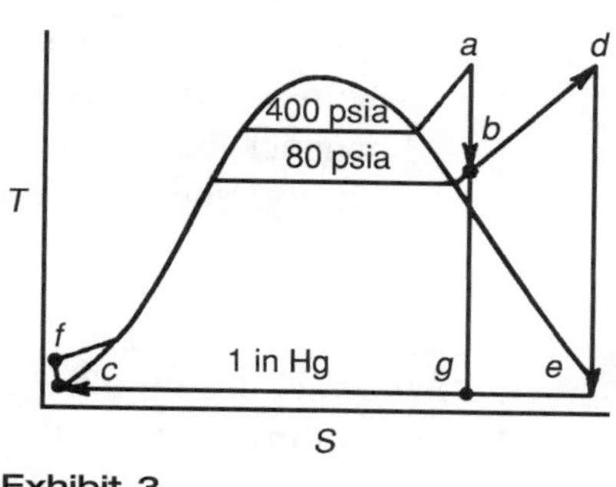

Exhibit 3

Quality at point *e* may be obtained in the usual manner for mixture of liquid and vapor. This is found to be 89.2 percent. In likemanner, H_e is found to be equal to 981.06 Btu/lb. Also, H_c is equal to 47.06 Btu/lb.

$$\text{Work of pump} = \int p\,dv$$

$$F = (P_j - P_c)(144)\bar{V}_c$$
$$= (1/778)(400 - 0.4912)(144)(0.01607) = 1.19 \text{ Btu/lb}$$

(a) The *ideal cycle efficiency* is found in accordance with the following formula:

$$\frac{H_a - H_b + H_d - H_e - F}{H_a - H_c + H_d - H_b - F} \times 100$$

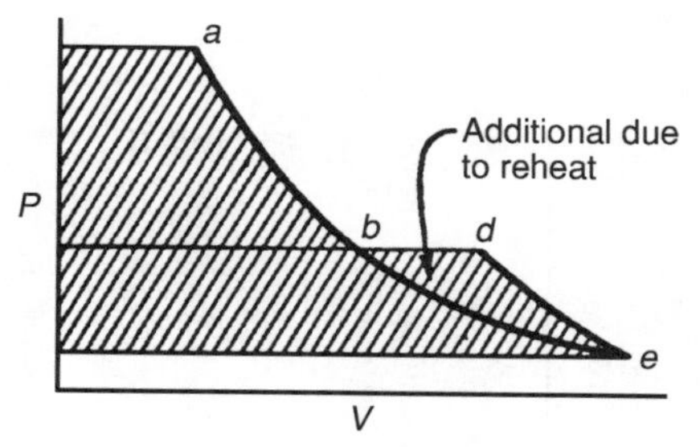

Exhibit 4

By substitution in this formula, the efficiency is found to be 37.4 percent.

(b) The Rankine cycle efficiency is found from this relation:

$$\frac{H_a - H_g}{H_a - H_c} \times 100.$$

By substitution in this formula, the *Rankine efficiency* is found to be equal to 36.5 percent, with the help of the following relations:

$$S_a = S_g = 1.6396 = S_f + x_g \times 1.9452$$

from which x_g is calculated out as 0.7959. Note that H_g is equal to 881.06 Btu/lb.

(c) The chief advantage due to reheating is in the increased quality at the lower pressures, improving the engine efficiency markedly. But an optimum must be reached where too high a quality must not be given to the steam. In this manner too much energy is not given to the condenser (increasing unavailability) and blade erosion is reduced in the turbine if the quality is too low.

Example 9.14

An actual test on a turbogenerator gave the following data: 29,760 kw delivered with a throttle flow of 307,590 lb steam per hour under the following conditions: throttle pressure 245 psia, superheat at throttle 252°F, exhaust pressure 0.964 in. Hg abs., absolute pressure at the one bleeding point 28.73 in. Hg, temperature of the feed water leaving the bleeder heater 163°F. For the corresponding ideal unit find (a) percent throttle steam bled at the one stage, (b) net work for each pound of throttle steam, (c) ideal steam rate, (d) cycle efficiency. For the actual unit find (e) the combined steam rate, (f) combined thermal efficiency, and (g) combined engine efficiency.

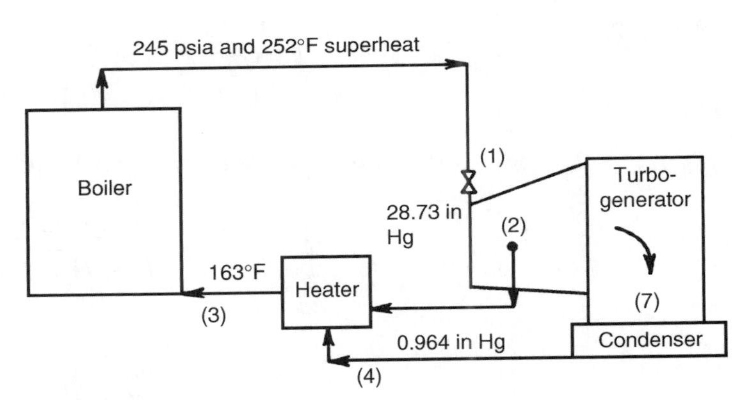

Exhibit 5

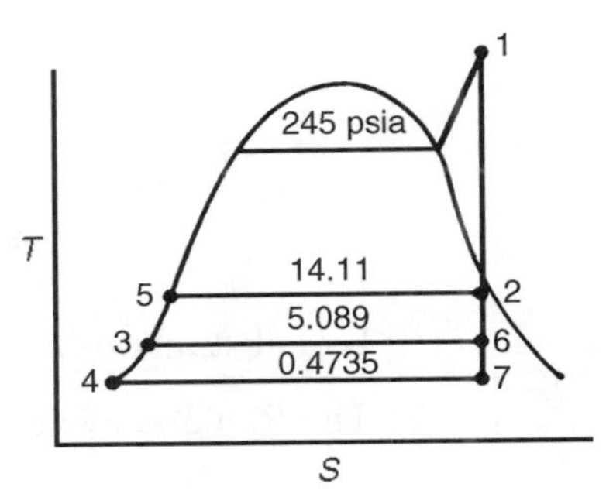

Exhibit 6

Solution

Refer to the steam flow sheet Exhibit 5 and the *TS* diagram Exhibit 6. Now with the use of the steam tables and Mollier diagram determine the following:

$S_1 = 1.676$
$H_1 = 1366$ Btu per lb
$H_2 = 1106$ Btu per lb
$P_2 = 14.11$ psia
$H_3 = 130.85$ Btu per lb
$P_3 = 5.089$ psia
$H_4 = 46.92$ Btu per lb
$P_4 = 0.4735$ psia
$H_5 = 178$ Btu/lb

(a) Percent throttle steam bled:

$$\frac{H_5 - H_4}{H_2 - H_4} \times 100 = 12.41 \text{ percent}$$

(b) Heat converted to work:

$$H_1 - H_2 + (1 - m_2)(H_2 - H_7) = 419.1 \text{ Btu/lb}$$

where $m_2 = 0.1241$.

(c) Ideal steam rate:

$$(3413/419.1) = 8.14 \text{ lb steam per kwhr}$$

(d) Cycle efficiency (heat converted into work/heat supplied):

$$\frac{419.1}{H_1 - H_3} = \frac{419.1}{1366 - 130.85} \times 100 = 36 \text{ percent}$$

(e) Combined steam rate: lb steam consumed/kwhr generated

$$307{,}590/29{,}760 \equiv 10.34 \text{ lb per kwhr}$$

(f) Combined thermal efficiency:

$$\frac{3413}{\text{heat supplied}} = \frac{3413}{10.34 \times (H_1 - H_3)}$$

$$\frac{3413}{10.34 \times 1235.15} \times 100 = 27.3 \text{ percent}$$

(g) Combined engine efficiency $= (27.3/36) \times 100 = 75.7$ percent

Example 9.15

A turbogenerator is operated on the reheating-regenerative cycle with one reheat and one regenerative feed-water heater. Throttle steam at 400 psia and 700°F total steam temperature is used. Exhaust is at 2 in. Hg abs. Steam is taken from the turbine at a pressure of 63 psia for both reheating and feedwater heating. Reheat is to 700°F. For the ideal turbine working under these conditions find: (a) percentage of throttle steam bled for feed-water heating, (b) heat converted to work per pound of throttle steam, (c) heat supplied per pound of throttle steam, (d) ideal thermal efficiency, (e) TS diagram and layout showing boiler, turbine, condenser, feed-water heater, and piping and using the same letters to designate corresponding points on the two diagrams.

Solution

Refer to Exhibits 7 and 8. Use the steam tables and Mollier diagram. Values obtained by the student may differ slightly because of the steam tables used and interpolations within the Mollier diagram, but this is not to be disturbing in any way.

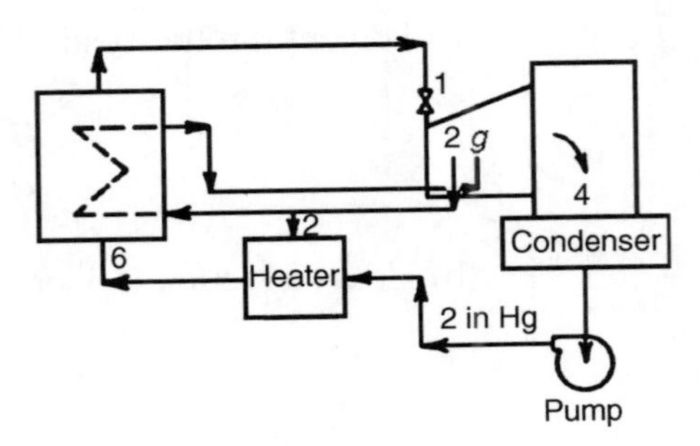

Exhibit 7

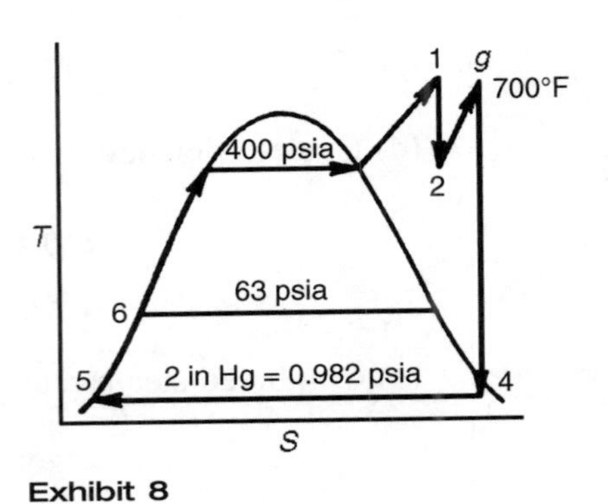

Exhibit 8

$P_1 = 400$ psia	$H_2 = 1178$ Btu per lb
$t_1 = 700°F$	$H_g = 1380.1$ (from steam tables)
$H_1 = 1362.3$ Btu per lb	$S_g = 1.8543$ (from steam tables)
$S_1 = 1.6396$	$H_6 = 265.3$ Btu per lb
	$H_5 = 69.1$ Btu per lb
	$H_4 = 1036$ Btu per lb

(a) Percent throttle steam bled:

$$\frac{H_6 - H_5}{H_2 - H_5} = \frac{196.17}{1107.9} = 0.1771, \quad \text{or} \quad 17.71 \text{ percent}$$

(b) Heat converted to work per pound throttle steam:

$$(H_1 - H_2) + (0.8229)(H_g - H_4) = 467.3 \text{ Btu/lb}$$

(c) Heat supplied per pound throttle steam:

$$(H_1 - H_6) + (H_g - H_2) = 1299.1 \text{ Btu/lb}$$

(d) Ideal thermal efficiency:

$$(467.3)/(1299.13) \times 100 = 36.1 \text{ percent}$$

In Exhibit 5, the factor m_2 is the decimal part of the original bleed at point 2. If we let w_1 be the original weight of steam and sufficient steam is bled so that it is just condensed, the heat removed from the bleed steam is given by

$$m_2 w_1 = (H_2 - H_f)$$

where H_f is the enthalpy of liquid at the corresponding pressure. For more penetrating coverage see standard handbooks and texts on the subject.

Example 9.16

A cogeneration steam power plant supplies some 20,000 lbm/min of steam at 1000 lbf/in.^{2} and 1200°F to the high pressure turbine. The steam is exhausted from the turbine at 100 psia and 10,000 lbm/min is supplied to the food processing plant and the rest is either re-heated to 900°F for supply to the low pressure turbine or tapped off to heat the feed water in the closed feed-water heater. The steam is exhausted from the low pressure turbine at 10 psia, where it is mixed with the 60°F water returning from the process in a condenser. The water leaves the condenser as saturated liquid and enters the boiler feed pump. Exhibit 9 fills in the rest of the details. Determine

(a) The mass flow rate of the 100 psia steam injected into the closed feed-water heater to bring the termperatue of the exiting feed water equal to 100 psia saturated water temperature.

(b) The flow rate of water required to cool condenser water to saturated liquid at 10 psia

(c) The work produced by the turbines

(d) The utilization factor.

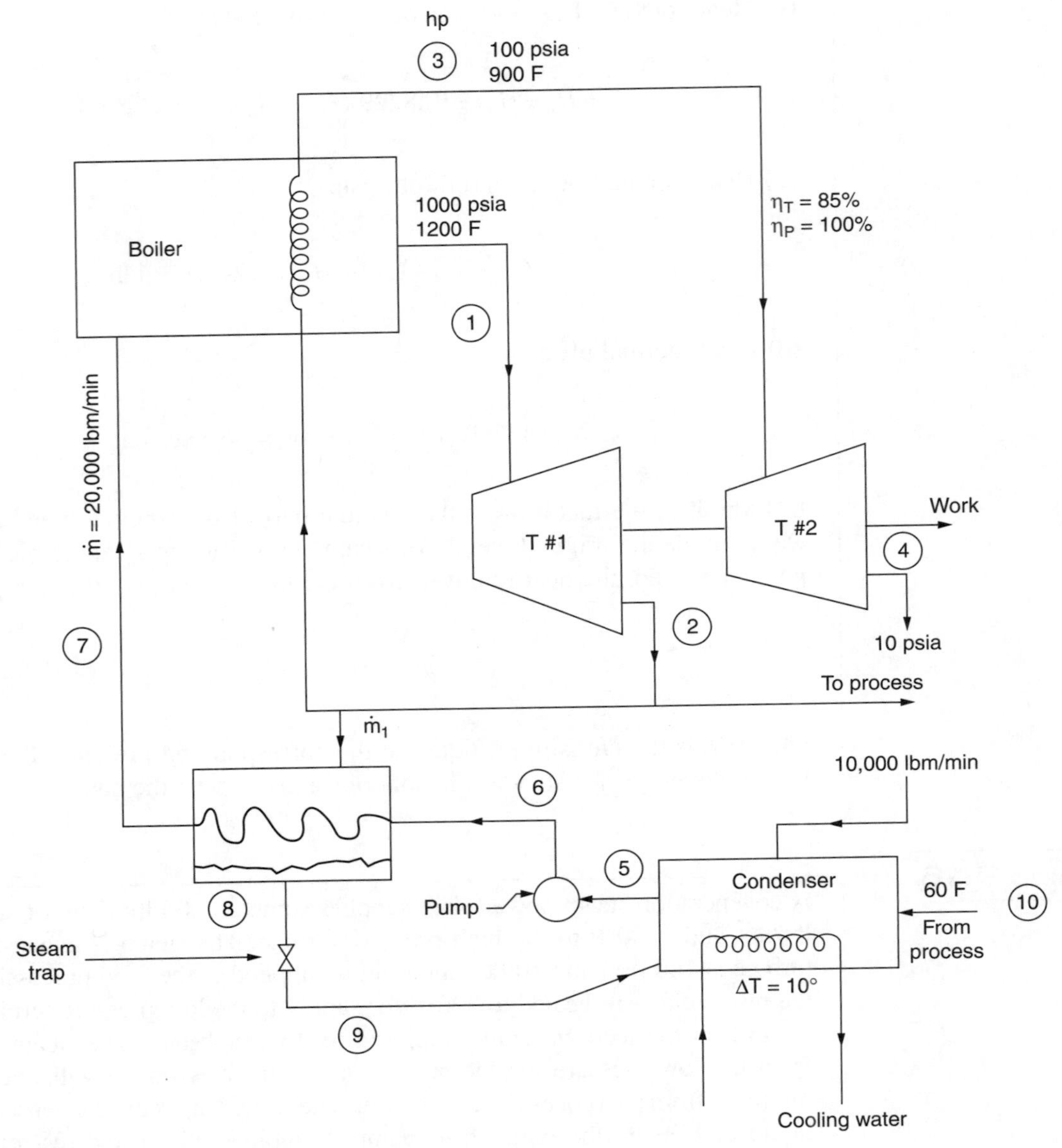
hp
3
100 psia
900 F
Boiler
1000 psia
1200 F
1
η_T = 85%
η_P = 100%
ṁ = 20,000 lbm/min
T #1
T #2
Work
4
2
10 psia
7
To process
ṁ1
10,000 lbm/min
6
5
Condenser
60 F
10
From
process
Pump
8
Steam
trap
ΔT = 10°
9
Cooling water

Exhibit 9

Solution

Table of values

	1	2 S	2	3	4S	4	5	6	7	8	9	10
P	1000	100	⟶		10	⟶		1000	→	100	10	→
T	1200			900					←	327.9		60
h	1619.7	1296	1344.6	1480.9	1212.0	1252.3	161.2	164.2	301.6	298.6	→	28.08
s	1.7261	→		1.883								
v							0.01659			0.01773		

To find mass flow rate of steam exiting the low pressure turbine we need to find the enthalpy exiting the turbines.

$$\eta T_1 = 0.85 = \frac{h_1 - h_2}{h_1 - h_{2S}} = \frac{1619.7 - h_2}{1619.7 - 1296}$$

$$h_2 = 1344.6$$

$$\eta T_2 = 0.85 = \frac{h_3 - h_4}{h_3 - h_{4S}} = \frac{1480.9 - h_4}{1480.9 - 1212}$$

$$h_4 = 1252.3$$

$$h_6 = h_5 + \int_5^6 vdp = \frac{161.2 + (0.1659)(144)(990)}{778} = 161.2 + 3.04$$

$$h_6 = 164.2$$

(a) Closed FWH

$$m_1(h_2) + (10,000 - m_1)h_6 = m_1 h_8 + (10,000 - m_1)h_7$$

$$h_7 \approx h_8 + \int vdp = \frac{298.6 + (.01773)(1000 - 100)(144)}{778} = 301.6$$

$$\dot{m}_1(1344.6) + (10,000 - \dot{m}_1)(164.2) = \dot{m}_1(298.6) + (10,000 - \dot{m}_1)(301.6)$$

$$\dot{m}_1 = 1166 \text{ lbm/min.}$$

(b) This flow rate required to cool condenser water.

$$(10{,}000\ \text{lbm/min})h_{10} + (10{,}000 - 1160)h_4 + 1166(h_9)$$
$$= 20{,}000(h_5) + Q$$
$$(10{,}000)(28.08) + (8834)(1252.3) + 1166(298.6)$$
$$= 20{,}000(161.2) + \dot{m}_{\text{cooling}}(10)$$
$$\dot{m} = 847{,}000\ \text{lbm/min}$$

(c) Work produced by turbines.

$$\text{Turbine } \#1 = \dot{m}(h_1 - h_2) = 20{,}000(1619.7 - 1344.6)$$
$$= \frac{5{,}502{,}000\ \text{Btu/min}}{42.41\ \text{Btu/Hp min}} = 1.297 \times 10^5\ \text{hp}$$
$$\text{Turbine } \#2 = 8834(h_3 - h_4) = \frac{8834(1480.9 - 1252.3)}{42.41}$$
$$= 4.76 \times 10^4\,\text{hp}$$

(d) Utilization factor =

$$t_u = \frac{\text{Network out } + \text{process HT}}{\text{Total HT. In.}}$$
$$t_u = \frac{W_{T\#1} + w_{T\#2} - w_P + \text{process HT}}{\text{Total HT}}$$
$$t_u = \frac{5.5 \times 10^6 + 2.019 \times 10^6 - 20{,}000(3.04) + 10{,}000(1344.6 - 28.08)}{20{,}000(1619.7 - 301.6) + 8834(1480.9 - 1344.6)}$$
$$t_u = 74.8\ \text{percent}$$

THE MERCURY-VAPOR CYCLE

In the mercury-vapor-steam cycle the basis is the effects of the difference in thermodynamic properties of the two pure fluids. We know that steam works under relatively high pressures with an attendant relative low temperature. Mercury, on the other hand, has its vapor characteristic as operating under low pressures with attendant high temperature.

In the cycle the pressures are so selected that the mercury vapor condenses at a temperature higher than that at which steam evaporates. The processes of mercury vapor condensation and steam evaporation take place in a common vessel called the *condenser-boiler*, which is the heart of the cycle. In the steam portion of the cycle, condenser water carries away the heat of steam condensation; in the mercury portion of the cycle, steam picks up the heat of condensation of the mercury vapor. Thus, there is a great saving in heat and the economies effected reflect consequent improvement in cycle efficiency.

Refer to Figure 9.1 for the flow sheet hookup and to Figure 9.2 for the process shown on the *TS* diagram.

The same furnace serves the mercury boiler and the steam superheater. Mercury vapor is only condensed, not superheated. Now, if the condenser-boiler is high enough above the mercury boiler, the head of mercury is great enough to return

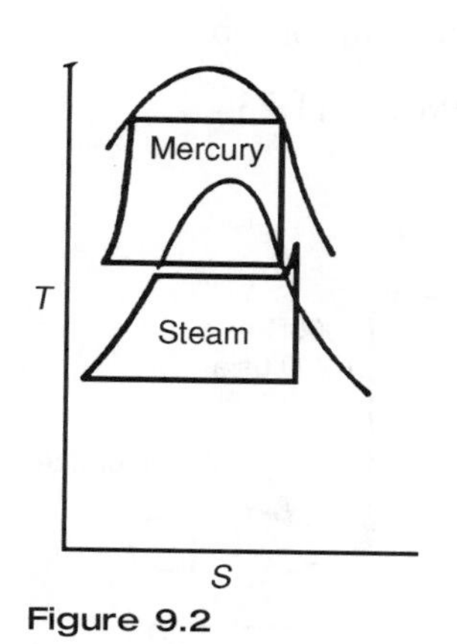

Figure 9.2

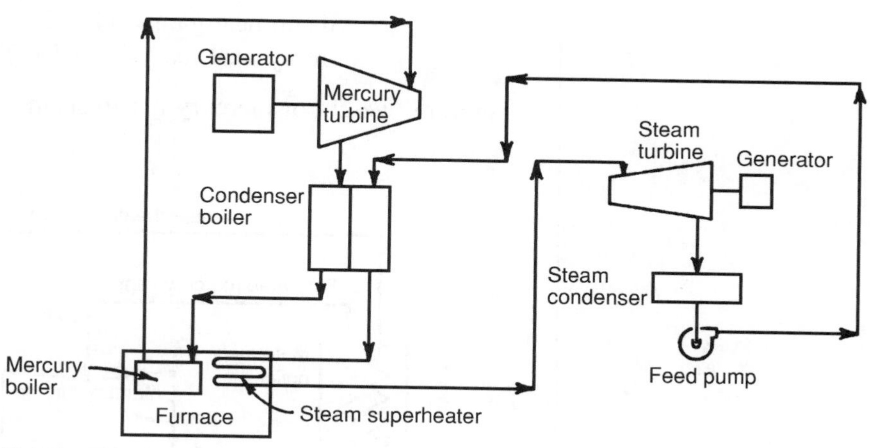

Figure 9.1

the liquid mercury to the boiler by gravity, making the use of a mercury feed pump unnecessary.

Example 9.17

A binary cycle steam and mercury plant has a maximum temperature of 1000°F for both. The mercury is condensed in the steam boiler at 10 psia and the steam pressure is 1200 psia. Condenser pressure is 1 psia. Expansions in both turbines are assumed to be constant entropy. Steam cycle has superheat, but no reheat or regeneration. Find the efficiency of the cycle as described. Find its efficiency without the mercury.

Solution

Refer to Exhibit 10 for flow sheet. Now set up two columns.

Mercury cycle

$$H_{m1} = 151.1 \text{ Btu per lb}$$
$$S_{m1} = 0.1194$$
At 10 psia, $S_{me} = 0.1194$

Quality x is found as follows:

$$0.1194 = 0.0299 + x(0.1121)$$
$$x_m = 0.798$$
$$H_{me} = 22.6 + 0.798(123)$$
$$= 120.7$$

Steam cycle

$$H_{s1} = 1499.2 \text{ Btu per lb}$$
$$S_{s1} = 1.6293$$
At 1 psia, $S_{se} = 1.6293$

Quality x is found as follows:

$$1.6293 = 0.1326 + x(1.8456)$$
$$x_s = 0.81$$
$$H_{se} = 69.7 + 0.81(1,036.3)$$
$$= 69.7 + 839 = 908.7$$

Mercury cycle

$$H_{mf} = 22.6$$

Steam cycle

$$H_{sj} = 69.7$$

Assume 98 percent quality steam leaving the mercury condenser. Then

$$H_{sw} = 571.7 + 0.98(611.7) = 1,171.7$$

Balance around the mercury condenser:

$$\text{Steam heat gain} = 1171.7 - 69.7 = 1102 \text{ Btu per lb}$$
$$\text{Mercury heat loss} = 120.7 - 22.6 = 98.1 \text{ Btu per lb}$$

Therefore, weight of mercury per lb steam = 1102/98.1 = 11.23.

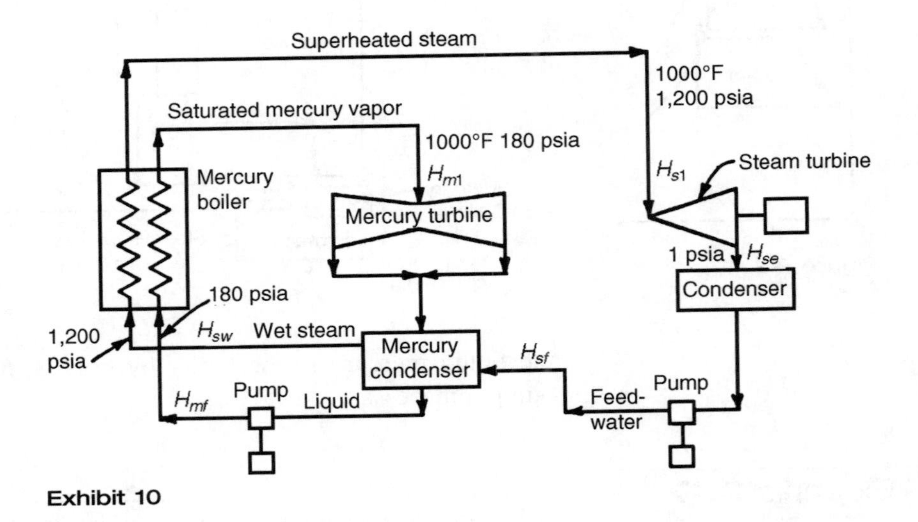

Exhibit 10

Heat input per pound of steam:

$$\text{For mercury} = 11.23 \text{ Hg } (H_{m1} - H_{mf}) = 11.23(151.1 - 22.6) = 1442$$

$$\text{For steam} = (H_{s1} - H_{sw}) = (1499.2 - 1171.7) = 327.5$$

Work done per pound of steam:

$$\text{For mercury} = 11.2316 \text{ Hg } (H_{m1} - H_{me}) = 11.23(151.1 - 120.7) = 342$$

$$\text{For steam} = (H_{s1} - H_{se}) = (1499.2 - 908.7) = 590.5$$

$$\text{work total} = 932.5 \text{ Btu}$$

Binary cycle efficiency

$$(932.5/1769.5)100 = 52.7 \text{ percent}$$

Steam cycle efficiency without mercury topping turbine

$$(590.5)/(1499.2 - 69.7) \times 100 = 41.3 \text{ percent}$$

Symbols used:
Mercury cycle:

H_{m1} = enthalpy at turbine inlet
S_{m1} = entropy at turbine inlet
S_{me} = entropy of exhaust
H_{me} = enthalpy of exhaust
H_{mf} = enthalpy of condensed mercury

Steam cycle:

H_{s1} = enthalpy at turbine inlet
S_{s1} = entropy at turbine inlet
S_{se} = entropy of exhaust
H_{se} = enthalpy of exhaust
H_{sf} = enthalpy of condensed steam
H_{sw} = enthalpy of wet steam leaving mercury condenser

Example **9.18**

(a) Draw a diagram showing the principal paths of fluid flow for a mercury-steam power plant. Indicate on the diagram the state of the fluid, whether liquid or saturated vapor or superheated vapor. (b) What overall station efficiency would you expect for a 1000°F mercury steam cycle? (c) Approximately how many pounds of mercury per pound of steam per hour are circulated? Why?

Solution

(a) See Exhibit 10.

(b) Refer to Example 9.17, and note that for the steam cycle alone without the mercury topping turbine the efficiency is about 40 percent, for the binary cycle, between 50 and 55 percent.

(c) About 10 lb of mercury are circulated for each pound of steam vaporized. Heat of condensation of mercury is about 100 Btu per lb while heat of vaporization of water is about 1000 Btu per lb. Thus to vaporize 1 lb of steam, 1000/100 = 10 lb of mercury must be circulated.

CHAPTER 10

Internal-Combustion Engines and Cycles

OUTLINE

ACTUAL INDICATOR CARDS—OTTO CYCLE

Although the compression and expansion curves of the actual *PV* or indicator cards approximate the polytropic curve $PV^n = C$, they are not truly adiabatic. This is largely because there is a heat transfer between the gases and the cooled cylinder walls, but is further complicated by a variation in the specific heat as the temperature of the gases changes. For an engine using gasoline as fuel, the polytropic n has a value of 1.3 for compression and expansion curves.

The compression pressures for engines operating on this cycle vary from 50 to 250 psi, depending on type of fuel for which the engine is designed. For automobile engines pressures range from 80 to 120 psi. Lower pressures are required for kerosene; blast furnace gas, 150 psi; alcohol, 250 psi. Figure 10.1 shows an actual indicator card for the four-cycle process.

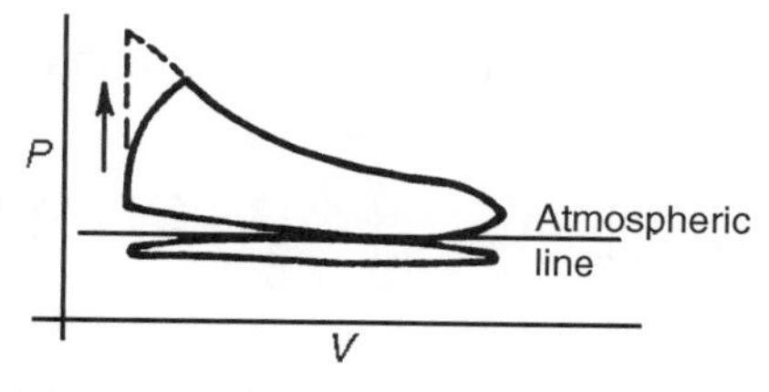

Figure 10.1

The thermal efficiency of the Otto cycle is determined by

$$\eta_t = 1 - \left(\frac{V_2}{V_1}\right)^{k-1} \quad \textbf{(10.1)}$$

Example 10.1

Calculate the compression ratio necessary to produce a compression pressure of 150 psia with a compression curve index of 1.30 and an initial compression of 13.7 psia.

Solution

This is a polytropic process and the following holds:

$$\frac{V_1}{V_2} = \left(\frac{P_2}{P_1}\right)^{1/n} = r_c = \left(\frac{150}{13.7}\right)^{1/1.3} = 6.29$$

Example 10.2

Calculate the area of the intake port of one of the cylinders of an automobile motor so that the velocity through it is 100 fps. The engine bore is 3.06 in. and the stroke 3.75 in., engine speed is 3800 rpm. The intake port opens at top dead center and closes at 40° past bottom center, or 220° of crank travel. Volumetric efficiency is 60 percent.

Solution

Volumetric efficiency is obtained from the following:

$$\eta_v = \frac{\text{actual intake charge (cu ft of mixture)}}{\text{piston displacement (cu ft)}}$$

$$\text{Piston displacement} = 0.785D^2L = 0.785 \times 3.06^2 \times 3.75 = 27.5 \text{ cu in.} = V$$

$$\begin{aligned}\text{Actual intake charge } &= \text{vol efficiency} \times PD \\ &= 0.60 \times (27.5/1{,}728) = 0.0096 \text{ cu ft} = Q\end{aligned}$$

Q = velocity through port × port area × time intake is open

Time intake valve open = time interval for receiving charge = t

$$t = \text{time for one revolution} \times 220/360$$

$$t = \frac{60}{3800} \times \frac{220}{360} = \frac{\text{sec}}{\text{rev}} \times \text{rev} = 0.0096 \text{ sec.}$$

During the next $(140 + 360)$ degrees, the intake port is inactive, so that the acceptance of the charge has nothing to do with this period.

Therefore

$Q = 0.0096 = 100$ fps × port area × 0.0096 sec

from which port area $A = \dfrac{0.0096}{100 \times 0.0096} = 0.01$ sq ft, or 1.44 sq in.

Note that A is the average port area. The full port area is greater. The average port area is only correct if we assume instantaneous full port opening and closing.

Example 10.3

A 10- by 18-in. single-acting gas engine runs 200 rpm and makes 96 explosions per minute. The gross weight on a Prony brake arm was 140 lb, tare of 20 lb, and brake arm was 60 in. long. Indicator card area was 1.18 sq in. and the length of the card was 3 in. Scale on the spring was 200 psia per in. Find (a) indicated horsepower, (b) brake horsepower, (c) friction horsepower, and (d) mechanical efficiency.

Solution

(a)
$$\text{ihp} = \frac{PLAN}{33{,}000} = \frac{78.7 \times 18/12 \times 78.5 \times 96}{33{,}000} = 27 \text{ ihp}$$

The value of 78.7 in the above equation is the mean effective pressure MEP determined so that 1.18/3 × 200 = 78.7 psi. The value of A is $0.785 \times 10^2 =$ 78.5 sq in.

(b)
$$\text{bhp} = \frac{2\pi NT}{33{,}000} = \frac{2 \times 3.1416 \times 200 \times (60/12 \times 140 - 20)}{33{,}000} = 22.85$$

(c) Friction hp = ihp – bhp = 27 – 22.85 = 4.15

(d) Mechanical efficiency = 22.85/27 × 100 = 84.6 percent

GASOLINE ENGINE—COMPRESSION RATIO AND FUEL ECONOMY

The thermal efficiency increases with compression ratio. Because the cycle efficiency is an index of the work done on the piston by the air-fuel mixture for each Btu of fuel burned, it represents the fuel economy of the engine. Thus, an increase in efficiency due to a higher compression ratio improves the economy by lowering the fuel consumption per unit of power output.

GASOLINE ENGINE—COMPRESSION RATIO AND EXHAUST TEMPERATURE

From the technical literature it may be seen that the cold-air standard of the Otto cycle is indicated in terms of thermal efficiency to be

$$\eta_t = 1 - \left(\frac{1}{r_c}\right)^{0.4} \qquad \textbf{(10.2)}$$

It is possible to derive an expression from Equation 10.2 to take the form of

$$T_e = \frac{k}{r_c^{0.4}} + T_i \qquad \textbf{(10.3)}$$

where T_e is exhaust temperature in degrees Rankine, k is a factor mainly dependent on the heat supplied to the air during combustion, and T_i is intake temperature in degrees Rankine. Thus we see that with k constant and T_i constant, *an increase in the compression ratio tends to lower the exhaust temperature* T_e.

The effect of opening the throttle is to speed up the engine. Therefore, the time required to burn the fuel in each cylinder is less, and partial combustion may be occurring at point of exhaust, with consequent raising of the exhaust temperature. Furthermore, the loss of heat through the cylinder walls is a small fraction of the heat input, because the gases are in contact with the cylinder for a shorter time. This fact further tends to increase the temperature of the exhaust.

WORK FOR OTTO CYCLE

Below is the work formula for the Otto cycle. This is derived from the equations for work from the isentropic and isobaric processes for gases. Heat is supplied in the action from *B* to *C* in the cycle of operation.

$$\text{Work} = \frac{(P_3V_3 - P_4V_4) - (P_2V_2 - P_1V_1)}{k-1} \text{ ft-lb} \tag{10.4}$$

In terms of temperatures

$$\text{Work} = \frac{WR}{k-1}[(T_3 - T_4) - (T_2 - T_1)] \text{ ft-lb.} \tag{10.5}$$

Heat supplied in the Otto Cycle is given by the already-familiar expression

$$Q = Wc_v(T_3 - T_2) \text{ Btu.} \tag{10.6}$$

Example 10.4

Calculate the volumetric efficiency of a six-cylinder automobile motor of $3\frac{5}{16}$–in. bore and $3\frac{3}{4}$–in. stroke when running at 2000 rpm, with 60 cfm of entering air.

Solution

$$PD = \text{Piston displacement} = \frac{\pi D^2 L}{4} = \left(\frac{\pi}{4}\right)\frac{(3.3125)^2 \text{ in}^2(3.75)\text{in}}{1728 \text{ in}^3/\text{ft}^3}$$

$$= 0.0187 \text{ cu ft}$$

$$\text{No. of suction strokes per min} = N = 2000/2$$

$$\text{Volume displaced per min} = V \times N \times \text{number of cylinders}$$

$$= 0.0187 \times 1000 \times 6 = 112.2 \text{ cu ft}$$

$$\text{Volumetric efficiency} = 60/112.2 \times 100 = 53.5 \text{ percent}$$

Example 10.5

A cold air standard Otto cycle (constant specific heat at room temperature) has air inlet at 14.7 psia and 540° R. The compression ratio is 10 and the heat addition to the air during combustion is 700 Btu/lbm. Determine the following:

(a) temperature of the air at TDC after compression

(b) air temperature after combustion

(c) maximum pressure in the cylinder

(d) thermal efficiency of cycle

(e) net work of cycle

Solution

First sketch the *PV* and *TS* diagrams for the Otto cycle, as shown in Exhibits 1 and 2.

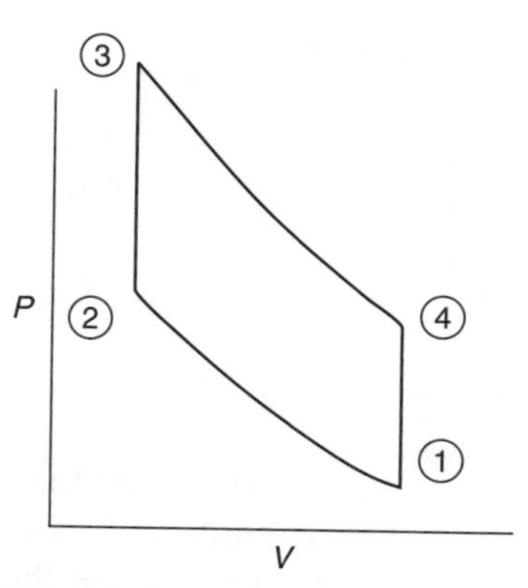

Exhibit 1

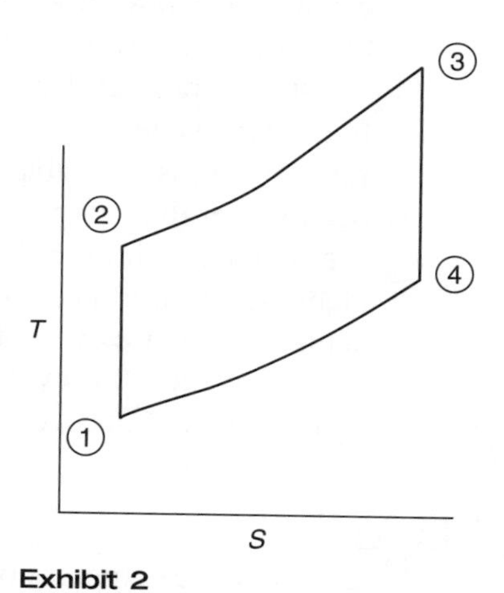

Exhibit 2

(a) $T_2 = T_1\left(\dfrac{V_1}{V_2}\right)^{k-1} = T_1(r_c)^{k-1} = 540(10)^{1.4-1} = 1356°\text{R}$

(b) Combustion process:

$$Q = C_N d_T$$
$$700 \text{ Btu/lbm} = 0.171 \text{ Btu/lbm °R}(T_3 - 1356)$$
$$T_3 = 5450°\text{R}$$

(c) $p_2 = p_1(r_c)^k = 14.7(10)^{1.4} = 369$ psia

Process (2) – (3) is constant volume:

Therefore $\dfrac{p_2 v_2}{T_2} = \dfrac{p_3 v_2}{T_3}$

$$p_3 = \frac{T_3}{T_2}(p_2) = \frac{5450}{1356}(369)$$

$$p_3 = 1483 \text{ psia.}$$

(d) $\eta_t = 1 - \frac{1}{r_c^{k-1}} = 1 - \frac{1}{10^{0.4}} = 60$ percent

(e) $\eta_T = \frac{w_{net}}{Q_{in}}$

$$0.60 = \frac{W_{net}}{700}$$

$$w_{net} = 420 \text{ Btu/lbm.}$$

THE DIESEL CYCLE

After proving mathematically that such an engine was practical, Dr. Rudolph Diesel built the first engine of its kind in 1892. When he tried to start it, the first explosion wrecked the engine. Unfortunately, while in the midst of a successful career, Dr. Diesel mysteriously disappeared from a cross-channel steamer while on a trip from Antwerp to London in 1913.

It may be seen from an inspection of Equation 10.2 that as r_c increases, the bracketed factor increases and the efficiency decreases. Therefore, a low cutoff fuel ratio r_0 is desirable for best thermal efficiency. The point of cutoff seldom occurs later than 10 percent of the stroke or r_0 equal to 2.4, and is usually earlier. We may also observe that a working substance with a high value of k is advantageously helpful because efficiency increases with k. However, k for a real gas actually decreases with an increase in temperature.

As in the Otto cycle, the value of k in the cold-air standard is 1.4. Lower values, about 1.35, would be used in the hot-air standard.

Example **10.6**

Calculate the compression ratio required in a diesel engine to obtain a compression pressure of 450 psia, assuming the air at the beginning of compression to exist at 13 psia and 150°F. Assume a compression curve exponent of 1.35; calculate the temperature at the end of compression.

Solution

Assume 1 lb of air and apply the perfect-gas law. But before proceeding let us set up nomenclature as follows:

P_2 = initial pressure, psia
P_3 = final pressure, psia
V_2 = initial volume, cu ft
V_3 = final volume, cu ft
R = gas constant for air, 53.3
n = compression-curve exponent
T_2 = initial temperature, R
T_3 = final temperature, R
w = weight of air, lb

$$V_2 = \frac{wRT_2}{P_2 \times 144} = \frac{1 \times 53.3 \times (150 + 460)}{13 \times 144} = 17.4 \text{ cu ft}$$

Refer to the polytropic process of gases. Then

$$V_3 = V_2\left(\frac{P_2}{P_3}\right)^{1/n} = 17.4\left(\frac{13}{450}\right)^{1/1.35} = 1.255 \text{ cu ft.}$$

Compression ratio $r_c = V_2/V_3 = 17.4/1.255 = 13.9$. Proceeding to the next part of the question, the determination of the final temperature, set up the equation

$$T_2 = T_3\left(\frac{P_2}{P_3}\right)^{(1.35-1)/n} = T_3\left(\frac{13}{450}\right)^{0.35/1.35} = (150+460) = T_3 \times 0.398$$

from which $T_3 = 610/0.398 = 1530$ R, or the equivalent of 1070°F.

Example **10.7**

A four-cycle six-cylinder diesel engine of $4\frac{1}{4}$ – in. bore and 6-in. stroke running at 1200 rpm, has 9 percent CO_2 present in the exhaust gases. The fuel consumption is 28 lb per hr. Assuming that 13.7 percent CO_2 indicates an air-fuel ratio of 15 lb of air to 1 lb of fuel, calculate the volumetric efficiency of the engine. The intake air temperature is 60°F and the barometric pressure is 29.80 in. Hg.

Solution

Atmospheric air contains 76.9 percent nitrogen by weight. If an analysis of the fuel oil shows zero nitrogen before combustion, all of the nitrogen in the exhaust gases must come from the air. Therefore, with 13.7 percent CO_2 by volume in the dry exhaust the nitrogen content is

$$N_2 = 76.9/100 \times 15 = 11.53 \text{ lb } N_2 \text{ per lb of fuel oil}$$

or $$11.53/28 = 0.412 \text{ mole } N_2 \text{ per lb fuel oil}$$

$$\text{Percentage of } CO_2 \text{ in exhaust gases} = \frac{CO_2}{N_2 + CO_2} \text{ moles.}$$

Then we can say

$$\frac{13.7}{100} = \frac{CO_2}{CO_2 + 0.412}.$$

Solving for CO_2, we obtain $CO_2 = 0.0654$ mole. Because mole percent is equal to volume percent, for 9 percent CO_2 in the exhaust gases

$$0.9 = \frac{CO_2}{CO_2 + N_2} = \frac{0.0654}{0.0654 + N_2}.$$

Solve for N_2, which is found to be equal to 0.661 mole. The weight of N_2 is determined in the usual manner to be 0.661 × 28 = 18.5 lb. The air for combustion is

$$\frac{N_2}{0.769} = \frac{18.5}{0.769} = 24.1 \text{ lb air per lb fuel oil.}$$

The specific volume of air at 60°F and 29.8 in. Hg is 13.02 cu ft per lb. Thus, the actual charge drawn into the cylinder is found in accord with

$$24.1 \times \frac{13.02}{3600} \times 28 = 2.45 \text{ cu ft per sec.}$$

We remember that the volumetric efficiency is the ratio of actual charge drawn into the cylinder divided by the piston displacement. Now the piston displacement for a single cylinder is

$$0.785(4.25)^2 \times 6 \times 1/1728 = 0.0492 \text{ cu ft.}$$

For the six cylinders the volume displaced is

$$\frac{6\text{N}}{60} \times \text{PD} = 6 \times \frac{1200}{2} \times \frac{1}{60} \times 0.0492 = 2.95 \text{ cu ft per sec.}$$

Finally, volumetric efficiency is 2.45/2.95 × 100 = 83 percent.

Example **10.8**

A four-stroke cold air standard Diesel cycle has air inlet at 14.5 psia and 540° R. The compression ratio is 18 and the heat addition is 800 Btu/lbm. Determine

(a) Pressure at the end of compression

(b) Temperature after ignition

(c) Cut off ratio

(d) Thermal efficiency of cycle

(e) Work produced in Btu/lbm

Solution

Sketch the *PV* and *TS* diagrams of the cycle as shown in Exhibits 3 and 4.

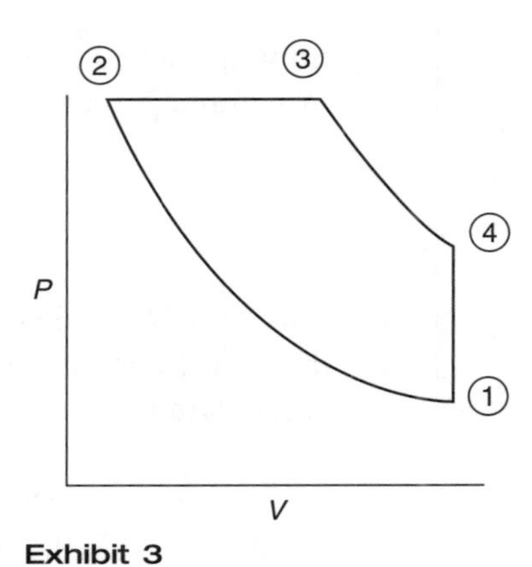

Exhibit 3

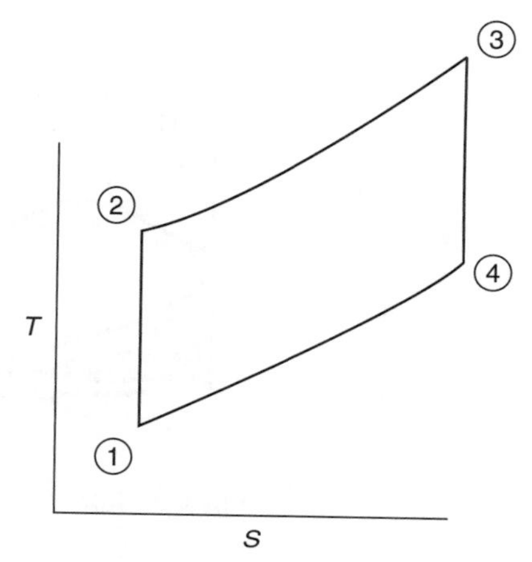

Exhibit 4

(a)
$$P_2 = P_1\left(\frac{V_1}{V_2}\right)^k = 14.5(18)^{1.4} = 829 \text{ psia}$$

(b) Combustion $= Q = c_p d_t$

$$800 \text{ btu/lbm} = 0.24 \text{ btu/lbm°R}(\Delta T)$$
$$\Delta T = 3333$$
$$T_2 = T_1\left(\frac{v_1}{v_2}\right)^{k-1} = 540(18)^{1.4-1} = 1716°\text{R}$$

Therefore $T_3 = T_2 + \Delta T$

$$T_3 = 1716 + 3333 = 5049°\text{R}$$

(c)
$$r_{cut\ off} = \frac{v_3}{v_2} = \frac{T_3}{T_2} = \frac{3333}{1716} = 1.94$$

(d)
$$\eta_t = 1 - \frac{1}{r^{k-1}}\left[\frac{r_{cut}^k - 1}{k(r_{cut} - 1)}\right]$$
$$\eta_t = 1 - \frac{1}{18^{1.4-1}}\left[\frac{1.94^{1.4} - 1}{1.4(1.94 - 1)}\right] = 63.4 \text{ percent}$$

(e)
$$\eta_t = \frac{w_{\text{net}}}{Q_{\text{in}}} = \frac{w_{\text{net}}}{h_3 - h_2}$$
$$0.634 = \frac{w_{\text{net}}}{124(5049 - 1716)}$$
$$w_{\text{net}} = 507 \text{ btu/lbm}$$

Figures 10.2 through 10.5 are actual *PV* diagrams of the Otto and diesel cycles at various load conditions.

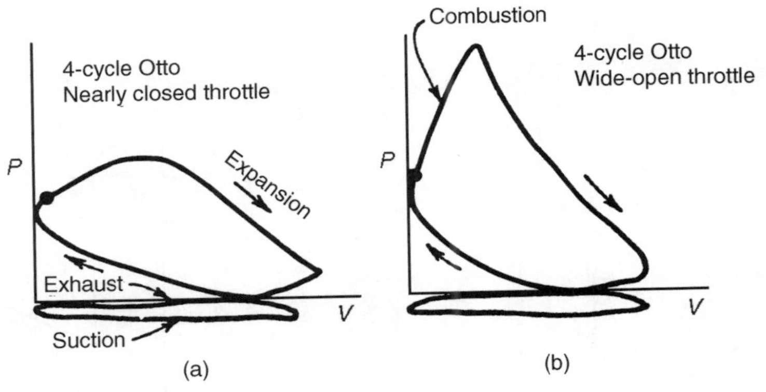

Figure 10.2

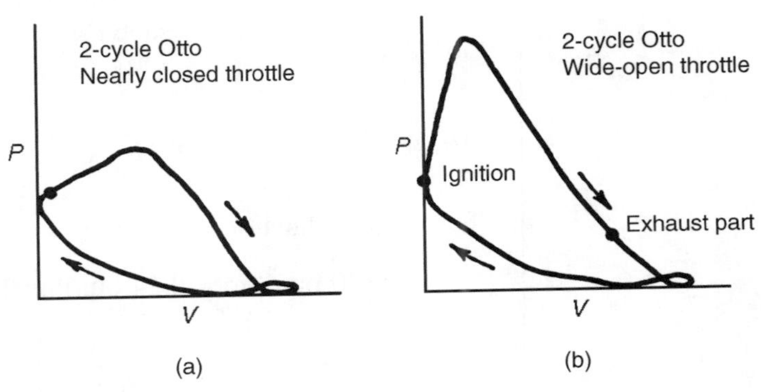

Figure 10.3

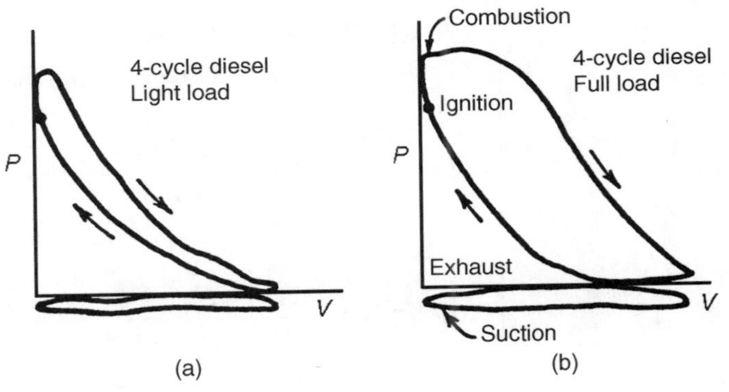

Figure 10.4

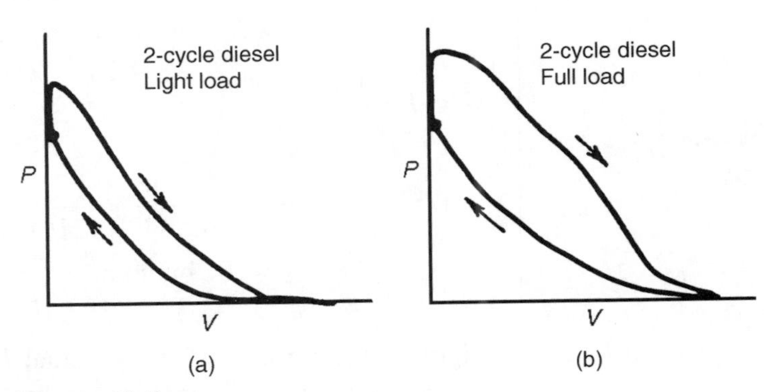

Figure 10.5

Example 10.9

A diesel engine of the air-cell type has a cylinder bore of 4.25 in. and a stroke of 6 in. Assuming flat surfaces for both cylinder head and piston face, calculate the distance in inches between these two surfaces under the following conditions: pressure at the end of compression at 500 psia, pressure at the beginning of compression at 13.7 psia, volume between piston and head is 30 percent of the total clearance volume with 70 percent being in the air cell. Refer to Exhibit 5.

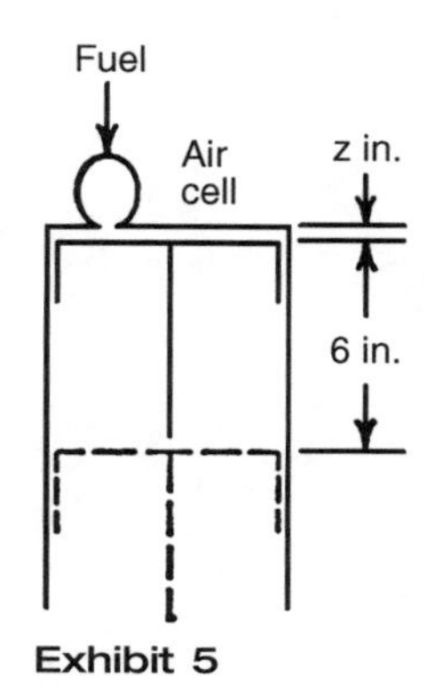

Exhibit 5

Solution

$$\text{Displacement } V_2 - V_3 = \frac{\pi}{4} D^2 L$$

$$V_2 - V_3 = 0.785 \times 4.25^2 \times 6 = 85 \text{ cu in.}$$

Refer to Figure 10.3 and assume compression curve exponent equal to 1.35.

$$r_c = \frac{V_2}{V_3} = \left(\frac{P_3}{P_2}\right)^{1/n} = \left(\frac{500}{13.7}\right)^{1/1.35} = 14.32$$

Now, because $V_2 - V_3 = 85$ cu in. and $V_2 = 14.32 \times V_2$, by substitution

$$14.32 \times V_3 - V_3 = 85$$

from which $V_3 = 85/(14.32 - 1) = 6.4$ cu in. From the problem, the volume at end of compression is 30 percent of total clearance volume. Then

$$6.4 \times 0.3 = 0.785 \times (4.25)^2 \times l.$$

The distance l is calculated from the above as

$$l = \frac{6.4 \times 0.3}{0.785(4.25)^2} = 0.135 \text{ in.}$$

Percent clearance may be determined although not required here. With c designated as clearance fraction and relation between c and r_c being $r_c = (1 + c)/c$,

$$r_c = 14.32 = \frac{1+c}{c}$$

from which $c = 1/13.32 = 0.0752$, or 7.52 percent.

The ideal Diesel cycle is not a very good approximation of how an actual engine works. The fuel does not ignite instantaneously and the piston does not go down at the rate of the expansion of the gas in the cylinder. As a consequence, a fictitious cycle called the dual cycle is much closer to the actual diesel cycle. Example 10.10 demonstrates the use of this cycle.

Example 10.10

Consider an engine for which air intake is 60° F and 14.7 psia. The engine has a compression ratio of 17 and a cut-off ratio of 1.8 The pressure doubles during the constant volume heat addition process. Sketch the *PV* and *TS* diagrams and then determine the values at each of the states.

Solution

The *PV* and *TS* diagrams are shown in Exhibit 6 and Exhibit 7.

$$r = V_1/V_2 \quad r = V_4/V_3$$

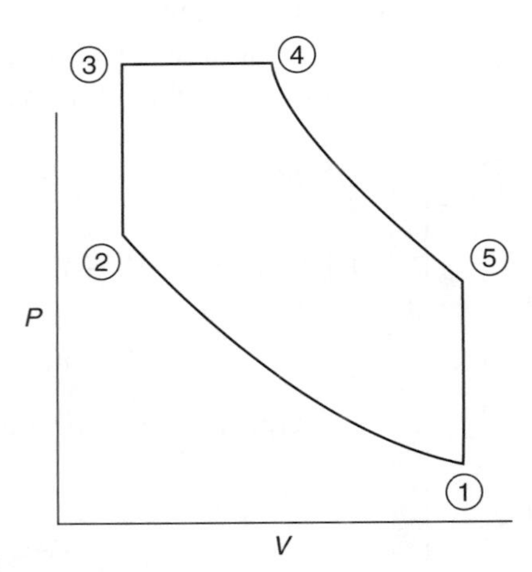

Exhibit 6

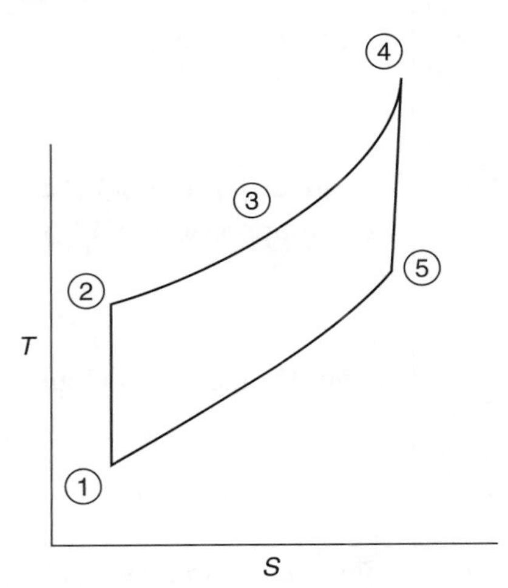

Exhibit 7

$$r_{\text{compression}} = \frac{V_1}{V_2} \qquad r_{\text{cut off}} = \frac{V_4}{V_3}$$

Determine values of properties at various states.

	1	2	3	4	5
p	14.7	776	1552		67.4
T_r	520	1615	3230	5814	2373

State 2

$$T_2 = T_1\left(\frac{v_1}{v_2}\right)^{k-1} = 520(17)^{1.4-1} = 1615°\text{R}$$

$$p_2 = p_1\left(\frac{v_1}{v_2}\right)^{k} = 14.7(17)^{1.4} = 776\text{ psia}$$

State 3

$$p_3 = 2p_2 = 2(776) = 1552 \text{ psia.}$$

$$\frac{p_2 v_2}{T_2} = \frac{p_3 v_3}{T_3}$$

$$T_3 = \frac{p_3}{p_2}(T_2) = 2(1615) = 3230°\text{R}$$

State 4

$$\frac{p_3 v_3}{t_3} = \frac{p_4 v_4}{T_4}$$

$$T_4 = T_3\left(\frac{v_4}{v_3}\right) = 3230(1.8) = 5814°\text{R}$$

State 5

$$\frac{T_4}{T_5} = \left(\frac{v_5}{v_4}\right)^{k-1} \quad \text{and} \quad \frac{v_1}{v_2} = \frac{v_5}{v_3} = 17\frac{v_4}{v_3} = 1.8$$

Combining we get $\frac{v_5}{v_4} = 9.4$

$$5814 - (9.4)^{k-1}T = 2373$$

$$p_5 = p_4\left(\frac{v_4}{v_5}\right)^k$$

$$p_5 = 1552\left(\frac{1}{9.4}\right)^{1.4} = 67.4 \text{ psia.}$$

Now with all the properties the work and thermal efficiency may easily be found,

$$\text{Work Net} = Q_{in} - Q_{out} = C_v(T_3 - T_2) + C_p(T_4 - T_3) - C_v(T_5 - T_1)$$
$$\text{Work Net} = 0.17(3230 - 1615) + 0.24(5814 - 3230) - 0.17(2373 - 520)$$
$$\text{Work Net} = 579.7 \text{ btu/lbm.}$$

$$\text{Thermal eff} = \frac{Q_{in} - Q_{out}}{Q_{in}}$$

$$\eta_t = \frac{C_v(T_3 - T_2) + C_p(T_4 - T_3) - C_v(T_5 - T_1)}{C_v(T_3 - T_2) + C_p(T_4 - T_3)}$$

$$\eta_t = \frac{5797}{0.17(3230 - 1615) + 0.24(5814 - 3230)}$$

$$\eta_t - \frac{579.7}{894.7} = 64.8 \text{ percent}$$

CHAPTER 11

Gas Turbines, Nozzles, and High Speed Flow

The gas turbine is a prime mover that operates on the Brayton cycle and includes a compressor, a combustion chamber, and a turbine as illustrated in Figure 11.1. Any combination of these elements working together is considered a gas turbine; however, the use of these components in conjunction with an Otto-cycle engine, a diesel engine, a steam power plant, or other common forms of prime movers is not considered a gas-turbine power plant. Other components such as regenerators, compressor interstage coolers, turbine interstage combustion chambers, and split compressors or split turbines may be added to the compressor, combustion chamber, and turbine, and the resulting power plant is still considered a gas turbine. The addition of these components merely improves the performance or mechanical operation of the gas turbine and does not change its fundamental principles.

In the internal-combustion-engine cycles, all processes occur in the same cylinder. In contrast, in the gas-turbine cycle, compression occurs in one machine (the compressor) and expansion occurs in another machine (the turbine). Thus, the *flow work* of getting the air into and out of the compressor and the burned gases into and out of the turbine must be considered.

The operation of the gas turbine and the principal factors affecting its power and efficiency may be understood by referring to Figure 11.2 and the enthalpy-entropy diagram in Figure 11.3. Air is taken into the compressor at point 1 and delivered at point 2. At C part of the compressed air mixes with the fuel, which burns in the combustion chamber, the remainder passing around the chamber. The mixture of hot gases at point 3 passes on and enters the turbine at A. After expanding to point 4, the gases pass out of the exhaust at E. The principal factors

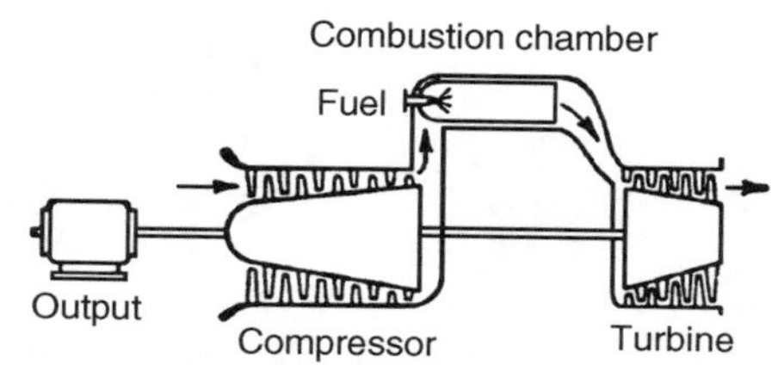

Figure 11.1 Schematic diagram of simple-cycle gas turbine

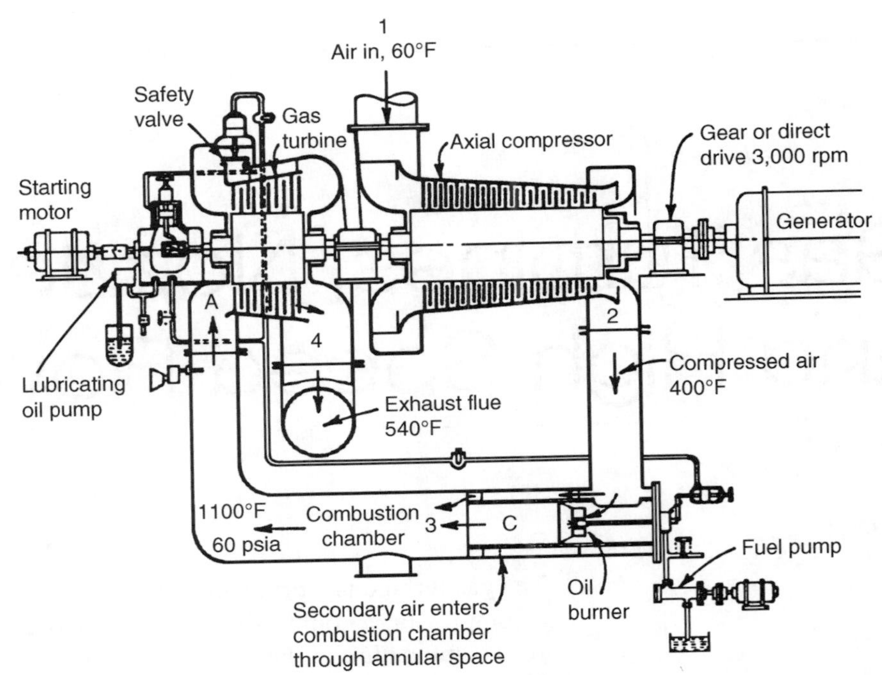

Figure 11.2 The continuous-combustion gas-turbine unit includes a high-efficiency air compressor, a combustion chamber, and a multistage turbine

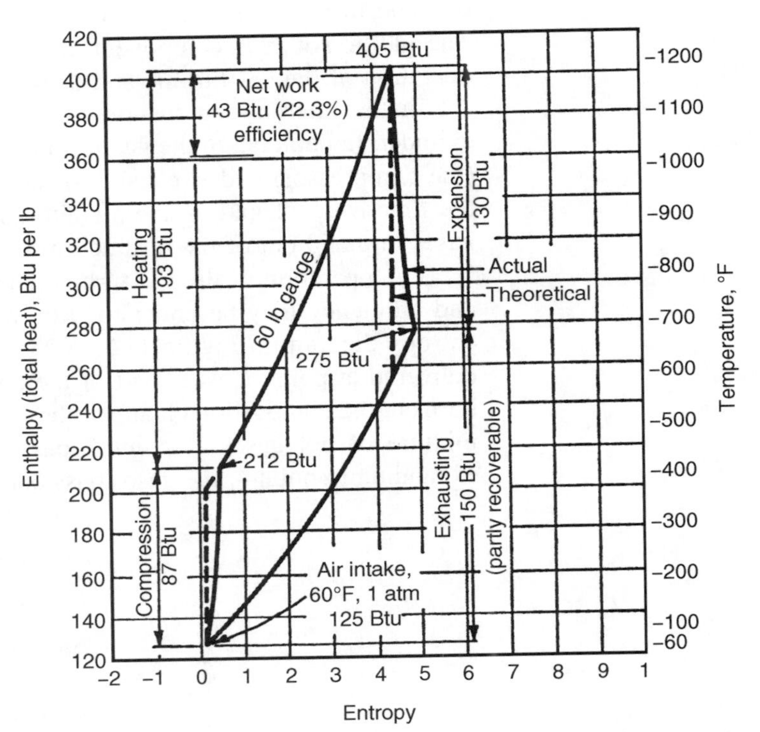

Figure 11.3 Heat-entropy diagram for gas-turbine cycle resembles that for a diesel cylinder except that compression ratio is only 5:1

affecting the gas turbine's efficiency are air temperatures, combustion process, and heat exchange.

The effects of altitude on performance are beyond the scope of this presentation and readers are referred to the technical literature for details.

Example 11.1

The components of an "open" type of gas-turbine cycle are a turbine-generator, a combustor, a compressor, and a regenerator.

(a) Make a schematic diagram of the apparatus. Label the equipment, and indicate the kind of working fluid and its direction of flow.

(b) Sketch pressure-volume and temperature-entropy diagrams of the equivalent ideal cycle. Number the corresponding thermodynamic states on the diagrams shown and the diagram of part (a). Draw the temperature-entropy diagram to conform to a regenerative effectiveness of 100 percent.

(c) Consider the working fluid of the ideal cycle of part (b) to have the properties of air at all times with c_p equal to 0.24 Btu/(lb)(°F) and k equal to 1.40. The pressure ratio is 4, atmospheric conditions of temperature is 60°F, combustor outlet temperature is 1200°F, and regenerator effectiveness is 100 percent. Compute the cycle thermal efficiency.

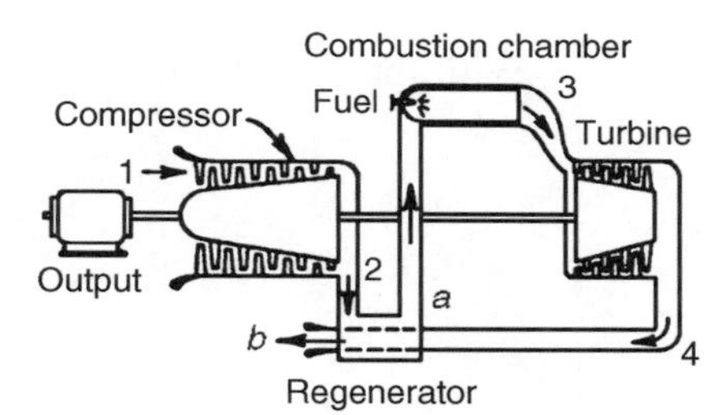

Exhibit 1 Schematic diagram of a regenerative-cycle gas turbine

Solution

(a) Refer to Exhibit 1. This is self-explanatory.

(b) Refer to Exhibit 2 for the *PV* diagram and to Exhibit 3 for the *TS* diagram. Dotted lines represent 100 percent regenerative effectiveness.

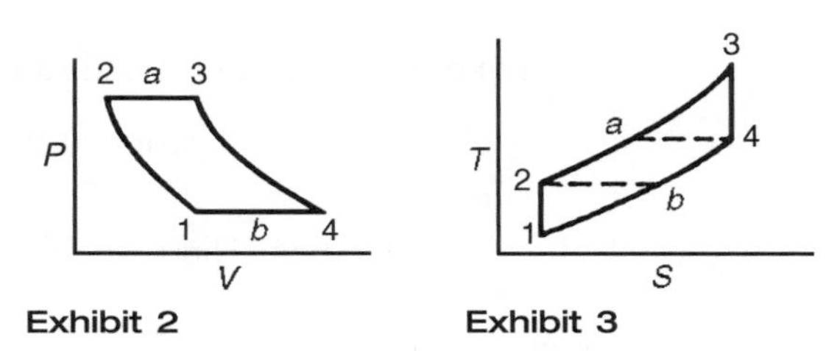

Exhibit 2 **Exhibit 3**

Where

Process 1 – 2 Isentropic compression in compressor
Process 2 – 3 Constant pressure heat addition in combustor
Process 3 – 4 Isentropic expansion in turbine
Process 4 – 1 Constant pressure heat rejection

(c) Net work is equal to turbine work minus compressor work. This is the same as for the simple cycle. Assuming constant specific heat with a 100% effective regenerator

$$\dot{m}Cp(T_3 - T_a) = \dot{m}Cp(T_3 - T_4)$$

$$\text{Efficiency} = \frac{(T_3 - T_4) - (T_2 - T_1)}{T_3 - T_4} = 1 - \frac{T_2 - T_1}{T_3 - T_4}$$

$$= 1 - \frac{T_1}{T_4} = 1 - \frac{T_2}{T_3}.$$

Assume the process of compression is adiabatic. Then

$$T_1 = 60 + 460 = 520°\text{R}$$
$$T_3 = 1200 + 460 = 1660°\text{R}.$$

With the use of the relationship for adiabatic compression

$$\frac{T_2}{T_1} = \left(\frac{P_2}{P_1}\right)^{(k-1)/k}$$

solve for T_2.

$$T_2 = 520(4)^{(1.4-1)/1.4} = 520 \times 1.486 = 772°\text{R}.$$

Likewise $$T_4 = \frac{1660}{1.486} = 1120°\text{R}.$$

Finally, the efficiency is found to be

$$\eta_T = 1 - \frac{520}{1120} = 1 - \frac{772}{1660} = 0.535, \text{ or } 53.5 \text{ percent.}$$

The net work is determined to be for 1 lb/s of working substance

work of turbine – work of compressor

$$1 \times 0.24[(1660 - 1120) - (772 - 520)] = 129.8 - 60.5 = 69.3 \text{ Btu}.$$

The work of the turbine is next found in accordance with

$$\dot{m}C_p(T_3 - T_4) = 1 \times 0.24(1660 - 1120) = 129.8 \text{ Btu/s.}$$

Finally the efficiency is determined as

$$\eta_T = \frac{\text{net work}}{\text{work turbine}} = \frac{69.3}{129.8} \times 100 = 53.5 \text{ percent.}$$

Example 11.2

A gas turbine consists of a compressor, a combustor, and an expander. Air enters the compressor at 60F and 14.0 psia and is compressed to 56 psia; the isentropic efficiency of the compressor is 82 percent. Sufficient fuel is injected to give the mixture of fuel vapor and air a heating value of 200 Btu per lb. Combustion may be assumed complete, and the weight of fuel may be neglected. The expander reduces the pressure to 14.9 psia, with an engine efficiency of 85 percent. Assume that the combustion products have the same thermodynamic properties as air, with c_p equal to 0.24 and constant. The isentropic exponent may be taken as 1.4.

(a) Sketch the complete cycle on both *PV* and *TS* diagrams.

(b) Find the temperature after compression, after combustion, and at exhaust.

(c) Determine Btu per lb of air supplied, the work required to drive the compressor, the work delivered by the expander, the net work produced by the gas turbine, and its thermal efficiency.

Solution

(a) Refer to Exhibits 4 and 5. The *ideal cycle* is given by 1-2-3-4-1. *Actual compression* takes place along 1-2′. *Actual heat added* lies along 2′-3′. The *ideal expansion* process path is 3′-4′. For the *actual expansion* see path along 3′-4″. *Ideal work* is equal to c_p × ideal temperature difference. *Actual work* is equal to c_p × actual temperature difference. See also Figure 11.2.

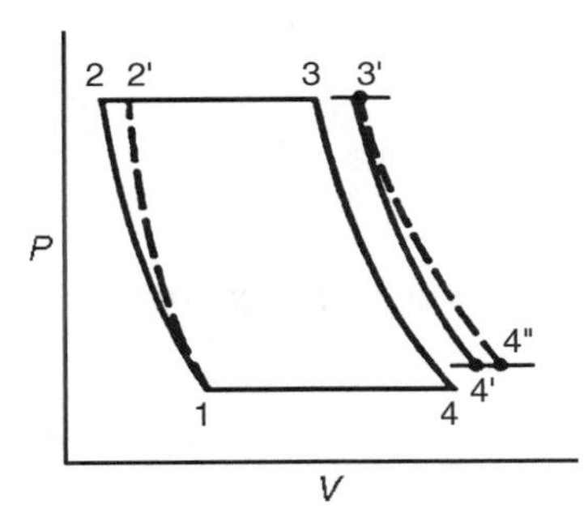

Exhibit 4

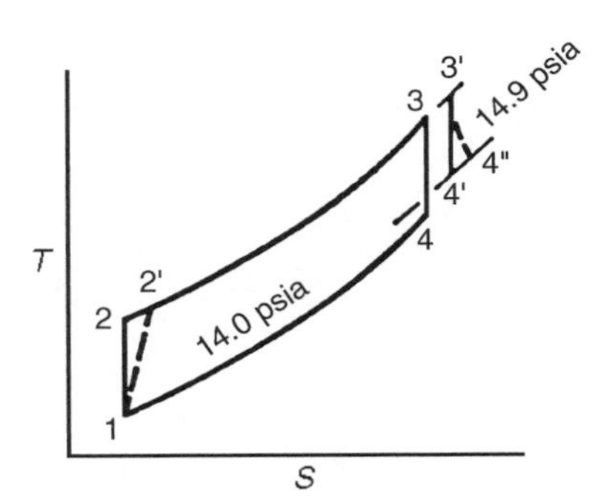

Exhibit 5

(b) $\text{Efficiency (isentropic)} = \dfrac{\text{ideal work of compression}}{\text{actual work of compression}}$

$$\text{Efficiency (isentropic)} = \frac{c_p(T_2 - T_1)}{c_p(T_{2'} - T_1)} = 0.82$$

Entering air temperature is 60°F or 520°R. Then, on the same basis as the previous problem, determine T_2.

$$T_2 = 520\left(\frac{56}{14}\right)^{1.4-1/1.4} = 773°\text{R}$$

Then insert the values of T_1 and T_2 in the equation for efficiency and solve for $T_{2'}$. This is found to be 829°R, or 369°F. This is temperature after compression.

$$\eta = 0.82 = \frac{C_p(T_2 - T_1)}{C_p(T_{2'} - T_1)}$$

$$0.82 = \frac{773 - 520}{T_{2'} - 520}$$

$$T_{2'} = 829$$

For temperature after combustion use the following approach.

$$Q = C_p(T_{3'} - T_{2'}) = 200 = 0.24(T_{3'} - 829),$$

$$T_{3'} = 1663°\text{R} = 1203°\text{F}$$

from which $T_{3'}$ is found to be 1663°R. This temperature is equivalent to 1203°F.

$$\text{Engine efficiency} = \frac{\text{ideal work of expansion}}{\text{actual work of expansion}}$$

$$\text{Engine efficiency} = \frac{C_p(T_{3'} - T_{4''})}{C_p(T_{3'} - T_{4'})} = 0.85$$

Using a similar approach as before, determine T_4 by use of the relation

$$\frac{T_{4'}}{T_{3'}} = \left(\frac{P_{4'}}{P_{3'}}\right)^{(k-1)/k}$$

from which by rearrangement

$$T_{4'} = 1663\left(\frac{14.9}{56}\right)^{(1.4-1)/1.4} = 1140°\text{R}.$$

Now insert the values of $T_{4'}$ and $T_{3'}$ in the equation for engine efficiency, and solve for $T_{4''}$.

$$\eta_e = 0.85 = \frac{Cp(1663 - T_{4''})}{Cp(1663 - 1140)}$$

$$T_{4'} = 1218°\text{R}$$

This is found to be 1218°R, or 758°F. This is temperature after expansion, i.e., at exhaust.

(c) Work of compression $= C_p(T_{2'} - T_1) = 0.24\,(829 - 520) = 74.2$ Btu

$$\text{Work delivered by expander} = C_p(T_{3'} - T_{4''}) = 0.24\,(1663 - 1218) = 106.9 \text{ Btu}$$

Net work = 106.9 – 74.2 = 32.7 Btu. And the thermal efficiency is

$$\eta = \frac{\text{net work}}{\text{heat supplied}} = \frac{32.7}{200} \times 100 = 16.4 \text{ percent.}$$

Example **11.3**

A cold-air standard regenerative gas turbine develops some 6000 hp net. Air enters the compressor at 15 psia and is compressed to 100 psia, at which time it passes through a constant pressure regenerator. The air leaves the regenerator at 700°F and passes into the combustion chamber, where some 650 Btu/lbm is added to

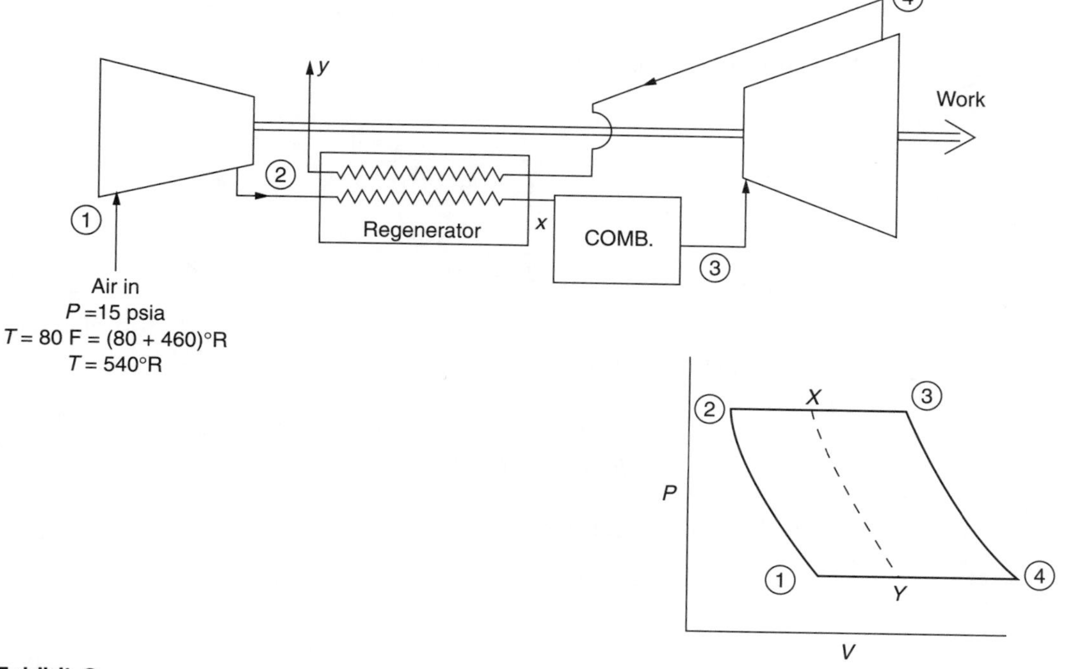

Exhibit 6

the gas. Exhibit 6 illustrates the turbine and its *PV* characteristics.

Determine (a) regenerator effectiveness, (b) mass flow in lbm/min (neglecting fuel), (c) thermal efficiency.

Solution

First we must find the values of the various states. It is useful to organize the data in a table, as shown in Exhibit 7. Calculations showing how these values were determined follow.

Exhibit 7

	1	2	*x*	3	4	*y*
P	15	80	→	→	15→	→
T	540	871	1160	3868	2398	

The process from 1 to 2 is isentropic.

$$T_2 = T_1\left(\frac{P_2}{P_1}\right)^{(k-1)/k} = 540\left(\frac{80}{15}\right)^{(1.4-1)/1.4} = 871\ ^\circ\text{R}$$

Combustion chamber:

$$Q = C_p(T_3 - T_x)$$
$$650 \text{ Btu/lbm} = 0.24 \text{ Btu/lbm}^\circ\text{R}(\Delta T)$$
$$\Delta T = 2708$$
$$T_3 = T_x + \Delta T = 1160 + 2708$$
$$T_3 = 3868^\circ\text{R}$$

The process from 3 to 4 is isentropic.

$$T_4 = T_3\left(\frac{P_4}{P_3}\right)^{(k-1)/k}$$
$$T_4 = 3868\left(\frac{15}{80}\right)^{(1.4-1)/1.4} = 2398^\circ\text{R}$$

(a) Regenerator effectiveness:

$$E_R = \frac{h_x - h_2}{h_4 - h_2} = \frac{T_x - T_2}{T_4 - T_2} = \frac{1160 - 871}{2398 - 871} = 18.9 \text{ percent}$$

(b) Mass flow rate of air:

$$\text{Pow} = m^{\bullet}\,(\text{Work net})$$

$$m^{\bullet} = \frac{\text{Power}}{\text{Work net}}$$

$$m^{\bullet} = \frac{(6000\ hp)(42.42)\ \text{Btu/Hp min}}{[(h_3 - h_4) - (h_2 - h_1)]\ \text{Btu/lbm}}$$

$$m^{\bullet} = \frac{(6000)(42.42)}{0.24[(13868 - 2398) - (871 - 540)]} = 931\ \text{lbm/min}$$

(c) Thermal efficiency:

$$\eta_t = \frac{\text{Energy wanted}}{\text{Energy that costs \$}} = \frac{w_T - w_c}{650\ \text{Btu/lbm}}$$

$$\eta_t = \frac{273.4}{650} = 42\ \text{percent}$$

Often we want to change the velocity of flow of gases and vapors. The device most convenient to use for this is the nozzle. The nozzle may be designed to either increase or decrease the velocity. When high speed is required, as with military aircraft where conservation of momentum is used, the exit speed of the exhaust products needs to be very high.

Example **11.4**

Air (constant specific heat) enters the isentropic diffuser of a turbojet engine at 1100 ft/sec, pressure of 6 psia and temperature of 400°R. The air decelerates to a low velocity as it enters an 80 percent efficient compressor with a pressure ratio of 12. Fuel is injected into the combustion chamber, producing heat at a rate of 800 Btu/lbm of air flowing through the turbine. The turbine is 85 percent efficient and is connected to an isentropic nozzle. Exhibit 8 shows the engine and its *TS* characteristics.

Determine (a) the velocity of the gas leaving the met and (b) thermal efficiency of the engine.

Solution

The first step is to determine the values at each of the states. The table in Exhibit 9 organizes these data. Calculations showing how these values were determined follow.

First law through diffuser:

$$\frac{Va^2}{29\ cJ} + h_a = h_1 + \frac{v_{1^2}}{29\ cJ}$$

$$\frac{(1100)^2}{2(32.2)(778)} + 96 = h_1 \quad h_1 = 120\ \text{Btu/lb.}$$

$$P_1 = P_a\left(\frac{T_1}{T_a}\right)^{k/(k-1)} = 6\left(\frac{500}{400}\right)^{1.4/(1.4-1)} = 13\ \text{psia.}$$

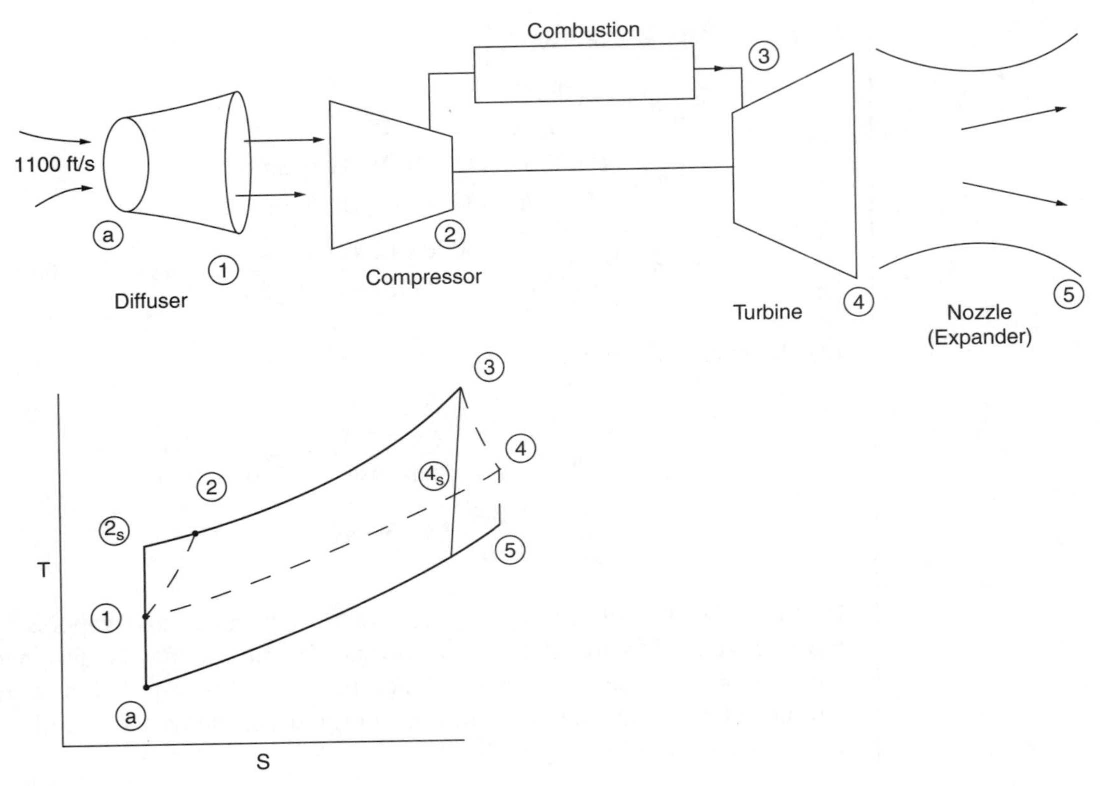

Exhibit 8

Exhibit 9

	A	1	2*s*	2	3	4*s*	4	5
P	6	13	156→	→	→	81→	→	6
T	400	500	1017	1146	4479	3719	3833	1822
h	96	120						

Now analyze the compressor:

Pressure ratio = 12

$$T_2 = T_1\left(\frac{P_2}{P_1}\right)^{(k-1)/k} = 500\left(\frac{156}{13}\right)^{(1.4-1)/1.4} = 1017°\text{R}$$

$$\eta_c = 0.80 = \frac{h_{2S} - h_1}{h_2 - h_1} = \frac{c_p}{c_p}\left(\frac{1017-500}{T_2 - 500}\right)$$

$$T_2 = 1146.$$

Consider the combustion chamber:

$$h_3 = h_2 + 800$$
$$C_P T_3 = C_P T_2 + 800 = 0.24(1146) + 800$$
$$T_3 = 4479°\text{R}$$

Note that with a turbojet $W_c = W_t$; that is, all of the turbine work is used to power the compressor.

$$h_2 - h_1 = h_3 - h_4$$
$$C_P(1146 - 500) = C_P(4479 - T_4)$$
$$T_4 = 3833$$

To get P_4 we need to get T_{4S}

$$\eta_T = .85 = \frac{T_3 - T_4}{T_3 - T_{4S}} = \frac{4479 - 3833}{4479 - T_{4S}}$$

$$T_{4S} = 3719$$

$$P_4 = P_3\left(\frac{T_4}{T_3}\right)^{k/(k-1)} = 156\left(\frac{3719}{4479}\right)^{1.4/(1.4-1)} = 81 \text{ psia}$$

Exit nozzle (isentropic):

$$T_5 = T_4\left(\frac{P_5}{P_4}\right)^{(1.4-1)/1.4} = 3833\left(\frac{6}{81}\right)^{0.4/1.4} = 1822°\text{R}$$

(a) First law:

$$h_4 = h_5 + \frac{v_5^{\ 2}}{29 \text{ cJ}}$$

$$0.24(3833) = 0.24(1822) + \frac{v_5^{\ 2}}{2(32.2)(778)}$$

$$v_5 = 4918 \text{ ft/s.}$$

(b) $$\eta_T = \frac{Q_{\text{in}} - Q_0}{Q_{m'}} = \frac{800 - 0.24(T_5 - T_A)}{800} = \frac{800 - 0.24(1822 - 400)}{800}$$

$$\eta_T = 57.3 \text{ percent}$$

The velocity of sound and Mach numbers are terms commonly used when working with nozzles. We find that if the back pressure at the exit of the nozzleis low enough, the gas flowing in the nozzle increases in speed as the nozzle area is reduced, until the velocity is the speed of sound or Mach equal to one. If the area is then increased in the direction of flow, the speed of the gas goes beyond the speed of sound. This is typical flow in a converging/diverging nozzle as shown in Figure 11.4.

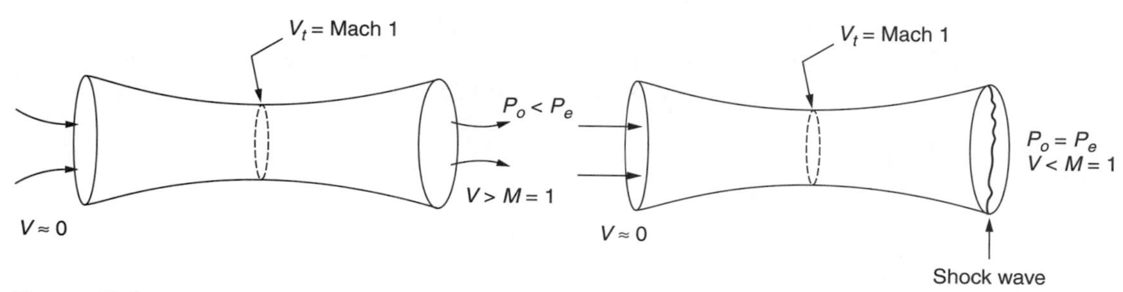

Figure 11.4

If the back pressure is increased to the exit pressure of the nozzle, a standing shock wave is produced. The characteristics of both nozzles in Figure 11.4 can be determined through the use of Table 11.1.

Table 11.1 One-dimensional compressible flow functions for an ideal gas with $k = 1.4$

(a) Isentropic Flow Functions				(b) Normal Shock Functions				
M	T/T_o	p/p_o	A/A^*	M_x	M_y	p_y/p_x	T_y/T_x	p_{oy}/p_{ox}
0	1.000 00	1.000 00	∞	1.00	1.000 00	1.0000	1.0000	1.000 00
0.10	0.998 00	0.993 0	5.8218	1.10	0.911 77	1.2450	1.0649	0.998 92
0.20	0.992 06	0.972 50	2.9635	1.20	0.842 17	1.5133	1.1280	0.992 80
0.30	0.982 32	0.939 47	2.0351	1.30	0.785 96	1.8050	1.1909	0.979 35
0.40	0.968 99	0.895 62	1.5901	1.40	0.739 71	2.1200	1.2547	0.958 19
0.50	0.952 38	0.843 02	1.3398	1.50	0.701 09	2.4583	1.3202	0.929 78
0.60	0.932 84	0.784 00	1.1882	1.60	0.668 44	2.8201	1.3880	0.895 20
0.70	0.910 75	0.720 92	1.094 37	1.70	0.640 55	3.2050	1.4583	0.855 73
0.80	0.886 52	0.656 02	1.038 23	1.80	0.616 50	3.6133	1.5316	0.812 68
0.90	0.860 58	0.591 26	1.008 86	1.90	0.595 62	4.0450	1.6079	0.767 35
1.00	0.833 33	0.528 28	1.000 00	2.00	0.577 35	4.5000	1.6875	0.720 88
1.10	0.805 15	0.468 35	1.007 93	2.10	0.561 28	4.9784	1.7704	0.674 22
1.20	0.776 40	0.412 38	1.030 44	2.20	0.547 06	5.4800	1.8569	0.628 12
1.30	0.747 38	0.360 92	1.066 31	2.30	0.534 41	6.0050	1.9468	0.583 31
1.40	0.718 39	0.314 24	1.1149	2.40	0.523 12	6.5533	2.0403	0.540 15
1.50	0.689 65	0.272 40	1.1762	2.50	0.512 99	7.1250	2.1375	0.499 02
1.60	0.661 38	0.235 27	1.2502	2.60	0.503 87	7.7200	2.2383	0.460 12
1.70	0.633 72	0.202 59	1.3376	2.70	0.495 63	8.3383	2.3429	0.423 59
1.80	0.606 80	0.174 04	1.4390	2.80	0.488 17	8.9800	2.4512	0.389 46
1.90	0.580 72	0.149 24	1.5552	2.90	0.481 38	9.6450	2.5632	0.357 73
2.00	0.555 56	0.127 80	1.6875	3.00	0.475 19	10.333	2.6790	0.328 34
2.10	0.531 35	0.109 35	1.8369	4.00	0.434 96	18.500	4.0469	0.138 76
2.20	0.508 13	0.093 52	2.0050	5.00	0.415 23	29.000	5.8000	0.061 72
2.30	0.485 91	0.079 97	2.1931	10.00	0.387 57	116.50	20.388	0.003 04
2.40	0.464 68	0.068 40	2.4031	∞	0.377 96	∞	∞	0.0

Source: Fundamentals of Thermodynamics, 3rd Ed., Moran and Shapiro, © 1996 John Wiley & Sons. Reprinted with permission of John Wiley & Sons, Inc.

Example 11.5

A converging-diverging nozzle at steady state has an entering area of 20 in.^{2}. Air acting as an ideal gas ($K = 1.4$) flows through the nozzle. Entering air is $T = 540°$R, $P = 100$ psia and the nozzle is choked.

If entering velocity is $M = 0.1$, determine (a) A_t and (b) $m^\bullet$

If exit area is 8 in.^{2}, find (c) M_e and (d) P_e.

Solution

(a)

$$A/A^* = 5.8218$$

$$A^* = A_t = 20 \text{ in.}^2/5.8218 = 3.44 \text{ in.}^2$$

(b) $C = \sqrt{krt} = \sqrt{(1.4)\left(\frac{1545}{29}\right)\frac{\text{ft lbf}}{\text{lbm }^\circ R}(540\ ^\circ R)\left(\frac{32.2\text{ ft lbm}}{\text{lbf } s^2}\right)}$

$C = 1139$

and entering mach number = 0.1

$$M = \frac{V}{C}$$
$$V = MC = 0.1(1139) = 113.9\text{ ft/s.}$$

$$m^\bullet = \rho\, Av \quad \rho = \frac{\rho}{RT} = \frac{(100\text{ lbf}/\text{in}^2)(144\text{ in}^2/\text{ft})}{\frac{1545}{29}\frac{\text{ft lbf}}{\text{lbm }^\circ\text{R}}(540\ ^\circ\text{R})}$$

$$\rho = 0.50\text{ lbm/ft}^3$$

$$m^\bullet = \frac{(0.50\text{ lbm/ft}^3)(20\text{ in}^2)(113.9\text{ ft/s})}{(144\text{ in}^2/\text{ft}^2)}$$

$$m^\bullet = 7.9\text{ lbm/s}$$

(c) $A/A^* = 8/3.44 = 2.33$

$$M_e = 2.37$$

(d) $P_e/P_o = 0.06840$

$$P_e = 100(0.06840) = 6.8\text{ psia}$$

Example 11.6

Air as an ideal gas ($K = 1.4$) flows through a converging-diverging nozzle. It enters at 150 psia and 500°R at a low velocity. A normal standing shockwave is at the exit where $M = 2.4$ (see Exhibit 10).

Determine the following properties after the shock: (a) M_y; (b) P_y; (c) T_y

(a) From Table 11.1, at $M_x = 2.4$, $M_y = 0.523$

(b) $P_x/P_o = 0.0684$

$$P_x = 150(0.0684) = 10.20\text{ psia}$$
$$P_y/P_x = 6.5533$$
$$P_y = (10.2)(6.5533) = 66.8\text{ psia}$$

(c) $T_x/T_o = 0.46468$

$$T_x = 500(0.46468) = 232.3^\circ\text{R}$$
$$T_y/T_x = 2.0403$$
$$T_y = T_x(2.0403) = 232.3(2.0403) = 474^\circ\text{R.}$$

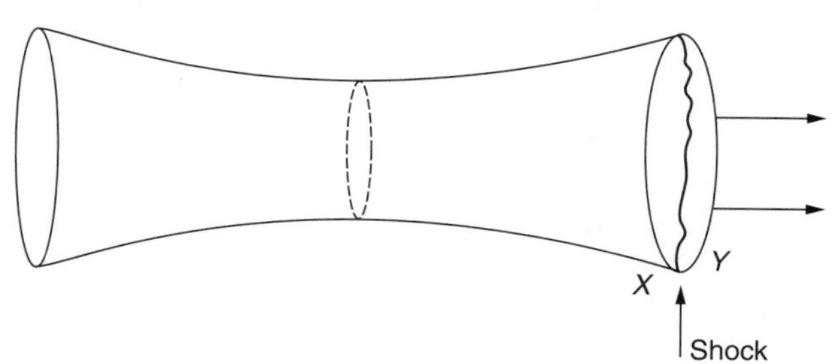

Exhibit 10

CHAPTER 12

Refrigeration

OUTLINE

The natural flow of heat from a hot body to a cold body will be rapid if the difference in temperature is great, if the cooling or heating surface is large, and if the resistance to heat flow is small. When these conditions are not favorable, the flow will be correspondingly slow. We know that heat will not flow from a cold to a hot body, nor between bodies of equal temperature.

Refrigeration involves the artificial means for removing heat when conditions are unfavorable to natural or rapid flow. It is concerned not only with producing low temperatures where desired, but also with accelerating the natural flow of heat at normal temperatures. Refrigeration is accomplished by providing a substance that is colder than the substance to be refrigerated.

REFRIGERANTS

Many liquids boil at temperatures low enough for refrigeration, but comparatively few are suitable for refrigeration purposes. Those that have practical usefulness are called refrigerants. Increased pressure on any of these liquids raises its boiling point. Decreased pressure has the reverse effect and lowers the boiling temperature. The term "boiling point" is generally understood to mean the temperature at which vaporization takes place under atmospheric pressure at sea level. In this presentation and in engineering, the term "boiling temperature" will be used in referring to temperatures of vaporization at pressures other than atmospheric. The candidate for licensure should take into the examination tables of refrigerants, such as ammonia and HFC 134a, usually appearing in handbooks or other sources.

REFRIGERATING TERMS

Understanding of the following terms is essential to facile handling of the problems appearing on the examination.

Refrigerating Effect

This is the amount of heat absorbed in the evaporator. This is also the amount of heat removed from space to be cooled. It is measured by subtracting heat content (enthalpy) of one pound of liquid refrigerant as it enters the expansion valve from the heat content of the same pound as it enters the compressor (refer to Figs. 12.1 and 12.2 from heat content at F or F' subtract heat content at either condition D' (saturated liquid) or D (subcooled liquid)).

Ton of Refrigeration

When the boiling refrigerant removes sensible heat from the environment of the evaporator at a rate equivalent to the melting of 1 ton (2000 lb) of water ice in 24 hr, the rate of heat removal is a ton of refrigeration. This is equivalent to the removal at a rate equal to

$$2000 \times \frac{144}{24} = 12{,}000 \text{ Btu per hr}$$

where 144 Btu per lb is the heat of melting 1 lb of water ice. Heat of sublimation of 1 lb of dry ice (CO_2) is equal to 275 Btu. To say that a refrigeration machine has a capacity of 10 tons is to say that the rate of refrigeration is $10 \times 200 = 2000$ Btu per min. Note that 1 ton of refrigeration is equal to a rate of 200 Btu per min.

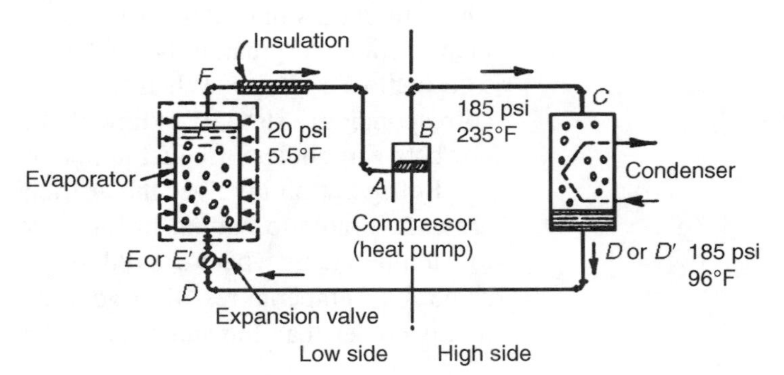

Figure 12.1

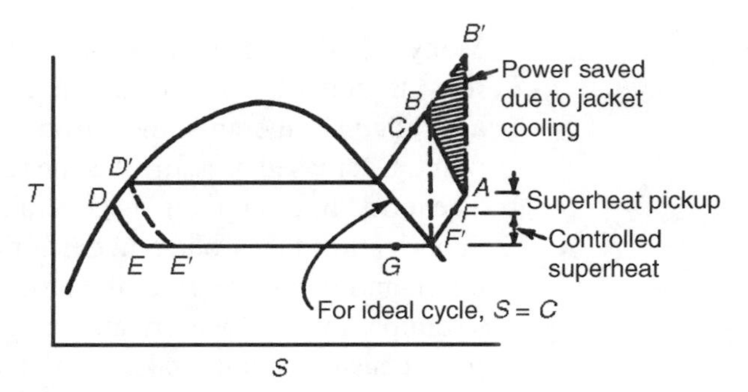

Figure 12.2

Refrigerant Circulated

Dividing 200 Btu per min by the refrigerating effect, in Btu per lb of refrigerant, gives pounds of refrigerant circulated each minute.

Work of Compression

This is the amount of heat added to the refrigerant in the compressor cylinder. It is measured by subtracting heat content of one pound of refrigerant at compressor suction conditions (point F', F, or A in Fig. 12.2) from heat content of the same pound at compressor discharge conditions (point B or B' in Fig. 12.2).

Theoretical Horsepower Requirements

Multiplying work of compression in Btu per lb by pounds of refrigerant circulated in an hour, and dividing this product by 2545 Btu per hp-hr gives theoretical horsepower requirements.

$$\text{Hp} = \frac{\text{Work of compression} \times \text{refrigerant circulated}}{2545} \qquad \textbf{(12.1)}$$

Coefficient of Performance

This is the ratio of refrigerating effect to work of compression. A high coefficient of performance (COP) means high efficiency. Theoretical COP ranges from about 2.5 to more than 5.

$$\text{C.O.P.} = \frac{\text{Refrigerating effect}}{\text{Work of compression}} \qquad \textbf{(12.2)}$$

Horsepower Per Ton of Refrigeration

The mechanical input in horsepower divided by tons of refrigeration effect produced provides the answer to this quantity. If COP is known, horsepower per ton of refrigeration (TR) can be figured directly.

$$\text{hp} = \frac{12{,}000 \text{ Btu per hr}}{2545 \times \text{COP}} \qquad \textbf{(12.3)}$$

Standard-Ton Conditions

An evaporating temperature of 5°F, a condensing temperature of 86°F, liquid before expansion valve at 77°F, and suction-gas temperature at 14°F are the necessary conditions for the standard TR. Refrigerating machines are often rated under these conditions.

Head Pressure

This is the pressure at the discharge of the compressor or in the condenser. This is also known as "high-side" pressure.

Suction Pressure

This is the pressure at the compressor suction or at the outlet of the evaporator. This is also known as "low-side" pressure.

COMPRESSION REFRIGERATION CYCLE

Refer to Figures 12.1 and 12.2. In Figure 12.1 the pipelines are considered short so that pressure drop becomes negligible. The cycle shown is for ammonia. Check the conditions for pressure and temperature. If the condenser is clean and there is just enough surface, then the liquid refrigerant will leave the condenser at condition D' as saturated liquid. If the surface is clean and a large amount of cooling water at low temperature is used, the liquid is subcooled to condition D. The pipe loses sensible heat from B to C. The refrigerant picks up heat from F to A through pipe insulation. Saturated vapor at F' inside the evaporator is superheated in the evaporator from F to F' to ensure dry vapor to the compressor to ensure against wet compression and resulting compressor destruction.

Starting with liquid at D, the refrigerant expands through the expansion valve along the constant-enthalpy line with a large pressure drop and a small increase in volume, as indicated by line DE. The saturated liquid that has not already flashed into vapor now boils in the evaporator, changing its state from a saturated liquid to a saturated vapor at the downstream pressure condition within the evaporator. This refrigerating effect is accompanied by a relatively small pressure drop and a large increase in volume, as along path EF'. This is where each pound of refrigerant does its work of cooling. Vapor is superheated and reaches the compressor suction, where its volume is reduced and pressure increased along path $F'FAB$ under normal operating conditions. In the compressor, work is done on the refrigerant to raise its pressure to cause flow and also to raise its temperature above the cooling medium (water or air) so that heat can leave the refrigerant and cool it.

In the condenser the refrigerant loses its heat and superheat of compression and condenses out as liquid along path $BCD'D$. Then the cycle starts anew. Note that the shaded area in the compression cycle is power saved due to jacket cooling of the compressor. If the compression were adiabatic and reversible, then the path of compression would follow AB'. But because there is real cooling either by air or water, the compression path would lie along AB with the shaded area in Figure 12.2 indicative of some degree of isothermal cooling. Cool and large quantities of water increase shaded area and reduce power consumption accordingly.

Refrigerant Circulated per Ton of Refrigeration

First determine how much refrigerating effect we can expect from each pound of refrigerant. This is the difference between the enthalpy at D or D' and F'. It can also be expressed as the difference between the heat removed by the condenser water and the work done by the compressor. Refer to Figure 12.2. Enthalpy of saturated liquid at 185 psi is 150.5 Btu per lb and enthalpy of saturated vapor at 20 psi is 613.5 Btu per lb. The difference is 462.6 Btu and each pound of circulating ammonia removes that amount of heat in the evaporator. Then

$$200/462.6 = 0.432 \text{ lb ammonia per min per TR.}$$

Horsepower per Ton of Refrigeration

This is a common method of expressing the actual efficiency of a compression system. Its relation to coefficient of performance is

$$\frac{\text{hp}}{\text{TR}} = \frac{4.71}{\text{COP}} \tag{12.4}$$

This expression may be applied to either actual or ideal cycles.

Wet and Dry Compression

The state point of the refrigerant vapor as it enters the compressor suction may be in the wet region (point G, Fig. 12.2), saturated vapor point (point F'), or in the superheat region (point F or A). If compression occurs with the vapor wet (point G), the process is known as *wet compression*. If compression occurs at either points F', F or A, it is known as *dry compression*. You will note that under wet compression the cycle approaches the ideal or Carnot cycle. Naturally we would expect wet compression to be more efficient than gas refrigeration. This is in accord with the Joule-Thomson effect.

For the ideal refrigeration machine wet compression would give a large coefficient of performance between two temperature limits. However, wet compression in the actual machine produces low volumetric efficiencies due to vaporization of liquid particles in the compressor cylinder. Actually, the mechanical efficiency of the machine for dry compression is much better all around, and dry compression is used universally in order to protect the machine against damage.

DISPLACEMENT OF THE VAPOR-COMPRESSION MACHINE

For a particular capacity the size of the compressor depends on the number of pounds of refrigerant that are removed from the evaporator and circulated per unit time in the closed cycle.

If a refrigeration plant is to have a capacity of N tons of refrigeration, then the rate is $N \times 200$ Btu per min. From this and the refrigerating effect per pound, the weight of refrigerant circulated per minute may be determined. For 100 percent volumetric efficiency the displacement V_d is

$$V_d = (\text{sp vol at suction})(200 \times N/\text{refrigerating effect}) = \text{cfm} \qquad \textbf{(12.5)}$$

The displacement at volumetric efficiency E_v is obtained by dividing the displacement at 100 percent by E_v as a decimal less than unity. Volumetric efficiencies will fall between 65 and 85 percent as limits in actual compressors.

Example 12.1

Calculate the displacement in an ammonia compressor for a 50-ton refrigeration machine when operating with ammonia at 0°F in the expansion coils (evaporator). At this temperature the heat absorbed by the evaporation of 1 lb of ammonia is 500 Btu (refrigerating effect), and the specific volume is 9 cu ft per lb. The vapor enters the compressor saturated. If speed is 180 rpm and stroke is 1.2 × bore, what are bore and stroke, the compressor being single-acting?

Solution

The heat to be absorbed is equal to (50 ton)(200 Btu/min-ton) = 10,000 Btu per min. The compressor displacement is

$$\left(\frac{10000 \text{ Btu/min}}{500 \text{ Btu/lb}}\right)(9 \text{ ft}^3/\text{lb}) = 180 \text{ cfm}$$

Now let N be compressor speed (rpm), D be bore (ft), L be stroke (ft), V be piston displacement (cu ft per stroke). Then

$$\text{Displacement } (V) = \frac{\pi}{4}D^2LN = \frac{\pi}{4}(D^2)\text{ ft}^2(1.2D)\text{ ft}(180\text{ rpm}) = 180\text{ cfm}$$

$$0.942D^3\text{ft}^3(180\text{ rpm}) = 180\text{ cfm}$$

$$169.6D^3 = 180\text{ cfm}$$

Solving for D

$$D = 1.02\text{ ft}$$

also

$$L = 1.2\text{D} = 1.2 \times 1.02 = 1.224\text{ ft.}$$

Example 12.2

Consider a Carnot Refrigerator Using R134a entering the compressor at 10 psia and leaving as a saturated vapor at 100 psia and entering the condensor where it leaves as a saturated liquid.

Determine
a) TS Diagram of cycle
b) Coefficient of performance (COP)
c) If mass flow rate is 10 lb/s, how many tons of refrigeration can the unit produce?

Solution

a) $T_H = 79.17°\text{F}$ $h_f = 36.99$
$h_g = 112.46$ $s_f = .0768$
$s_g = .2169$ $T_L = -29.71$
$s_f = .0068$ $s_g = .2265$
$h_f = 2.91$ $h_g = 97.37$

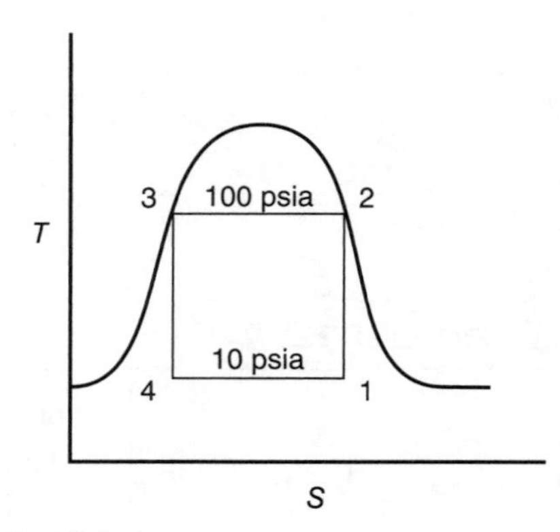

Exhibit 1

b)
$$\text{COP} = \frac{T_L}{T_H - T_L} = \frac{430.29}{29.17 + 29.71}$$

$$\text{COP} = 3.95$$

c)
$$Q = \dot{m}(h_1 - h_4)$$

Determine quality entering compressor.

$$S_1 = S_2 = .2169 = xS_f + (1 - x)S_g$$

$$.2169 = x\,(.0068) + (1 - x)(.2265)$$

$$x = .0437$$

Now find quality entering evaporator

$$S_3 = S_4 = .0768 = xS_f + (1 - x)S_g$$

$$0.0768 = x\,(.0068) + (1 - x)(.2265)$$

$$x = .68$$

To find $Q_{\text{EVAP}} = \dot{m}(h_1 - h_4)$

$$\text{h}_1 = .0437\,(2.91) + (1 - .0437)(97.37) = 93.24 \text{ Btu/lb}$$

$$\text{h}_4 = .68\,(2.91) + (1 - .68)97.37 = 33.14 \text{ Btu/lb}$$

$$Q_{\text{REF}} = \frac{(10\text{lb/s})(60\text{s/min})(93.24 - 33.14)\text{Btu/lb}}{200\text{Btu/min-ton}}$$

$$Q_{\text{REF}} = 100.3 \text{ Tons Refrigeration}$$

Example **12.3**

A single-stage ammonia compressor is producing 10 tons of refrigeration and the power consumed is 15 bhp. Suction pressure is 25 psi, condensing pressure 180 psi. Brine temperature is 20°F off brine cooler. Determine actual coefficient of performance (COP) and amount of ammonia circulated.

Solution

$P_{\text{EVAP}} = 25 \text{ psig} + 15 \text{ atm}$
$P_{\text{EVAP}} = 40 \text{ psia}$
$h_g = 615 \text{ Btu/lb}$
$P_{\text{COND}} = 180 + 15 = 195 \text{ psia}$
$h_f = 149 \text{ Btu/lb}$

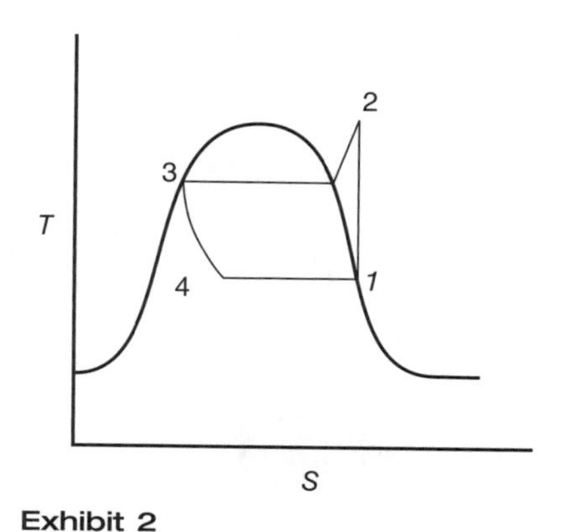

Exhibit 2

$$\text{COP} = \frac{\text{refrigerating effect}}{\text{power consumed}}$$

$$= \frac{(12{,}000 \text{ Btu/hr ton})(10 \text{ tons})(24 \text{ hr})}{(15 \text{ hp})(2545 \text{ Btu/hp hr})(24 \text{ hr})} = 3.15$$

Assume no subcooling and that saturated vapor enters compressor. Then

$$h_3 = h_4$$

$$\left(\frac{288{,}000 \text{ Btu/ton ice in 24 hrs}}{h_1 - h_4}\right) = 620 \text{ lb per TR}$$

CARNOT CYCLE FOR REFRIGERATION

This is a reversed cycle of the Carnot type, as shown in Figure 12.3. The refrigerant is isentropically compressed along *ab* from a cold temperature T_1 in the evaporator to T_2 above that of some naturally available heat sink. The refrigerant then discharges heat at constant temperature T_2 along *bc*. Heat-sink temperature is T_0. At some point *c*, an isentropic expansion *cd* lowers the temperature to T_1, which is below that of the environment, and heat flows from the environment into the evaporator at T_1 along *da* and the cycle is repeated.

The COP of this cycle may now be expressed as for the Carnot cycle

$$\frac{T_1}{T_2 - T_1} \qquad \textbf{(12.6)}$$

This is the highest possible COP for all cycles operating between the temperature limits T_1 and T_2 and serves as the standard of comparison for other cycles that more nearly approach the natural events.

Conclusions Drawn from the Carnot Cycle

It is desirable that work expended be kept to a minimum because it is paid for. Work will be reduced as T_2 is lowered. Steps taken to keep this temperature down are important. T_0 is the lowest temperature attainable by a natural coolant (well water, air). There is a definite limit of improvement here. Work will be reduced as the evaporator temperature T_1 is increased. There are limits, however. To freeze water, 32°F and below is necessary. To cool air, higher temperatures may be used, *i.e.,* 50°F.

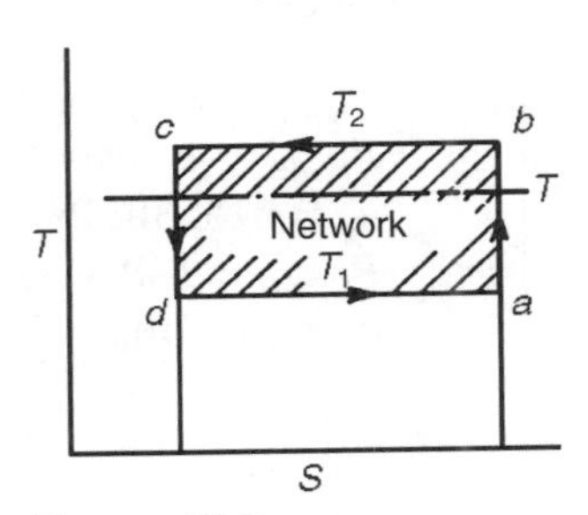

Figure 12.3

Example 12.4

In an ammonia condensing machine (compressor plus condenser) the water used for condensing is at 55°F and the evaporator is at 15°F. (a) Calculate the ideal COP. (b) If 1.5 hp per TR is required, what is actual COP? Mechanical efficiency may be taken as 90 percent.

Solution

(a) Ideal COP:

$$\frac{T_L}{T_H - T_L} = (460 + 15)/(55 - 15) = 11.885$$

(b) Actual COP:

$$\text{COP}_{\text{ACT}} = \frac{Q_L}{\text{work in}} = \frac{(12{,}000)(0.9)}{(1.5)(2545)} = 2.84$$

Example 12.5

A refrigeration cycle uses 134a and operates as an ideal vapor compression refrigerator. The cycle has a flow rate of 1 lb/s and operates between 5 psia and 50 psia. Determine

a) TS Diagram
b) Power to Compressor kW
c) Coefficient of performance (COP)
d) Tons refrigeration

Solution

a)

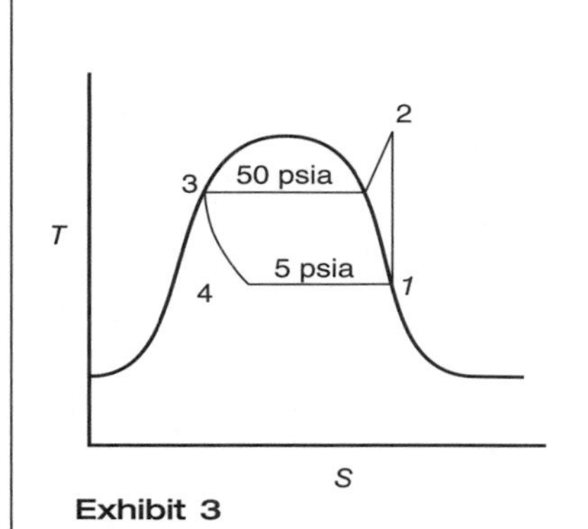

Exhibit 3

Determine values at states

	1	2	3	4
P	5	50	50	5
h	97.53	113.95	24.14	→
s	0.2311	→		

b) Power to comp = $(h_2 - h_1)\dot{m}$

$$\text{Power} = 1\text{lb/s}(113.95 - 97.53) = 16.42 \text{ Btu/s}$$

$$\text{Power} = \frac{(16.42\text{Btu/s})(3600\text{s/hr})}{3413\text{Btu/hr-kWhr}}$$

$$\text{Power} = 17.3 \text{ kW}$$

c)

$$\text{COP} = \frac{Q_L}{\text{work}} = \frac{h_1 - h_4}{h_2 - h_1} = \frac{97.53 - 24.14}{113.95 - 97.53} = 4.47$$

d)

$$\text{Tons Ref} = \frac{(\dot{m}\text{lb/s})(h_1 - h_4)\text{Btu/lb (60s/min)}}{200\text{Btu/min-ton}}$$

$$\text{Tons Ref} = \frac{(1)(97.53 - 24.14)(60)}{200} = 22.02 \text{ Tons}$$

Example 12.6

A vapor compression heat pump system using HFC 134a develops some 40,000 Btu/hr of heating. The condenser operates at 300 psia and the evaporator temperature is –10°F; both have negligible pressure drops. The refrigerant is a saturated vapor at the evaporator exit and is a liquid at 140°F at the exit of the condenser. The refrigerant exits the adiabatic compressor and enters the condenser at 220°F. Determine:

(a) Mass rate of flow of HFC 134a

(b) Compressor hp

(c) Compressor efficiency

(d) COP

Solution

The first step is to sketch the *TS* diagram of the device (Exhibit 4) and determine the values of the properties at the various states (Exhibit 5).

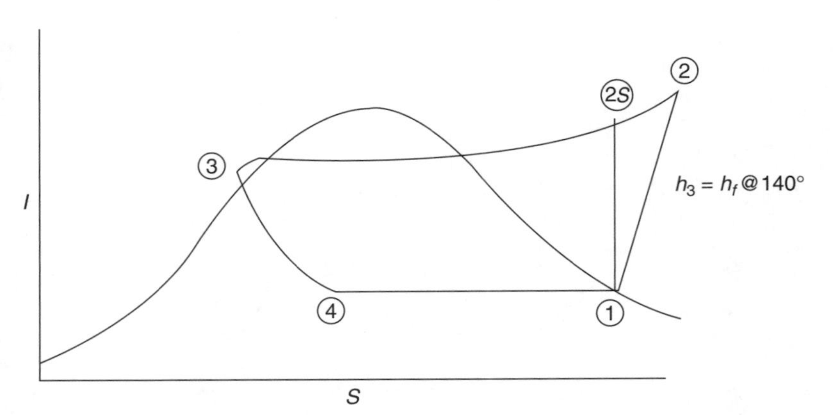

Exhibit 4

	1	2*S*	2	3	4
P		300→	→	→	→
T	−10		220	140	−10
h	100.29	126.17	140.36	59.08→	→
S	0.2236→	→			

Exhibit 5

(a) $\text{Mass rate of flow} = \dfrac{\text{heat capacity}}{(h_2 - h_3)} = \dfrac{40{,}000 \text{ Btu}}{(140.36 - 59.08) \times 60 \text{ min/hr}}$

Mass rate of flow = 8.2 lbm/min.

(b) Compressor hp = (mass rate of flow)$(h_2 - h_1)$

$$= \frac{(8.2 \text{ lbm/min})(140.36 - 100.29)\text{Btu/lbm}}{42.42 \text{ Btu/hp min}}$$

hp = 7.6 hp

(c) Efficiency of compressor

$$\eta_c = \frac{h_{2s} - h_1}{h_2 - h_1} = \frac{126.17 - 100.29}{140.36 - 100.29}$$

$$\eta_c = 64.6\%$$

(d) $\text{Actual COP} = \dfrac{\text{Energy wanted}}{\text{Energy costs}}$

$$\text{COP} = \frac{h_2 - h_3}{h_2 - h_1} = \frac{140.36 - 59.08}{140.36 - 100.29}$$

$$\text{COP} = 2.03$$

CHAPTER 13

Heating, Ventilating and Air Conditioning

OUTLINE

The concepts of heat transmission we reviewed in Chapter 4 may be applied to the study of the material in this chapter. Overall heat-transfer coefficients are determined in the same fashion in order to calculate building heat losses. The working substance is the air around us, which carries the heat or lower temperature into every corner of the building. The flow of heat or cool takes place at constant atmospheric pressure conditions.

Heating, ventilating, and air conditioning problems usually involve maintenance of conditions of temperature and humidity within a building. Essentially a building is a shell for keeping out the weather and permitting a satisfactory indoor climate to be maintained. How much heat must be added to or subtracted from an enclosure depends on the construction and tightness. Light construction increases heat losses; substantial construction, while costing more initially, keeps down the costs of operation and fuel. Flimsy and loose construction encourages air infiltration and results in high heat-load needs. Different types of construction are used in the northern and southern parts of the country, and difference in climate is the governing factor.

HEATING

Heat passes from a region of higher temperature to lower temperature. In cold weather heat is lost by a warm building by radiation, by convection, and by the escape of warm air replaced by the colder outside air.

Overall Coefficient of Heat Transmission

In actual calculations, standard handbooks list overall coefficients of heat transmission for actual thickness of walls, the material involved, and the surface resistances (film resistances). Such an overall coefficient is called U and represents the number of Btus passing through a square foot of surface of wall construction in one hour, for one degree temperature difference between inside and outside air temperatures. There is assumed to be still air on the inside and 15 mph wind on the outside. Overall coefficients can be found by directly testing complete walls or by calculating from individual coefficients. For a simple brick wall the formula used in practice is

$$\frac{1}{U} = \frac{1}{f_i} + \frac{x}{k} + \frac{1}{f_0} \tag{13.1}$$

where

f_i = film conductance for still inside air equal to 1.65
f_0 = film conductance for outside 15-mph wind on wall surface equal to 6.0
k = conductivity of the brick (or any other material)
x = thickness, in.

Example 13.1

Calculate the overall coefficient of heat transmission through an 8-in. common brick wall with still air inside and 15-mph wind outside.

Solution

From Equation 13.1,

$$\frac{1}{U} = \frac{1}{1.65} + \frac{8}{5} + \frac{1}{6.0} = 2.37$$

$$U = 1/2.37 = .421 \text{ Btu/(sq ft)(hr)(°F)}.$$

Under actual conditions walls are more complicated, having any number of materials in series. In many cases an air space is included for its insulating value. For single air space and a simple wall,

$$U = \frac{1}{1/f_i + x/k_1 + 1/a + x_2/k_2 + 1/f_0}. \tag{13.2}$$

For a simple wall with several air spaces, add $1/a$ for each space. For more detailed data and instructions the student is referred to the *Guide of the American Society of Heating, Refrigerating and Air-Conditioning Engineers*. This excellent reference source and other handbooks give conductivity values for various materials, and overall coefficients for common wall, floor, and roof constructions.

Example 13.2

Compute the coefficient of heat transmission for a wood frame wall with 1 in. fir sheathing, building paper and yellow pine siding on outside of studs, air space and wood lath and plaster on inside the studs.

Solution

Refer to the ASHRAE *Guide* for conductivities. Then

$$U = \frac{1}{\frac{1}{1.65} + \frac{1}{1.10} + \frac{1}{0.50} + \frac{1}{2.50} + \frac{1}{6}} = 0.25.$$

Wind Velocity Effects on *U*

As we know, heat-transmission coefficients used in heating and air-conditioning load calculations are based on outside wind velocity of 15 mph and still air (50 fpm) on the inside surface. Equations 13.1 and 13.2 give us the U value for these conditions. For any one wall type the only change will occur in f_0. For low values of U occurring in multiple and air-space types of construction, a change in f_0 is not as effective as for higher values of U occurring in simple walls or single wall materials.

In order to give a quick visual indication of what happens when wind velocity changes, Figure 13.1 has been worked. Coefficients U based on an average wind velocity of 15 mph are plotted against correction factors for various wind velocities designated in the *Guide*. Above an average wind velocity of 30 mph, infiltration takes on a more important role in heating and air-conditioning calculations. Once knowing the 15-mph coefficient U, read up to the proper wind-velocity curve, and then left to read the multiplier. To use it,

$$\text{Corrected } U = \text{multiplier (from chart)} \times U. \tag{13.3}$$

Tables giving conductivities and overall coefficients for heating, ventilating, and air conditioning are readily available in the *Guide*.

Heat flows through an air space by conduction, radiation, and by convection currents. It has been shown that there is always a certain amount of air movement even in "theoretical dead-air spaces." The resulting mechanism is rather complex, but experimental work shows that from about 3/4 to $1\frac{1}{2}$ in. the conductance of an air space is more or less constant; below 3/4 in., the conductance increases rapidly as the space gets narrower. For air spaces wider than $1\frac{1}{2}$ in., figure the surface conductance on each side of the space instead of using a single conductance value for the air space.

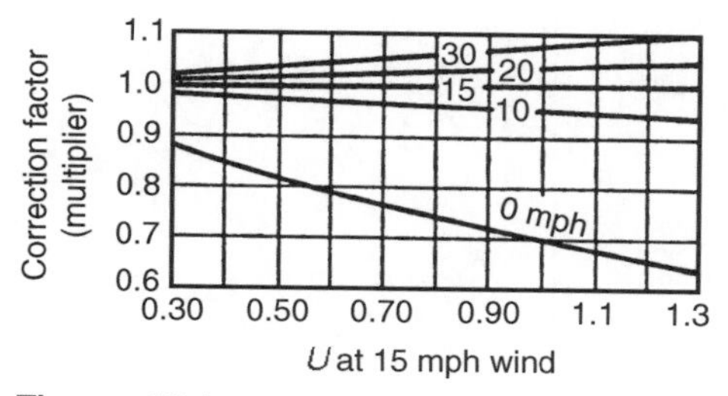

Figure 13.1

Insulation

The heat loss through walls, floors, and roofs can be reduced by using insulating materials having high heat-flow resistance. Most insulating materials depend on extremely low conductivity for their effectiveness due to the retention of air within tiny pores. Reflective insulations, such as aluminum foil, feature low emissivity and high reflectivity, and effectively stop transfer by radiation.

Choice of insulation for a particular installation depends on many factors, including insulating value, cost, ease of installation, chemical and physical stability, resistance to fire, vermin, etc. The *Guide* lists conductivities for various insulating materials. Suppliers of insulating materials also include overall coefficients for their products, but care should be exercised in evaluating the *U*'s published in their literature.

Infiltration Losses

Heat losses through infiltration of cold outside air displacing warm inside air through cracks around doors and windows can be very telling. In some cases, infiltration losses are much greater than transmission losses. This movement of air and displacement is due to wind pressure and temperature differences between indoors and outdoors.

Figuring Heating Loads

A series of such calculations is rather simple, but does involve experienced judgment in the choice of coefficients for certain conditions.

There are two kinds of heat-loss calculations. The first kind is made to determine the maximum heat loss with which to size the heating plant. The second kind is made to determine the total heat loss for a heating season or part of a season, so as to check fuel consumption or to estimate heating costs. In both bases, they begin with measurement of wall, floor, and roof surfaces, and determination of infiltration flows. The same coefficients of heat transmission are used in both cases. Deviation takes place when temperature differences are introduced. For design calculations, the maximum design temperature difference is used, while for seasonal figures the average difference (say January average) is used.

Example 13.3

Find the heat in Btu per hr that is to be supplied to a room 18 by 22 by 9 ft high to maintain a temperature of 70°F with the outside temperature of 0°F. One outside wall is of brick 12 in. thick, furred and plastered on wood lath, with two single-glazed weather-stripped windows each 4 by 6 ft. Rooms above, below, and on three sides are also maintained at 70°F. Wind velocity is 15 mph. Refer to Exhibit 1.

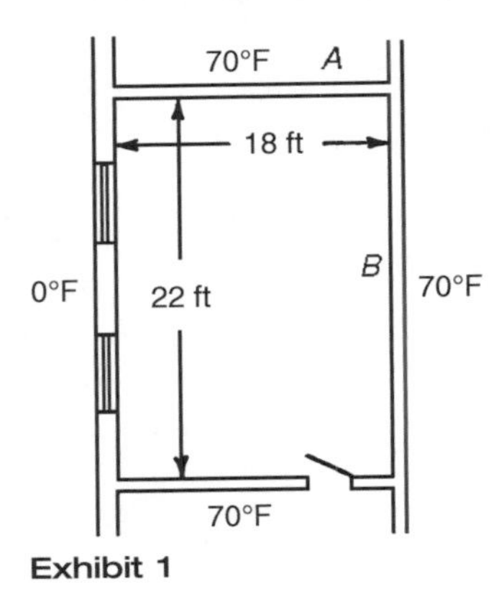

Exhibit 1

Solution

Areas. Window glass: 2 × 4 × 6 = 48 sq ft. Exposed wall (gross): 9 × 22 = 198 sq ft. Exposed wall (net): 198 – 48 = 150 sq ft.

Coefficients. From tables here or in the *Guide* for 12-in. brick walls furred and plastered on wood lath, $U = 0.24$. U for single-glazed window = 1.13.

Transmission losses. Use the equation

$$Q = U \times A \times (t_i - t_0). \qquad \textbf{(13.4)}$$

Through wall: $Q = 150 \times 0.24 \times 70 = 2520$ Btu per hr

Through glass: $Q = 48 \times 1.13 \times 70 = 3797$ Btu per hr

Infiltration losses.

$$Q = \frac{\text{cu ft per hr} \times (t_i - t_0)}{55.2}. \qquad (13.5)$$

Length of crack; two windows = 2[(3 × 4) + (2 × 6)] = 48. For 15-mph wind, from the *Guide*, infiltration = 22.9 cu ft per hr per ft of crack. Thus,

$$Q_{\text{infiltration}} = \frac{(\text{Crack Length ft})(\text{Infiltration ft}^3/\text{hr ft})(\Delta T)}{55.2} = \frac{(48)(22.9)(70)}{55.2} = 1394 \text{ Btu/hr}$$

Total losses from room and total heat required to balance these losses = 2520 + 3797 + 1394 = 7711 Btu per hr. For heating with steam, the equivalent direct radiation is 7711/240 = 32.1 EDR. If heating with 180°F hot water, the equivalent is 7711/150 = 51.4 EDR.

If the rooms adjoining were unheated, say at 35°F and a cellar below at 32°F, U for the partitions = 0.34 and for the floor = 0.24. The additional losses would then be

$$\text{Partition } A = 162 \times 0.34 \times (70 - 35) = 1928 \text{ Btu per hr}$$
$$\text{Partition } B = 198 \times 0.34 \times (70 - 35) = 2356 \text{ Btu per hr}$$
$$\text{Floor} = 396 \times 0.24 \times (70 - 35) = 3612 \text{ Btu per hr.}$$

Many designers add 15 to 20 percent of total losses for rooms exposed to prevailing winds.

Example **13.4**

By means of insulation, the loss in heat through a roof per square foot is reduced from 0.40 to 0.18 Btu per hr for each degree Fahrenheit difference between inside and outside temperatures. The area of the roof is 10,000 sq ft and the average difference between inside and outside temperatures is 35°F during the heating season of 5000 hr. If the heating value of coal is 13,000 Btu per lb and the efficiency of the heating plant is 60 percent, find the value of the coal saved per heating season at $15 per ton of 2000 lb.

Solution

$$(0.4 - 0.18) \times 10{,}000 \times 35 \times 5000 \times \frac{1}{0.6} \times \frac{1}{13{,}000} \times \frac{1}{2000} \times 15 = \$370.13$$

Example 13.5

A building having a volume of 500,000 cu ft is to be heated in zero weather to 70°F. The wall and roof surfaces aggregate 28,000 sq ft and the glass surface aggregates 7000 sq ft. The air is changed three times every hour (a 20-min. air change). Allowing transmission coefficients of 0.25 for the wall and roof surfaces and 1.13 for the single-glass-paned windows, calculate the square feet of steam radiation required if each square foot emits 240 Btu per hour.

Solution

The heating load consists of

$$\text{Wall and roof losses} = 0.25 \times 28,000 \times 70 = 490,000 \text{ Btu per hr}$$
$$\text{Glass loss} = 1.13 \times 7000 \times 70 = 553,700 \text{ Btu per hr}$$
$$\text{Ventilation load} = 500,000 \times 3 \times 0.075 = 112,500 \text{ lb per hr}$$
$$112,500 \times 0.24 \times 70 = 1.89 \times 10^6 \text{ Btu per hr}$$
$$\frac{(490 + 554 + 1.89 \times 10^3) \times 1000}{240} = 12,225 \text{ sq ft EDR.}$$

If hot-water radiation were used, 150 instead of 240 would be the emission in Btu per hr, and more radiation would be required to do the same job. The fuel-rate consumption may be determined by dividing the total heating load by the system efficiency and the result again by the gross heating value of the fuel as fired. It is convenient to remember that 4 sq ft of steam radiation are equivalent to the condensation rate of 1 lb steam per hour for low-pressure heating steam.

Example 13.6

A control room for an oil refinery unit is to be heated and ventilated by means of a central duct system. Ventilation is to be at a rate of 3 cfm of outside air per square foot of floor area. Room to be ventilated measures 40 by 60 ft, and is to be pressurized to keep out hazardous gases. Outside design temperature is −10°F and inside is to be maintained at 75°F. Determine steam consumption rate for maximum design conditions with the use of 5 psig saturated steam.

Solution

The rate of outside air to be handled by the heating and ventilating unit is 40 × 60 × 3 = 7200 cfm. The heating load is found to be

$$Q = 7200 \times 1.08 \times (75 + 10) = 660,000 \text{ Btu per hr}$$

where factor 1.08 = sp ht air (0.24) × min per hr (60)/sp vol air (13.3). Saturated steam at a pressure of 5 psig has heat of condensation equal to 960 Btu per lb. Then the steam rate required is simply

$$660,000/960 = 690 \text{ lb per hr.}$$

VENTILATING

Ventilating Terms and Definitions

The following terms are commonly used in ventilation work.

Cubical contents is the contents of the space to be ventilated expressed in cubic feet. Length × width × height = cubical contents. No deduction is made for equipment, tables, etc., within the space.

Cubic feet per minute (cfm) is the rate of air flow.

Capacity is the volume of air handled by a fan, group of fans, or by a ventilation system, usually expressed in cubic feet per minute.

Fan rating is a statement of fan performance for one condition of operation and includes fan size, speed, capacity, pressure, and horsepower.

Fan performance is a statement of capacity, pressure, speed, and horsepower input.

Fan characteristic is a graphical presentation of fan performance throughout the full range from free delivery to no delivery at constant speed for any given fan.

Resistance pressure (RP). The resistance pressure of any ventilating system is the total of the various resistance factors that oppose the flow of air in the system stated in inches water gauge (WG).

Static pressure (SP) is the force exerted by the fan to force air through the ventilating system. If exerted on the discharge side, it is said to be positive and if on the inlet side, it is negative or suction pressure. The total static pressure the fan exerts is the sum of these two pressure readings. In any system the SP exerted by the fan is equal to the RP of the system. The SP is measured by an inclined draft gauge in inches of water.

Velocity is the speed at which the air is traveling expressed in lineal feet per minute. It is measured by use of a velometer or anemometer. Average velocities may be calculated for a fan by dividing the cfm by the square-foot area of the fan

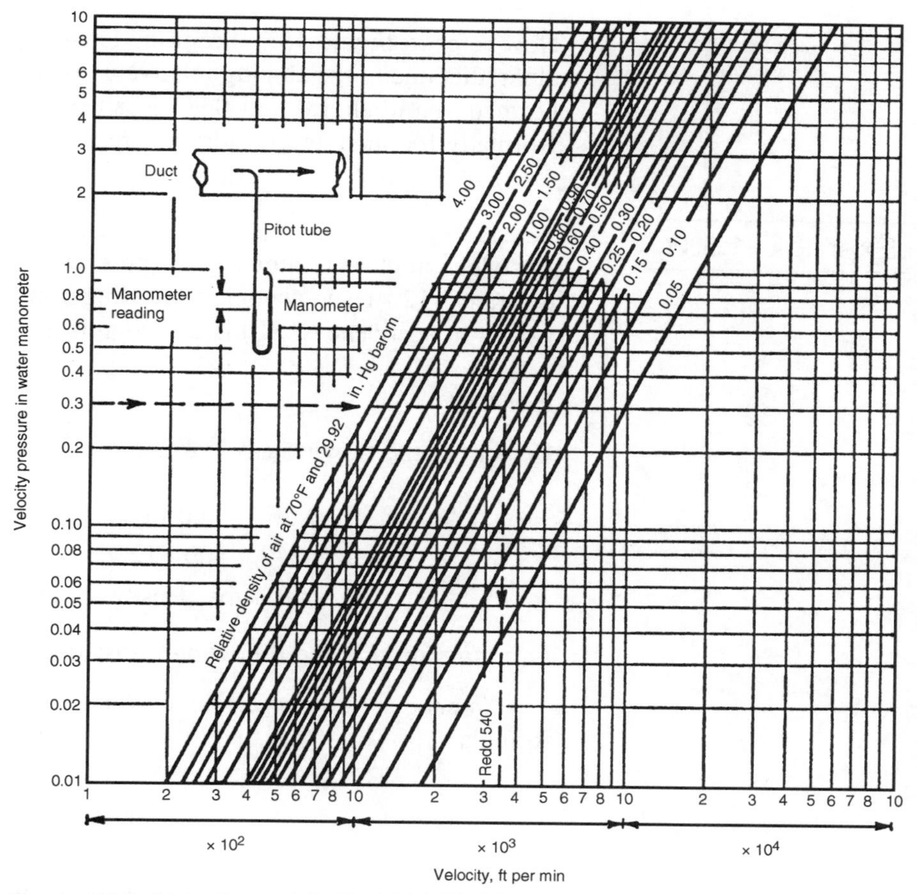

Figure 13.2 Velocity vs. velocity pressure, gases at various densities

discharge. Average velocities in any part of the system may be calculated by dividing the cfm flowing by the cross-sectional area of the duct.

Velocity pressure (VP) is a measure of the kinetic energy of horsepower in the moving air. It is measured directly by the use of a Pitot tube and a draft gauge. It can be calculated from the velocity from the formula

$$\text{VP} = (\text{velocity, fpm}/4005)^2. \tag{13.6}$$

Figure 13.2 shows a chart for determining the velocity of any gas within a duct from monometer readings in inches of water. First determine relative density of the flowing gas, remembering that density of standard air at 70°F and atmospheric pressure is 0.075 lb per cu ft. Then use manometer reading and read right to relative density line and down to actual velocity in feet per minute.

Example **13.7**

Air at 100°F is flowing in a duct. Pitot tube reading for velocity pressure is 1/2 in. WG. What is the actual velocity of flow?

Solution

Air at 100°F has a density of 0.075 × (460 + 70)/(460 + 100), or 0.071 lb per cu ft. Relative density is

$$0.071/0.075 = 0.946.$$

Now refer to Figure 13.2. From the left-hand side start at 1/2 in. and read right along this line to the sloping relative density line of 0.946. An interpolation must be made between 1.0 and 0.90. Read down to velocity of close to 2900 fpm. Normally for such temperatures no real deviation is made.

Example **13.8**

Acetylene at close to atmospheric pressure is flowing in a pipeline. Flowing temperature is 120°F and manometer reading from Pitot tube reads 1 in. WG. Molecular weight of acetylene (C_2H_2) is 26. What is the flow velocity?

Solution

Density of the flowing gas is found to be as follows:

$$\frac{26}{379} \times \frac{460+60}{460+120} = 0.0615 \text{ lb per cu ft}$$

$$\text{Relative density} = 0.0615/0.075 = 0.82.$$

From Figure 13.2 and following instructions given, find velocity to be 4600 fpm.

Horsepower output (air hp) of a fan, or air horsepower, is calculated from the formula

$$\text{Air hp} = \frac{\text{cfm} \times \text{TP}}{6{,}356} \tag{13.7}$$

where cfm is capacity in cubic feet per minute; TP is total pressure of water or static pressure + velocity pressure.

Brake horsepower (bhp) is the horsepower required to drive the fan. Brake horsepower is the input to the fan shaft required to produce the output air hp. Look at Figure 7.4 for a quick graphical presentation of air-flow-resistance pressure-air-horsepower relations.

Example 13.9

Air flows through a duct system at the rate of 20,000 cfm. Resistance pressure (static and velocity pressure) is 1 in. WG. Estimate brake horsepower of the fan assuming 70 percent for fan efficiency.

Solution

$$\text{Air hp (output hp)} = \frac{(cfm)(TP)}{6356}$$

$$\text{Air hp} = \frac{(20{,}000)(1\,\text{in H}_2\text{O})}{6356} = 3.15\ \text{hp}$$

$$\text{Mechanical eff.} = \frac{\text{Air hp}}{\text{bhp}}$$

$$\eta_{\text{fair}} = \frac{3.15\ \text{hp}}{\text{bhp}} = 70\%$$

$$\text{bhp} = 4.5\ \text{hp} \quad \text{suggest 5 hp fan.}$$

Mechanical efficiency of a fan is the ratio of horsepower output to horsepower input. Therefore,

$$\text{Mechanical efficiency} = \frac{\text{air hp}}{\text{bhp}}, \text{ expressed as a percentage.} \qquad \textbf{(13.8)}$$

Fan discharge or outlet is the place provided for receiving a duct through which air leaves the fan.

Fan inlet is the place provided for receiving a duct through which air enters the fan.

General Ventilation

Where little or no ductwork is required to ventilate a space, the application is known as general ventilation. In most cases, exhaust fans high in the side walls or in the roof are used with general movement of air through windows or louvers across the space and out the fans. The air movement is caused to flow throughout the space to remove smoke, fumes, gases, excess moisture, heat, odors, or dust or simply to provide a constant inflow of fresh, outside air by the removal of foul, stale air.

To select the fans properly for the solution of general ventilation problems, the cubical contents of the space to be ventilated should be determined by multiplying the length by the width by the average height. All dimensions used must be in feet to give volume (cubage) in cubic feet.

The next step is to select the rate of air change required to give satisfactory ventilation. The ASHRAE *Guide* or manufacturers' catalogues contain much useful data and should be used to solve examination questions.

Specific or Local Ventilation

Throughout industry there is hardly an industrial plant that does not have at least one and usually many operations, machines, or processes that require special or, as it is known in industry, specific attention. By this we mean a system of ductwork and fans designed to prevent release or spread of smoke, fumes, odors, dusts, vapors, or excess heat into other working areas. This means the control of the problem at its source. This local ventilation requires well-designed hoods or special collecting systems and a thorough knowledge of air-flow principles and laws governing the behavior of gases and vapors. Because it is beyond the scope of this presentation to cover all the detailed data required by the infinite variety of local exhaust problems, we shall review with the help of fundamental charts and tables that which has been of paramount value.

The best ventilating system can be rendered almost useless unless there is a well-selected balance between fan and duct system. Care should also be exercised to be sure that the fumes, gases, or vapors are collected as close to the source of generation as possible. The closer the better, and the less will be the ventilation needs with attendant reduction in heating load for outside air make-up. Follow these simple rules and no real trouble will ensue.

(a) Make all duct runs as short as possible with direct connections.

(b) Make area of inlet and outlet ducts equal to outside diameter of fan for lowest friction loss.

(c) Where hoods requiring a large area are required, use baffles to give higher edge velocities and reduce air volume required.

(d) Do not forget that you cannot take more air out of a room than you are putting into it. Location and sizing of even gravity air inlets are frequently as important as the design of the exhaust system.

For exhaust systems the quantity of air to be moved should be determined by the selection of a suitable face velocity at hood entrance or by other considerations such as heat or moisture absorption, cubic feet per minute of liquid surface, by duct velocities required to convey the material, or combinations of the above. For supply systems the air flow usually is already determined by the volume being exhausted or by other considerations similar to those listed for the exhaust system. But let us digress for a moment to look into the matter of evaporation of water from tanks.

AIR CONDITIONING

Equations of State

The water vapor present in the atmosphere is usually in the form of superheated steam as an invisible gas. The air is then "clear." If the atmosphere is cooled below the dew point, the excess vapor is condensed out in the form of minute drops of water or crystals of ice, so minute at first that they float as fog or cloud. If the droplets coalesce to form large drops, they fall to earth as rain. The maximum water vapor that can be held in the atmosphere increases greatly as the temperature is increased. At any temperature and pressure the quantity of water vapor can vary practically from none to maximum amount for that temperature. Absolute or "bone dry" air is never found in nature, but the amount of water vapor may be so low that it is difficult to measure it.

For practical air-conditioning purposes, it is frequently assumed that air and water vapor in the atmosphere follow the general gas equation for ideal gases, and that their mixture obeys Dalton's law of partial pressures.

To determine the maximum percent of water vapor for gauge pressure:

$$\frac{\text{Partial pressure of water (from steam tables), psi}}{14.7 + \text{gauge pressure, psi}}. \quad \textbf{(13.9)}$$

Gauge pressure in the denominator is for the entire system. Actual percent of water vapor by volume is

$$\text{Maximum percent of water vapor} \times \text{relative humidity.} \quad \textbf{(13.10)}$$

Example **13.10**

Find the mass of air contained in a room 25 by 30 by 10 ft at atmospheric pressure and 65°F.

Solution

Use equation of state: $PV = mRT$

$$m = \frac{PV}{RT} = \frac{(14.7 \times 144)\text{lbf/ft}^2(25 \times 30 \times 10)\text{ft}^3}{(525°\text{R})(53.3)(\frac{\text{ft lbf}}{\text{lbm°R}})}$$

$$m = 568 \text{ lbm}$$

Example **13.11**

Find relative humidity of air at 80°F dry bulb and 60°F dew point temp.

The relative humidity of an air mixture is the ratio of the water-vapor partial pressures at the two different temperatures.

$$\frac{\text{Partial pressure actual at dew-point temp.}}{\text{Partial pressure saturated at dry-bulb temp.}} \times 100 \quad \textbf{(13.11)}$$

Solution

From Equation 13.11 and the steam tables

$$\frac{0.2561 \text{ psia at } 60°\text{F}}{0.5067 \text{ psia at } 80°\text{F}} \times 100 = 50.6 \text{ percent.}$$

Of course, the relative humidity may be more quickly determined from the psychrometric chart.

The term relative humidity is often misused in a way to mean a feeling of dryness of the air. However, as a matter of fact, air with a relative humidity of 60 percent when the dry-bulb thermometer is 40°F is drier than air with a relative humidity of 10 percent and a dry-bulb temperature of 100°F. Relative humidity is the relationship of water vapor in the air at the dew point temperature to the amount that would be in the air if the air were saturated at the dry-bulb temperature. In reality it can be considered as a thermal condition of steam (or water vapor) at one pressure compared with steam (or water vapor) at a higher pressure.

Percentage humidity is not to be confused with relative humidity. Percentage humidity is defined as the percentage ratio of the existing weight of water vapor per unit weight of vapor-free air to the weight of water vapor which would exist per unit weight of vapor-free air if the mixture were saturated at the existing temperature and pressure. Percentage humidity and relative humidity approach equality when the vapor concentrations are low. *Percentage humidity is always somewhat smaller than relative humidity.*

Example 13.12

The temperature in a room is 72°F dry bulb and the relative humidity is 40 percent. Barometric pressure is 29.92 in. Hg. Find (a) the partial pressure of the water vapor in the air, (b) the humidity ratio.

Solution

(a) The pressure of saturated vapor at 72°F can be found in the steam tables as 0.38844 psia. For a relative humidity of 40 percent, we find the partial pressure of the water vapor to be

$$p_w = \phi \times p_s = 0.40 \times 0.38844 = 0.15538 \text{ psia}$$

(b) The humidity ratio is next found to be the same as previously discussed in the chapter on thermodynamics (Chapter 2).

$$\frac{p_w}{p_a} \times 0.622 = \frac{0.15538}{14.696 - 0.15538} \times 0.622$$
$$= 0.665 \times 10^{-2} \text{ lb water per lb dry air}$$

The Psychrometric Chart

This chart is well known and will not be reproduced here. Its use is of such importance that one should be taken along to the examination.

Latent Heat Load in Air Conditioning

Latent heat is defined as that which is added to or subtracted from a substance in changing its physical state from a solid to a liquid (latent heat of fusion) or from a liquid to a vapor (latent heat of vaporization). In air-conditioning work the latent heat refers to the change of liquid water to water vapor in humidifying or the change of water vapor to liquid water in dehumidifying.

The latent heat load in air-conditioning work is the sum of all items that produce water vapor. Such items as people, infiltration of outside air, coffee urns, and certain products give off water vapor that is latent heat. For a person at rest the latent heat given off is 180 Btu per hr. The average hourly total heat per person at rest is taken as 400 Btu per hr, of which 180 Btu is latent heat and 220 Btu is sensible heat. These values vary with degree of activity and environmental tem-

Table 13.1 Sources of internal heat btu per hr

1 Horsepower	2546.0	1 cu ft producer gas	150	60 gr moisture per cu ft
1 kilowatt	3415.0	1 cu ft illumin. gas	550–700	450 gr moisture per cu ft
100-watt lamp	341.5	1 cu ft natural gas	1000	675 gr moisture per cu ft

(Welsbach Burner averages 3 cu ft per hr; and Fishtail burner, 5 cu ft per hr)

	Heat Load of Person Normally Clothed at Rest, Btu per hr*								
Dry Bulb	**90°**	**85°**	**80°**	**75°**	**70°**	**65°**	**60°**	**55°**	**50°**
Latent heat	241	197	153	118	87	66	61	61	61
Sensible heat	96	144	193	232	267	293	324	354	385
Total heat	337	341	346	350	354	359	385	415	446
Gr. mois. evap. per hr	1589	1300	1011	779	578	433	405	405	405

*Add 25% for man at work.

perature and humidity. Reference should be made to the *Guide.*

The quantity of water vapor from infiltration of outside air is determined from the difference between grains of moisture at the outside conditions of dew point and the room dew point. The latent heat value for water vapor is taken at the outside dew point temperature. Manufacturers of air-conditioning equipment usually provide calculation forms complete with factors so that it becomes a simple matter to determine air-conditioning loads.

The quantity of water vapor (latent heat) given up in a space is more difficult to establish accurately than is the sensible heat load. The total sum of water vapor once established is converted from water vapor to heat by using an average value of 1040 Btu per lb water vapor. If the amount of water vapor is 50 lb per hr, the latent heat load would be 50 × 1040, or 52,000 Btu per hr.

Table 13.1 lists sources of internal heat. More elaborate tables are contained in the *Guide*.

Air Required to Pick Up Latent Heat

Because air is the medium of absorbing latent heat (moisture pickup), it is necessary to put the air in condition to be able to absorb moisture. This is accomplished similarly to the absorption of sensible heat in that the absolute humidity (dew point) is reduced sufficiently to make the air capable of taking on moisture. As previously explained, the pickup of sensible heat in the conditioned room is accomplished by the quantity of air circulated at a reduced dry-bulb temperature. The pickup of latent heat is accomplished by the same air circulated at a reduced entering dew-point temperature. Thus, if we circulate 9333 cfm to pick up 150,000 Btu sensible heat, this same quantity of air can have its dew point reduced sufficiently to accomplish the latent heat pickup. If the latent heat load is 52,000 Btu, this is equal to

$$\frac{52{,}000 \text{ Btu per hr}}{1040 \text{ Btu per lb}} = 50 \text{ lb per hr}$$

and

$$50 \times 7000 \text{ grains per lb} = 350{,}000 \text{ grains per hr.}$$

The quantity of water vapor to be picked up per cubic foot of air circulated is

$$\frac{350{,}000}{60 \text{ min} \times 9333} = 0.625 \text{ grains per cu ft.}$$

If the average room conditions are to be 80°F dry bulb and 50 percent relative humidity, the dew-point temperature is 59.5°F and the air contains 75.85 grains per lb, or 5.699 grains per cu ft, the air must be supplied to the room at a condition of 5.699 – 0.625, or 5.074 grains per cu ft. Referring to air-property tables, we find that 5.074 grains per cu ft corresponds to approximately 56°F dew point (5.06 grains per cu ft). Thus, the air entering dew point must not be higher than 56°F, or $3^1/_2$°F lower than the expected room dew point, so as to maintain the conditions established of 80°F dry bulb and 50 percent relative humidity.

Because the calculations were based on maintaining an average condition in the room, the return air is probably the best indication of the average conditions so that in most cases the thermostats and hygrostats are placed in the return air duct or near the return air grille. This is especially true if a number of rooms are connected to the air-conditioning apparatus. As we previously explained, the pickup of sensible heat in a conditioned room is accomplished by the quantity of air circulated at a reduced dry-bulb temperature and the pickup of latent heat is accomplished by the same air circulated at a reduced entering dew-point temperature.

Total Heat Load

For air-conditioning calculations, total heat is defined as the sum of all the sensible heat loads and all the latent heat loads, and is known as the *room total heat*. The room total heat does not establish the refrigeration load because it does not include the cooling load of outside air. The room total heat does, however, give us an easy means of establishing the dew point of the air entering the conditioned space.

Because the wet-bulb temperature determines the total heat in air, the total heat absorbed (sensible plus latent) by a quantity of air is measured by the difference between the respective wet-bulb temperatures. If H_1 is the enthalpy at the wet-bulb temperature of the air entering a space and H_2 at the wet-bulb temperature of the air leaving the space (return air), the total heat picked up in Btu per lb (W) of air is

$$\text{Total heat, Btu} = W(H_2 - H_1). \qquad \textbf{(13.12)}$$

If the sensible heat formula is divided by the total heat formula, the following relation exists:

$$\frac{\text{Sensible heat}}{\text{Total heat}} = \frac{W(T_2 - T_1)c_p}{W(H_2 - H_1)} = \frac{H_s}{H_t}. \qquad \textbf{(13.13)}$$

In the previous examples, the room condition was taken as 80°F dry-bulb temperature, 50 percent relative humidity, 59.5°F dew point, and 66.7°F wet-bulb temperature. H_s equals 150,000 Btu per hr, H_L equals latent heat equals 52,000 Btu per hr.

$$H_t = H_s + H_L = 150,000 + 52,000 = 202,000 \text{ Btu per hr} \qquad \textbf{(13.14)}$$

The dry-bulb temperature was taken as a differential of 15°F, i.e., $T_2 - T_1 = 80 - 65 = 15$°F, and the total heat corresponding to the room wet bulb of 66.7°F is 31.276 Btu per lb.

Substituting values in the formula, Equation 13.13,

$$\frac{150,000}{202,000} = \frac{15 \times 0.24}{31.276 - H_1} = 0.7426 = \frac{3.6}{31.276 - H_1}$$

from which H_1 is found to be equal to 26.42 Btu per lb.

By referring to tables of properties of air, 26.42 Btu per lb is seen to correspond to a wet-bulb temperature of about 60°F. Thus, the temperature of the air entering the conditioned area is 65°F dry bulb and 60°F wet bulb, which by referring to the psychrometric chart, gives a dew point of 57°F and a relative humidity of 75 percent. This is an easy method of determining the entering room dew point, which will necessarily vary with the quantity of air circulated.

A line drawn through the room conditions of the psychrometric chart and the air entering conditions and extended through the saturation curve will give a locus of all dew points for any given dry-bulb differential (see chart). Note that the lowest dew point of entering air would be about 55°F. Any lower dew point would cause lower relative humidities. This condition would not be objectionable unless it dropped low enough to cause the space to feel uncomfortably cool.

Example **13.13**

Ninety pounds of water vapor are released into the atmosphere of a room every hour. An air supply of 8250 cfm is provided. Find supply air dew point to maintain a room dew point of 59°F.

Solution

Moisture released to room is equal to 90 lb/hr. Required change in lbs per pound of air is

$$\frac{(90 \text{ lb/hr})(13.3 \text{ ft}^3/\text{lb})}{(8250 \text{ ft}^3/\text{min})(60 \text{ min/hr})} = 0.002418 \text{ lb}_{\text{water}}/\text{lb}_{\text{air}} \text{ to be added}$$

From psychrometric chart, at 59°F dew-point air has 0.0109 lb per lb. Then the required initial moisture content of supply is (0.0109 – 0.002418) = 0.00848. Dew point for this condition is found to be 52°F from chart.

Example **13.14**

The atmosphere of a room having a latent heat gain of 60,000 Btu per hr is to be maintained at a dew point of 58°F. If the air supply to the room is 8000 cfm, find the required dew point of the air supply.

Solution

Required change in latent heat content of air supply is

$$\frac{(60000\ \text{Btu/hr})(13.3\ \text{ft}^3/\text{lb})}{(8000\ \text{ft}^3/\text{min})(60\ \text{min/hr})} = 1.66\ \text{Btu/lb}$$

Latent heat content of room air from psychrometric chart is 0.0101 lbs per lb dry air and is equivalent to

$$0.0101 \times 1040 = 10.5\ \text{Btu/lb}$$

Latent heat content of air supply is 10.5 – 1.66 = 8.84 Btu. This is equivalent to (8.84 Btu)/(1040 Btu/lb) = 0.0085 lb water/lb dry air with an equivalent dew point from chart of 53.2°F.

Example **13.15**

A room having 80 percent of its total heat gain as sensible heat is to have its dry-bulb temperature maintained at 76°F and its wet-bulb temperature at 64°F. If air is to be supplied at a dry-bulb temperature of 55°F, find the required wet-bulb temperature of the air supply.

Solution

Air supply will heat up 76 – 55, or 21°F. Then, total heat × 0.80 is equal to 0.24 × 21, or

$$\text{Change in total heat} = 0.24 \times 21/0.80 = 6.3\ \text{Btu per lb.}$$

From psychrometric chart total heat of air (enthalpy) at 64°F wet-bulb temperature is 29.3. Required total heat of air supply is 29.3 – 6.3, or 23 Btu per lb. This is equivalent to a wet-bulb temperature of 54.4°F.

Example **13.16**

With outside conditions of 95°F dry-bulb and 78°F wet-bulb temperature and inside conditions of 75°F dry-bulb and 57.5°F wet-bulb temperature, determine the dew point, relative humidity, specific volume, grains of moisture per pound of dry air, and enthalpy, using the psychrometric chart.

Solution

Conditions	Outside Air	Inside Air
Dry bulb	95°F	75°F
Wet bulb	78°F	57.5°F
Dew point	71°F	43°F
Relative humidity	46 percent	32 percent
Specific volume	14.4 cu ft per lb	13.6 cu ft per lb
Grains per lb	114.4	40.9
Enthalpy (total heat)	40.64 Btu per lb	25.54 Btu per lb

Calculation of Cooling Load

Much of the work in calculating summer cooling load is identical with the calculation of winter heating load except that the heat flows into the building instead of out. There is one important new factor which has not been discussed, that of *solar heat* or radiant heat from the sun, which has a tremendous effect on cooling requirements, moreso when involving buildings than for process cooling.

Table 13.2

State	City	Design Dry Bulb, °F	Design Wet Bulb, °F
Alabama	Mobile	95	80
Arizona	Tucson	105	72
California	Los Angeles	90	70
Connecticut	New Haven	95	75
Delaware	Wilmington	95	78
District of Columbia	Washington	95	78
Florida	Miami	91	79
Illinois	Chicago	95	75
New Jersey	Trenton	95	78
New York	New York	95	75
Pennsylvania	Philadelphia	95	78
Texas	Houston	95	78
Utah	Salt Lake	95	65
Vermont	Burlington	90	73
Washington	Seattle	85	65
Wisconsin	Madison	95	75

Ordinary heat transmission takes place even when there is no sun effect. This transmission heat load is calculated as for heating. For the inside temperature, that of 80°F dry-bulb temperature and 50 percent relative humidity are most common although there may be slight deviations depending on a particular job. Outside design temperatures and wet-bulb temperatures for a number of locations are listed in Table 13.2. For a more complete list, refer to latest edition of the *Guide*.

You will recall that in calculating winter heating loads we choose an outdoor temperature 10 to 15°F above the lowest outdoor temperature ever recorded by the Weather Bureau. The choice of summer outdoor dry-bulb design temperature is not quite this simple. Wet-bulb temperature is just as important as dry-bulb temperature. The conditions given in Table 13.2 are those which will not be exceeded more than 5 to 8 percent of the time during an average summer cooling season (June through September) of 120 10-hr days, or 1200 hr. In choosing outdoor design conditions for a location where this information is not known, it is necessary to obtain Weather Bureau or other records of conditions during past years. Select values that have not been exceeded more than 5 percent of the time in previous average summers. Be sure to use daytime data and *not nighttime data*.

Cooling Load Due to Solar Radiation

The solar heat load is made up of that from sunlit walls and glass exposed to the rays of the sun. The *Guide* goes into this matter rather completely and the candidate is referred to that source for coverage. In its proper place we shall devote ourselves to the question and answer approach to focus attention.

Heat from Machinery and Appliances

When we calculate a winter heating load, we do not deduct the heat generated by all heat-producing equipment within the space to be heated because some of this equipment might be idle and thus not contributing heat. In summer, for which we are now calculating a cooling load, we have the opposite situation. We must consider any heat-generating machinery or appliance that is likely to be operating

Table 13.3 Heat gain from various sources

Source	Btu per hr		
	Sensible	**Latent**	**Total**
Electric-Heating Equipment			
Electrical equipment—dry heat—no evaporated water	100%	0%	100%
Electric oven—baking	80%	20%	100%
Electric equipment—heating water—stewing, boiling, etc.	50%	50%	100%
Electric lights and appliances per watt (dry heat)	3.4	0	3.4
Electric lights and appliances per kilowatt (dry heat)	3413	0	3413
Electric motors per horsepower	2546	0	2546
Electric toasters or electric griddles	90%	10%	100%
Coffee urn—large, 18-in. diameter—single drum	2000	2000	4000
Coffee urn—small, 12-in. diameter—single drum	1200	1200	2400
Coffee urn—approx. connected load per gallon of capacity	600	600	1200
Electric range—small burner	*	*	3400
Electric range—large burner	*	*	7500
Electric range—oven	8000	2000	10000
Electric range—warming compartment	1025	*	1025
Steam table—per square foot of top surface	300	800	1100
Plate warmer—per cubic foot of volume	850	0	850
Baker's oven—per cubic foot of volume	3200	1300	4500
Frying griddles—per square foot of top surface	*	*	4600
Hot plates—per square foot of top surface	*	*	9000
Hair dryer in beauty parlor—600 watts	2050	0	2050
Permanent-wave machine in beauty parlor—24-25 watt units	2050	0	2050
Gas-Burning Equipment			
Gas equipment—dry heat—no water evaporated	90%	10%	100%
Gas-heated oven—baking	67%	33%	100%
Gas equipment—heating water—stewing, boiling, etc.	50%	50%	100%
Stove, domestic type—no water evaporated—per medium-size burner	9000	1000	10000
Gas-heated oven—domestic type	12000	6000	18000
Stove, domestic type—heating water—per medium size burner	5000	5000	10000

Table 13.3 Heat gain from various sources (*Continued*)

Residence gas range—giant burner (about $5\frac{1}{2}$ in. diameter)	*	*	12000
Residence gas range—medium burner (about 4 in. diameter)	*	*	10000
Residence gas range—double oven (total size $18 \times 18 \times 22$ in. high)	*	*	18000
Residence gas range—pilot	*	*	250
Restaurant range—4 burners and oven	*	*	100000
Cast-iron burner—low flame—per hole	*	*	100
Cast-iron burner—high flame—per hole	*	*	250
Simmering burner	*	*	2500
Coffee urn—large, 18 in. diameter—single drum	5000	5000	10000
Coffee urn—small, 12 in. diameter—single drum	3000	3000	6000
Coffee urn—per gallon of rated capacity	500	500	1000
Egg boiler—per egg compartment	2500	2500	5000
Steam table or serving table—per square foot of top surface	400	900	1300
Dish warmer—per square foot of shelf	540	60	600
Cigar lighter—continuous flame type	2250	250	2500
Curling iron heater	2250	250	2500
Bunsen type burner—large—natural gas	*	*	5000
Bunsen type burner—large—artificial gas	*	*	3000
Bunsen type burner—small—natural gas	*	*	3000
Bunsen type burner—small—artificial gas	*	*	1800
Welsbach burner—natural gas	*	*	3000
Welsbach burner—artificial gas	*	*	1800
Fish-tail burner—natural gas	*	*	5000
Fish-tail burner—artificial gas	*	*	3000
Lighting fixture outlet—large, 3 mantle 480 c.p.	4500	500	5000
Lighting fixture outlet—small, 1 mantle 160 c.p.	2250	250	2500
One cu ft of natural gas generates	900	100	1000
One cu ft of artificial gas generates	540	60	600
One cu ft of producer gas generates	135	15	150
Steam-Heated Equipment			
Steam-heated surface not polished—per square foot of surface	330	0	330
Steam-heated surface polished—per square foot of surface	130	0	130
Insulated surface, per square foot	80	0	80
Bare pipes, not polished per square foot of surface	400	0	400
Bare pipes, polished per square foot of surface	220	0	220
Insulated pipes, per square foot	110	0	110
Coffee urn—large, 18-in. diameter—single drum	2000	2000	4000
Coffee urn—small, 12-in. diameter—single drum	1200	1200	2400
Egg boiler—per egg compartment	2500	2500	5000
Steam table—per square foot of top surface	300	800	1100
Miscellaneous			
Heat liberated by food per person, as in a restaurant	30	30	60
Heat liberated from hot water used direct and on towels per hour—barber shops	100	200	300

*Percent sensible and latent heat depends on use of equipment; dry heat, baking, or boiling.

at the same time that cooling is needed. If there is at all a question as to whether an appliance is operating, we shall consider that it will be operating. Table 13.3 shows the heat added by miscellaneous electric, gas, and steam-heated equipment. Considerable judgment factor is required in using values for equipment where the sensible and latent heat values in Table 13.3 are marked with an asterisk. Motor heat will be accounted for in the short form for figuring the cooling load. Electric lights are figured for full wattage whether incandescent or fluorescent.

Maximum Cooling Load

Table 13.4 Time lag in transmission of solar radiation through walls and roofs

Type and Thickness of Wall or Roof	Time Lag, hr
2-in. pine	1.5
6-in. concrete	3
4-in. gypsum	2.5
3-in. concrete and 1-in. cork	2
2-in. iron and cork	2.5
4-in. iron and cork	7.25
8-in. iron and cork	19
22-in. brick and tile wall	10

When does maximum cooling load occur? So far we have reviewed the procedure and the nature of the calculations involved in calculating summer cooling load. Let us see how we would go about determining the cooling load on a building of certain characteristics. If the top floor had a number of skylights and electric lighting, we could assume that on sunny days the lights would not be in use. We would only calculate the sun effect and exclude the lights. The sun effect on the east and west walls would not be added to the total sun load, for just one side or the other (whichever is the greater) would be included. At noon just the roof and skylight loads would be considered. Then the largest combination of them all is included. For parts of the country in the north latitude the maximum sun effect on the east wall occurs about 8 A.M., at which time there is zero sun effect on the west wall, and almost zero sun effect on the south wall. It may be shown that the sun effect on the east and west walls occurs at totally different times.

Time lag plays an important role in solar-heat gain. Sun will pass immediately through glass and will have an immediate effect on cooling load. However, the same sun impinging on walls will suffer a time lag in having its effect felt. Table 13.4 lists representative time lags for several materials.

As we can see from the table, the sun effect may not reach the inside and have its effect until several hours later. By the time this heat reaches the inside of the conditioned space, the sun effect on the south wall is at a maximum and the portion striking the windows of the south wall is entering the conditioned space immediately. Hence it is likely that the maximum solar-heat effect *inside* this room might occur sometime between 11 A.M. and 1 P.M. This maximum would *not* include the solar heat striking the solid portion of the south wall, because time lag would delay this heat and it would finally enter the conditioned space between

3 and 4 P.M. By this time most of the sun effect on the south windows would have disappeared, and because the room does not have an exposed west wall, the solar heat on the room would diminish as the afternoon progressed.

By combining the solar effects on glass and wall and by use of available extensive tables, a definite and orderly procedure may be established for determining the time of day at which the maximum effect of solar heat effect is actually entering the room.

Example 13.17

A building as illustrated in Exhibit 2 is to be maintained at 75°F dry-bulb and 64.4°F wet-bulb temperatures. This building is situated between two similar units that are not cooled. There is a second-floor office above and a basement below. The south wall, containing 45 sq ft of glass area, has a southern exposure. On the north side there are two show windows that are ventilated to the outside and are at outside conditions. Between the show windows is located the doorway. This

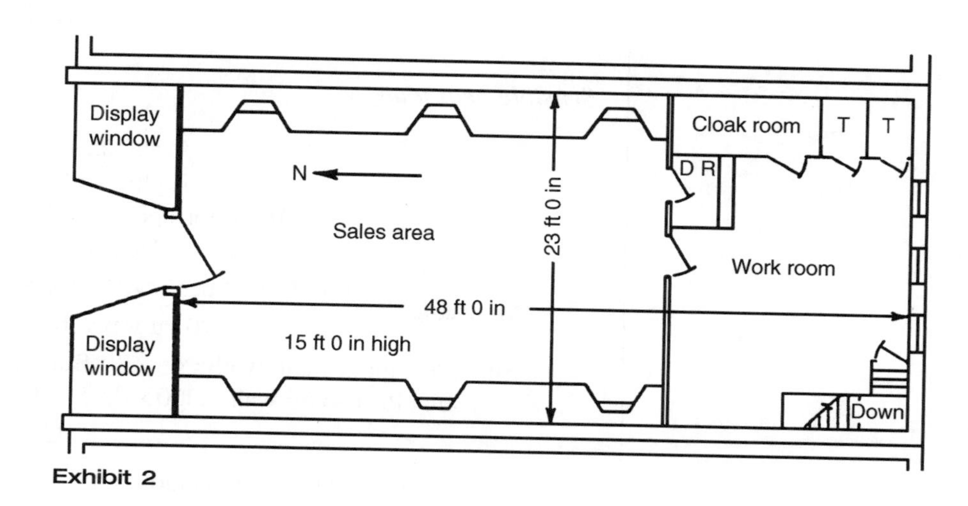

Exhibit 2

will be assumed as normally closed but it is opened quite frequently and allows an average of 600 cfm of outside air to be admitted. This will cause slightly more than two air changes per hour in the building. Number of persons is 35. Lighting is 1100 watts on a sunny day. Basement temperature is 80°F assumed. Maximum outside conditions for design purposes are 95°F dry-bulb and 78°F wet-bulb temperature. Ceiling fans equal 0.5 hp. Determine the refrigeration load.

Solution

Calculations of heat load using U coefficients:	U
East and west walls (24 in. brick, plaster one side)	0.16
North partition ($1\frac{1}{4}$ in. tongue-and-groove wood)	0.60
Plate-glass door	1.0
South wall (13 in. brick, plaster one side)	0.25

Windows (single thickness)	1.13
Floor (1 in. wood, paper, 1 in. wood over joists)	0.21
Ceiling (2 in. wood on joists, lath and plaster)	0.14

Temperature difference walls and ceiling: 95 – 75 = 20°F

Temperature difference floor: 80 – 75 = 5°F

Heat leakage:

$$\begin{aligned}
\text{East and west wall} &= 0.16\times 1440\times 20 = 4610 \text{ Btu per hr}\\
\text{North partition} &= 0.6\times 299\times 20 = 3590\\
\text{Glass door} &= 1.0\times 35\times 20 = 700\\
\text{South wall} &= 0.25\times 300\times 20 = 1500\\
\text{Windows } 45+46 &= 1.13\times 91\times 10 = 2060\\
\text{Floor} &= 0.21\times 1104\times 5 = 1160\\
\text{Ceiling} &= 0.14\times 1104\times 20 = \underline{3090}\\
\text{Total} &= 16{,}710 \text{ Btu per hr}
\end{aligned}$$

Sensible heat load

$$\begin{aligned}
\text{Occupants} &= 35\times 300 = 10{,}500 \text{ Btu per hr}\\
\text{Lights} &= 1100\times 3.413 = 3760\\
\text{Motor horsepower} &= 0.5\times 2546 = 1275\\
\text{Heat leakage above} &= 16{,}710\\
\text{Air leakage} &= \frac{36{,}000 \text{ cu ft per hr}\times 0.24\times(95-75)}{13.70 \text{ cu ft per lb}} = 12{,}600\\
\text{Sun effect glass south wall} &= 45\times 30 \text{ Btu/(hr)(sq ft)} = 1350\\
\text{Sun effect south wall} &= 300\times 0.25\times(120-95) = \underline{1875}\\
\text{Grand total} &= 48{,}070 \text{ Btu per hr}
\end{aligned}$$

$$\text{Dry tons of refrigeration} = 48{,}070/12{,}000 = 4.01 \text{ tons}$$

Latent heat load (moisture)

Air leakage

Conditions	Outside Air	Inside Air
Dry bulb	95°F	75°F
Wet bulb	78°F	64.4°F
Percent relative humidity	47	55
Dew point	71°F	58°F
Gr per lb	114.4	71.9
Sp vol		13.7 cu ft per lb

First determine the pounds per hour of air.

$$\frac{36{,}000\times(114.4-71.9)}{13.7\times 7000} = 15.95 \text{ lb per hr}$$

$$\begin{aligned}
\text{Air latent heat} &= 15.95\times 1040 = 16{,}900 \text{ Btu per hr}\\
\text{Person load} &= 35\times 100 = 3500 \text{ Btu per hr}\\
\text{Total latent} &= 16{,}900 + 3500 = 20{,}400 \text{ Btu per hr,}\\
&\quad \text{or } 20{,}400/12{,}000 = 1.70 \text{ tons}\\
\text{Total load} &= 4.01 \text{ sensible} + 1.70 \text{ latent} = 5.71 \text{ total tons}
\end{aligned}$$

Adiabatic Cooling

Let us examine a simple problem in cooling air adiabatically. Follow the process on the psychrometric chart. The water is merely recirculated and only a small amount of make-up is added due to loss by humidification; neither is water-heated or water-cooled before passing through the washer. For convenience let us superimpose temperature scales on this washer. The condition of entering air is 80°F dry bulb, 44°F dew point, 60°F wet bulb. With appropriate conditions and perfect internal mixing of water and air, the air leaves at 60°F and saturated (see Figure 13.3). Then, as we said before, all three temperatures become identical. Now, the wet bulb does not vary because enthalpy remains constant, so we have true adiabatic cooling, for the dry-bulb temperature has dropped. However, dew point has risen, for humidification has taken place. The heat released by the drop in dry bulb has been used in evaporating the water.

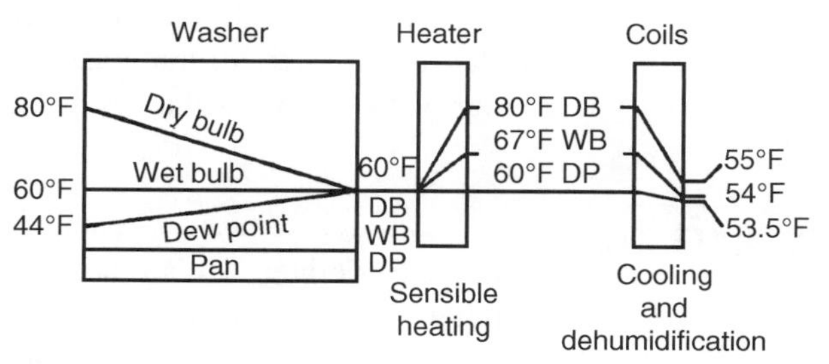

Figure 13.3

Now let us consider a heater placed in the air stream as shown. This adds only sensible heat to the air. With no moisture additions or subtractions, the dew point would remain the same as it is independent of sensible heat. But with the addition of sensible heat the total heat of the air and water vapor rises and thus the wet-bulb temperature is also raised. Also, there is a rise in dry-bulb temperature to some point above the wet-bulb temperature.

With the addition in line of a dehumidifier coil or washer, all three temperatures suffer a drop.

Example 13.18

A room is to be cooled by air from an adiabatic air washer, and has no moisture load, but does have a sensible heat load of 150 Btu per min. The conditions to be maintained are 80°F dry-bulb temperature and 60 percent relative humidity. Find the required entering wet-bulb temperature to the washer and the quantity of air that must be washed each minute.

Solution

From the psychrometric chart we see the wet-bulb temperature corresponding to the room conditions is 70°F. This is the entering wet-bulb temperature required. Then the air to be sensibly heated from 70 to 80°F with 150 Btu per min is found as follows:

$$\text{lb per min} = 150/(0.24 \times \text{temperature drop})$$
$$\text{lb per min} = 150/(0.24 \times 10) = 62.5 \text{ lb per min.}$$

Bypass System

From the nature of air-conditioning apparatus the air leaving the cooling coil or washer is practically saturated. The dew point so resulting is necessary to maintain proper conditions in the room. When the air-conditioning requirements are reduced due to lower outdoor temperatures, fewer people than calculated, less lighting, etc., the temperature leaving the apparatus is so low that the air entering the space will cause drafts and chilled environment. In order to avoid this condition, the leaving air must be sensibly heated to the proper dry-bulb temperature before being admitted to the space. Normally the difference (diffusion) between dry-bulb temperature to be maintained within the space and that of the entering air is from 15 to 20°F, depending on the manner of air distribution and kind of space.

In the bypass system, a portion of the return air is mixed with the outside air, which has passed through the conditioner together with the remainder of the return air. The quantity of return air being automatically controlled warms the saturated air leaving the conditioner to correct dry-bulb temperature. Because only a portion of the recirculated air that passes through the apparatus is refrigerated (the balance being passed around the apparatus), the refrigeration is reduced proportionately.

Example **13.19**

Air is to be delivered to a space at the outlet grille at 64°F dry bulb. The dehumidifier leaving air temperature (apparatus dew point) is 52°F. If the return air at 80°F dry bulb is available, what proportion of conditioned air and bypass air must be used?

Solution

Let x be proportion of conditioned air necessary and y be proportion of by-passed air necessary.

Then $$x + y = 1$$

$$52x + 80y = 64(x + y)$$

from which may be obtained

$$12x = 16y.$$

Therefore $x = 57.15$ percent conditioned air
$y = 42.85$ percent bypassed air.

Example **13.20**

A space to be conditioned has a sensible heat load of 10,000 Btu per min and a moisture load of 3.771 lb per min; 2300 lb of air are to be introduced each minute to this space for its conditioning to 80°F dry bulb and 50 percent relative humidity. Draw a sketch of this problem and work out the amount of air to be bypassed and the amount and temperature of air leaving the dehumidifier.

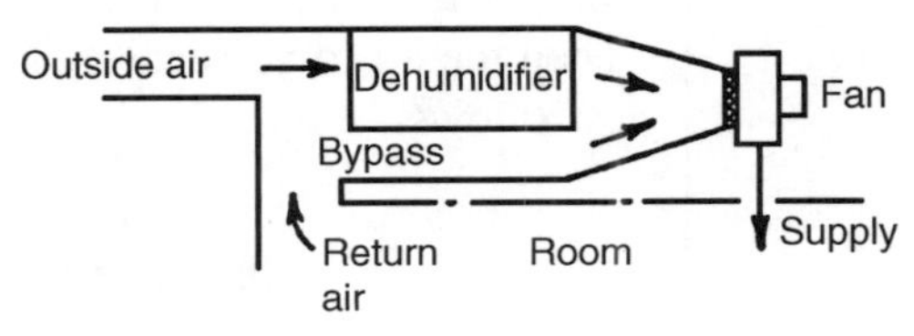

Exhibit 3

Solution

Refer to Exhibit 3. Now set up the following table of conditions.

	Room Conditions	**Air Leaving Dehumidifier**
Dry bulb	80°F	54°F
Wet bulb	67°F	54°F
Dew point	60°F	54°F
Relative humidity	50 percent	100 percent
Total heat	31.15	22.54
lb H_2O/lb air	0.0110	0.00887
Sp vol	13.84	13.13
Lb per min	2,300	1,610

Moisture load = 3.771 lb/min × 1040 Btu/lb = 3920 Btu/min
Sensible load = 10,000 Btu per min
Total load = 13,920 Btu per min

Solve by trial and error, assuming 53°F air leaving dehumidifier.

Total heat at 80°F and 50 percent rel. humidity = 31.15
Total heat at 53°F and 100 percent rel. humidity = 21.87
Difference is the pickup = 9.28 Btu

On the basis of the first trial, air circulated is 13,920/9.28 = 1500 lb per min. Check this figure as follows:

$$1500 \times 0.24 \times (80 - 53) = 9700 \text{ Btu per min.}$$

This is not enough. By trial and error again, assume 54°F air leaving dehumidifier.

Total heat at 80°F and 50 percent rel. humidity = 31.15
Total heat at 54°F and 100 percent rel. humidity = 22.54
Difference is the pickup = 8.70 Btu

Air circulated is now 13,920/8.70 = 1600 lb per min. Now check,

$$1600 \times 0.24 \times (80 - 54) = 9970 \text{ Btu per min.}$$

This figure is slightly short of the 10,000 Btu per min required, but will do. Thus,

Total air leaving dehumidifier = 1600 lb per min
Air bypassed = 2300 − 1600 = 700 lb per min

Temperature of air leaving dehumidifier = 54°F.

Example **13.21**

Air enters a drier at 70°F dbt and 20 percent relative humidity and leaves the drier at 180°F dbt and a relative humidity of 50 percent. The drier operates at atmospheric pressure of 14.3 psia. Determine the number of cubic feet of air entering per minute needed to evaporate 2.0 lb water per min from the material being processed.

Solution

At 70°F, saturation pressure is 0.3631 psia from steam tables. At 20 percent relative humidity, vapor pressure is 0.20 × 0.3631 = 0.073 psia.

$t_1 = 70°F \qquad RH_1 = 20$ percent

Using humidity ratio relations, pounds of water per pound of dry air is

$$w = (0.622)\left(\frac{P_v}{P - P_v}\right) = 0.622 \times (0.073/14.227) = 0.00319 \text{ lb per lb for entering air.}$$

Note that 14.3 – 0.073 = 14.227 psia. Then for leaving air,

$$w = (0.622)\left(\frac{(0.50)(7.5110)}{14.3 - (0.5)(7.511)}\right) = 0.622 \times (3.755/10.545) = 0.2215 \text{ lb } H_2O\text{/lb dry air.}$$

Thus water removed from system = 0.2215 – 0.00319 = 0.2183 lb water per lb dry air circulated.

$$\frac{2.0 \text{ lb water to be removed}}{0.2183 \text{ lb water removed per lb dry air}} = 9.17 \text{ lb per min air to be circulated}$$

To convert to cfm, first calculate air density at entering conditions.

$$\left(\frac{29}{379}\right) \times \left(\frac{14.227}{14.7}\right) \times \left(\frac{520}{530}\right) = 0.073 \text{ lb per cu ft}$$

Finally, air cfm = 9.17/0.073 = 125 cfm. Note application of humidity ratio.

Example **13.22**

An underground space 15 ft wide by 20 ft long by 10 ft high has a wall surface temperature of 55°F. Outside air conditions are 70°F dbt and 60 percent relative humidity, and the space is naturally ventilated to the outdoors. (a) Will surface condensation take place? (b) If so, what will be the dehumidification load if one air change per hour is assumed? The following data are to be used in the solution: moisture content in outside air is 4.788 grains per cu ft, moisture content to be maintained in conditioned space is 1.223 grains per cu ft. In the second part of the problem, to prevent condensation on the walls and ceilings, space air must be dehumidified to a low dry-bulb temperature of 50°F and a relative humidity of 30 percent.

Solution

Use the graph in Exhibit 4 to determine the possibility of condensation.

(a) Will surface condensation take place? Refer to Exhibit 4; 70°F dbt and 60 percent relative humidity intersect at a ground temperature of 55°F. Because this is surface temperature (dew point), condensation will take place.

(b) The dehumidification load is

$$\frac{\text{Room volume} \times (4.788 - 1.223)}{7000} = \frac{(15 \times 20 \times 10) \times (4.788 - 1.223)}{7000}$$
$$= 1.528 \text{ lb condensate per hour.}$$

Because the weight of 1 pint of water is about 1 lb, the dehumidification load is 1.528 × 24 = 37 pints per 24 hr.

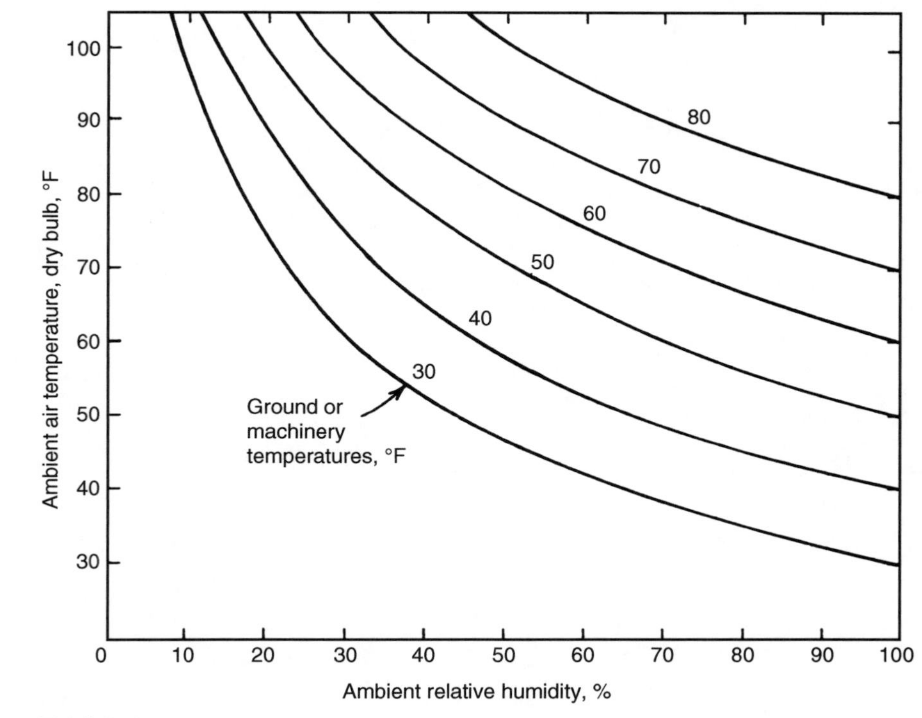
Ambient air temperature, dry bulb, °F
100
90
80
70
60
50
40
30
80
70
60
50
40
30
Ground or machinery temperatures, °F
0
10
20
30
40
50
60
70
80
90
100
Ambient relative humidity, %

Exhibit 4

APPENDIX A

Engineering Economics

Donald G. Newnan

OUTLINE

This is a review of the field known variously as *engineering economics, engineering economy*, or *engineering economic analysis*. Since engineering economics is straightforward and logical, even people who have not had a formal course should be able to gain sufficient knowledge from this chapter to successfully solve most engineering economics problems.

There are 35 example problems scattered throughout the chapter. These examples are an integral part of the review and should be examined as you come to them.

The field of engineering economics uses mathematical and economics techniques to systematically analyze situations which pose alternative courses of action. The initial step in engineering economics problems is to resolve a situation, or each alternative in a given situation, into its favorable and unfavorable consequences or factors. These are then measured in some common unit—usually money. Factors which cannot readily be equated to money are called intangible or irreducible factors. Such factors are considered in conjunction with the monetary analysis when making the final decision on proposed courses of action.

CASH FLOW

A cash flow table shows the "money consequences" of a situation and its timing. For example, a simple problem might be to list the year-by-year consequences of purchasing and owning a used car:

Year	Cash Flow	
Beginning of first year 0	−$4500	Car purchased "now" for $4500 cash. The minus sign indicates a disbursement.
End of year 1	−350	Maintenance costs are $350 per year.
End of year 2	−350	
End of year 3	−350	
End of year 4	−350 +2000	This car is sold at the end of the fourth year for $2000. The plus sign represents the receipt of money.

This same cash flow may be represented graphically, as shown in Fig. A.1. The upward arrow represents a receipt of money, and the downward arrows represent disbursements. The horizontal axis represents the passage of time.

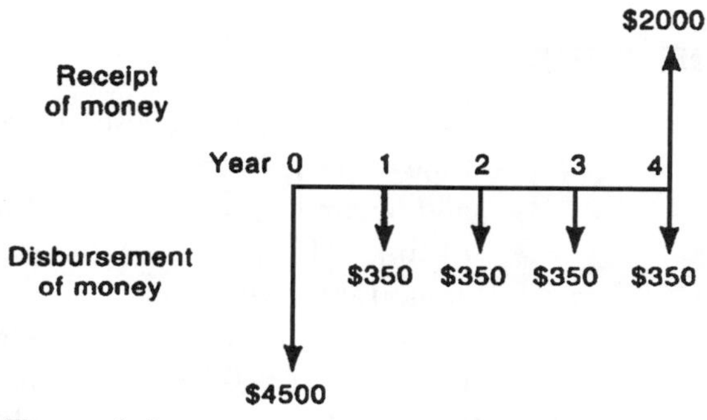

Figure A.1

Example A.1

In January 1993 a firm purchased a used typewriter for $500. Repairs cost nothing in 1993 or 1994. Repairs are $85 in 1995, $130 in 1996, and $140 in 1997. The machine is sold in 1997 for $300. Complete the cash flow table.

Solution

Unless otherwise stated, the customary assumption is a beginning-of-year purchase, followed by end-of-year receipts or disbursements, and an end-of-year resale or salvage value. Thus the typewriter repairs and the typewriter sale are assumed to occur at the end of the year. Letting a minus sign represent a disbursement of money and a plus sign a receipt of money, we are able to set up the cash flow table:

Year	Cash Flow
Beginning of 1993	–$500
End of 1993	0
End of 1994	0
End of 1995	–85
End of 1996	–130
End of 1997	+160

Notice that at the end of 1997 the cash flow table shows +160, which is the net sum of –140 and +300. If we define year 0 as the beginning of 1993, the cash flow table becomes

Year	Cash Flow
0	–$500
1	0
2	0
3	–85
4	–130
5	+160

From this cash flow table, the definitions of year 0 and year 1 become clear. Year 0 is defined as the *beginning* of year 1. Year 1 is the *end* of year 1, and so forth.

TIME VALUE OF MONEY

When the money consequences of an alternative occur in a short period of time—say, less than one year—we might simply add up the various sums of money and obtain the net result. But we cannot treat money this way over longer periods of time. This is because money today does not have the same value as money at some future time.

Consider this question: Which would you prefer, \$100 today or the assurance of receiving \$100 a year from now? Clearly, you would prefer the \$100 today. If you had the money today, rather than a year from now, you could use it for the year. And if you had no use for it, you could lend it to someone who would pay interest for the privilege of using your money for the year.

Simple Interest

Simple interest is interest that is computed on the original sum. Thus if one were to lend a present sum P to someone at a simple annual interest rate i, the future amount F due at the end of n years would be

$$F = P + Pin$$

Example A.2

How much will you receive back from a \$500 loan to a friend for three years at 10 percent simple annual interest?

Solution

$$F = P + Pin = 500 + 500 \times 0.10 \times 3 = \$650$$

In Example A.2 one observes that the amount owed, based on 10 percent simple interest at the end of one year, is $500 + 500 \times 0.10 \times 1 = \$550$. But at simple interest there is no interest charged on the \$50 interest, even though it is not paid until the end of the third year. Thus simple interest is not realistic and is seldom used. *Compound interest* charges interest on the principal owed plus the interest earned to date. This produces a charge of interest on interest, or compound interest. Engineering economics uses compound interest computations.

EQUIVALENCE

In the preceding section we saw that money at different points in time (for example, \$100 today or \$100 one year hence) may be equal in the sense that they both are \$100, but \$100 a year hence is *not* an acceptable substitute for \$100 today. When we have acceptable substitutes, we say they are *equivalent* to each other. Thus at 8 percent interest, \$108 a year hence is equivalent to \$100 today.

Example A.3

At a 10 percent per year (compound) interest rate, $500 now is *equivalent* to how much three years hence?

Solution

A value of $500 now will increase by 10 percent in each of the three years.

Now = $500.00
End of 1st year = 500 + 10%(500) = 550.00
End of 2nd year = 550 + 10%(550) = 605.00
End of 3rd year = 605 + 10%(605) = 665.50

Thus $500 now is *equivalent* to $665.50 at the end of three years. Note that interest is charged each year on the original $500 plus the unpaid interest. This compound interest computation gives an answer that is $15.50 higher than the simple-interest computation in Example A.2.

Equivalence is an essential factor in engineering economics. Suppose we wish to select the better of two alternatives. First, we must compute their cash flows. For example,

	Alternative	
Year	***A***	***B***
0	–$2000	–$2800
1	+800	+1100
2	+800	+1100
3	+800	+1100

The larger investment in alternative *B* results in larger subsequent benefits, but we have no direct way of knowing whether it is better than alternative *A*. So we do not know which to select. To make a decision, we must resolve the alternatives into *equivalent* sums so that they may be compared accurately.

COMPOUND INTEREST

To facilitate equivalence computations, a series of compound interest factors will be derived here, and their use will be illustrated in examples.

Symbols and Functional Notation

i = effective interest rate per interest period. In equations, the interest rate is stated as a decimal (that is, 8 percent interest is 0.08).

n = number of interest periods. Usually the interest period is one year, but it could be something else.

P = a present sum of money.

F = a future sum of money. The future sum F is an amount n interest periods from the present that is equivalent to P at interest rate i.

A = an end-of-period cash receipt or disbursement in a uniform series continuing for n periods. The entire series is equivalent to P or F at interest rate i.

Table A.1 Periodic compounding: Functional notation and formulas

Factor	Given	To Find	Functional Notation	Formula
Single payment				
Compound amount factor	P	F	$(F/P, i\%, n)$	$F = P(1+i)^n$
Present worth factor	F	P	$(P/F, i\%, n)$	$P = F(1+i)^{-n}$
Uniform payment series				
Sinking fund factor	F	A	$(A/F, i\%, n)$	$A = F\left[\frac{i}{(1+i)^n - 1}\right]$
Capital recovery factor	P	A	$(A/P, i\%, n)$	$A = P\left[\frac{i(1+i)^n}{(1+i)^n - 1}\right]$
Compound amount factor	A	F	$(F/A, i\%, n)$	$F = A\left[\frac{(1+i)^n - 1}{i}\right]$
Present worth factor	A	P	$(P/A, i\%, n)$	$P = A\left[\frac{(1+i)^n - 1}{i(1+i)^n}\right]$
Uniform gradient				
Gradient present worth	G	P	$(P/G, i\%, n)$	$P = G\left[\frac{(1+i)^n - 1}{i^2(1+i)^n} - \frac{n}{i(1+i)^n}\right]$
Gradient future worth	G	F	$(F/G, i\%, n)$	$F = G\left[\frac{(1+i)^n - 1}{i^2} - \frac{n}{1}\right]$
Gradient uniform series	G	A	$(A/G, i\%, n)$	$A = G\left[\frac{1}{i} - \frac{n}{(1+i)^n - 1}\right]$

G = uniform period-by-period increase in cash flows; the uniform gradient.

r = nominal annual interest rate.

From Table A.1 we can see that the functional notation scheme is based on writing (to find/given, i, n). Thus, if we wished to find the future sum F, given a uniform series of receipts A, the proper compound interest factor to use would be $(F/A, i, n)$.

Single-Payment Formulas

Suppose a present sum of money P is invested for one year at interest rate i. At the end of the year, the initial investment P is received together with interest equal to Pi, or a total amount $P + Pi$. Factoring P, the sum at the end of one year is $P(1 + i)$. If the investment is allowed to remain for subsequent years, the progression is as follows:

Amount at Beginning of the Period	+	Interest for the Period	=	Amount at End of the Period
1st year, P	+	Pi	=	$P(1+i)$
2nd year, $P(1+i)$	+	$Pi(1+i)$	=	$P(1+i)^2$
3rd year, $P(1+i)^2$	+	$Pi(1+i)^2$	=	$P(1+i)^3$
nth year, $P(1+i)^{n-1}$	+	$Pi(1+i)^{n-1}$	=	$P(1+i)^n$

The present sum P increases in n periods to $P(1 + i)^n$. This gives a relation between a present sum P and its equivalent future sum F:

$$\text{Future sum} = (\text{present sum})(1 + i)^n$$
$$F = P(1 + i)^n$$

This is the *single-payment compound amount formula*. In functional notation it is written

$$F = P(F/P, i, n).$$

The relationship may be rewritten as

$$\text{Present sum} = (\text{Future sum})\ (1 + i)^{-n}$$
$$P = F(1 + i)^{-n}.$$

This is the *single-payment present worth formula*. It is written

$$P = F(P/F, i, n).$$

Example A.4

At a 10 percent per year interest rate, \$500 now is *equivalent* to how much three years hence?

Solution

This problem was solved in Example A.3. Now it can be solved using a single-payment formula. P = \$500, n = 3 years, i = 10 percent, and F = unknown:

$$F = P(1 + i)^n = 500(1 + 0.10)^3 = \$665.50.$$

This problem also may be solved using a compound interest table:

$$F = P(F/P, i, n) = 500(F/P, 10\%, 3).$$

From the 10 percent compound interest table, read $(F/P, 10\%, 3) = 1.331$.

$$F = 500(F/P, 10\%, 3) = 500(1.331) = \$665.50$$

Example A.5

To raise money for a new business, a man asks you to lend him some money. He offers to pay you \$3000 at the end of four years. How much should you give him now if you want 12 percent interest per year?

Solution

P = unknown, F = \$3000, n = 4 years, and i = 12 percent:

$$P = F(1 + i)^{-n} = 3000(1 + 0.12)^{-4} = \$1906.55$$

Alternative computation using a compound interest table:

$$P = F(P/F, i, n) = 3000(P/F, 12\%, 4) = 3000(0.6355) = \$1906.50$$

Note that the solution based on the compound interest table is slightly different from the exact solution using a hand-held calculator. In engineering economics the compound interest tables are always considered to be sufficiently accurate.

Uniform Payment Series Formulas

Consider the situation shown in Fig. A.2. Using the single-payment compound amount factor, we can write an equation for F in terms of A:

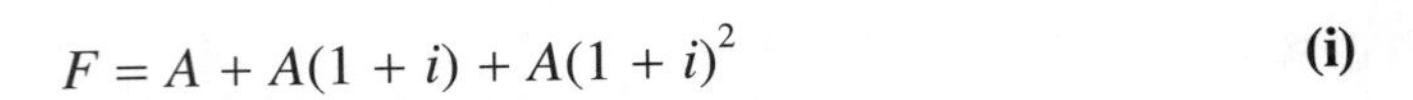

$$F = A + A(1 + i) + A(1 + i)^2 \quad \textbf{(i)}$$

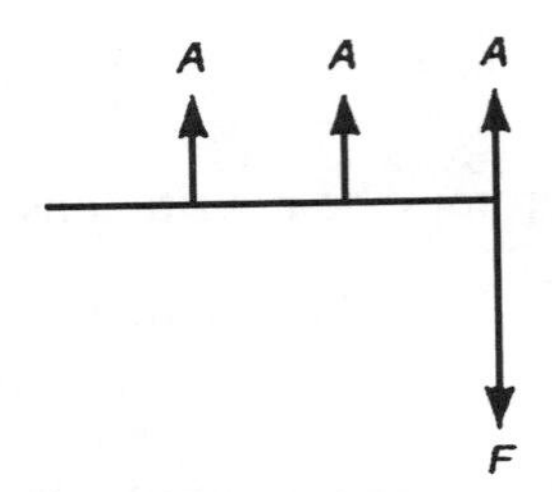

A = End-of-period cash receipt or disbursement in a uniform series continuing for n periods

F = A future sum of money

Figure A.2

In this situation, with $n = 3$, Eq. (i) may be written in a more general form:

$$F = A + A(1 + i) + A(1 + i)^{n-1} \quad \textbf{(ii)}$$

Multiply Eq. (ii) by $(1 + i)$ $\quad (1 + i)F = A(1 + i) + A(1 + i)^{n-1} + A(1 + i)^n \quad \textbf{(iii)}$

Subtract Eq. (ii): $\quad -F = A + A(1 + i) + A(1 + i)^{n-1} \quad \textbf{(ii)}$

$$iF = -A + A(1 + i)^n$$

This produces the *uniform series compound amount formula:*

$$F = A\left(\frac{(1+i)^n - 1}{i}\right)$$

Solving this equation for A produces the *uniform series sinking fund formula:*

$$A = F\left(\frac{i}{(1+i)^n - 1}\right)$$

Since $F = P(1 + i)^n$, we can substitute this expression for F in the equation and obtain the *uniform series capital recovery formula:*

$$A = P\left(\frac{i(1+i)^n}{(1+i)^n - 1}\right)$$

Solving the equation for P produces the *uniform series present worth formula:*

$$P = A\left(\frac{(1+i)^n - 1}{i(1+i)^n}\right)$$

In functional notation, the uniform series factors are

Compound amount $(F/A, i, n)$

Sinking fund $(A/F, i, n)$

Capital recovery $(A/P, i, n)$

Present worth $(P/A, i, n)$

Example **A.6**

If \$100 is deposited at the end of each year in a savings account that pays 6 percent interest per year, how much will be in the account at the end of five years?

Solution

$A = \$100$, $F =$ unknown, $n = 5$ years, and $i = 6$ percent:

$$F = A(F/A, i, n) = 100(F/A, 6\%, 5) = 100(5.637) = \$563.70$$

Example **A.7**

A fund established to produce a desired amount at the end of a given period, by means of a series of payments throughout the period, is called a *sinking fund.* A sinking fund is to be established to accumulate money to replace a \$10,000 machine. If the machine is to be replaced at the end of 12 years, how much should be deposited in the sinking fund each year? Assume the fund earns 10 percent annual interest.

Solution

$$\text{Annual sinking fund deposit } A = 10{,}000(A/F, 10\%, 12)$$
$$= 10{,}000(0.0468) = \$468$$

Example **A.8**

An individual is considering the purchase of a used automobile. The total price is \$6200. With \$1240 as a down payment, and the balance paid in 48 equal monthly payments with interest at 1 percent per month, compute the monthly payment. The payments are due at the end of each month.

Solution

The amount to be repaid by the 48 monthly payments is the cost of the automobile *minus* the \$1240 downpayment.

$P = \$4960$, $A =$ unknown, $n = 48$ monthly payments, and $i = 1$ percent per month:

$$A = P(A/P, 1\%, 48) = 4960(0.0263) = \$130.45$$

Example **A.9**

A couple sell their home. In addition to cash, they take a mortgage on the house. The mortgage will be paid off by monthly payments of \$450 for 50 months. The couple decides to sell the mortgage to a local bank. The bank will buy the mortgage, but it requires a 1 percent per month interest rate on its investment. How much will the bank pay for the mortgage?

Solution

$A = \$450$, $n = 50$ months, $i = 1$ percent per month, and $P =$ unknown:

$$P = A(P/A, i, n) = 450(P/A, 1\%, 50) = 450(39.196) = \$17{,}638.20$$

Uniform Gradient

At times one will encounter a situation where the cash flow series is not a constant amount A. Instead, it is an increasing series. The cash flow shown in Fig. A.3

Figure A.3

may be resolved into two components (Fig. A.4). We can compute the value of P^* as equal to P' plus P. And we already have the equation for P': $P' = A(P/A, i, n)$. The value for P in the right-hand diagram is

$$P = G\left[\frac{(1+i)^n - 1}{i^2(1+i)^n} - \frac{n}{i(1+i)^n}\right].$$

This is the *uniform gradient present worth formula.* In functional notation, the relationship is $P = G(P/G, i, n)$.

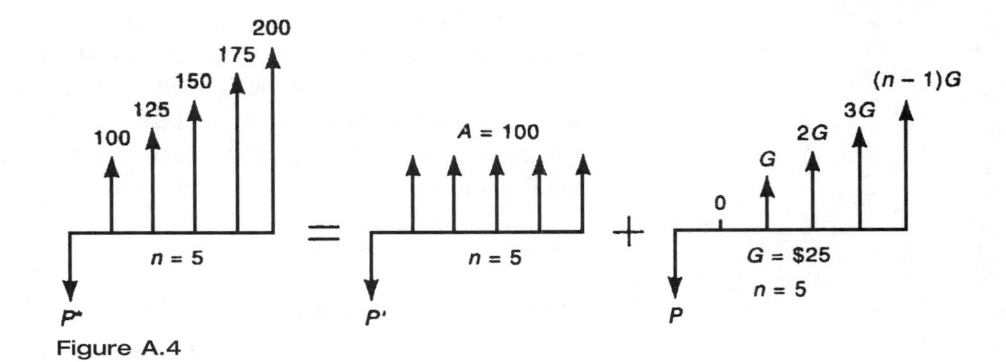

Figure A.4

Example A.10

The maintenance on a machine is expected to be $155 at the end of the first year, and it is expected to increase $35 each year for the following seven years (Exhibit 1). What sum of money should be set aside now to pay the maintenance for the eight-year period? Assume 6 percent interest.

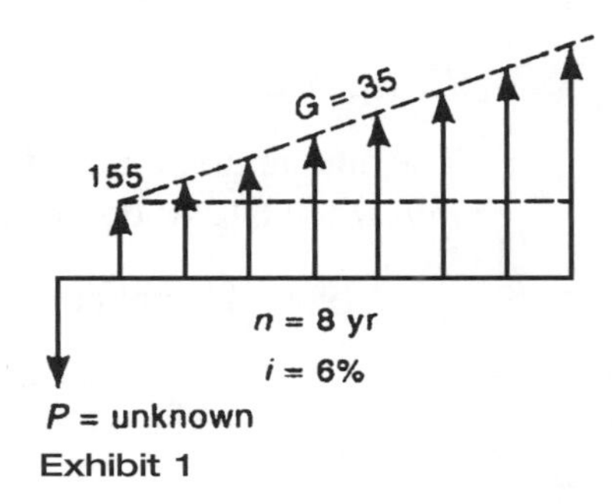

Exhibit 1

Solution

$$P = 155(P/A, 6\%, 8) + 35(P/G, 6\%, 8)$$
$$= 155(6.210) + 35(19.841) = \$1656.99$$

In the gradient series, if—instead of the present sum, P—an equivalent uniform series A is desired, the problem might appear as shown in Fig. A.5. The relationship between A' and G in the right-hand diagram is

$$A' = G\left[\frac{1}{i} - \frac{n}{(1+i)^n - 1)}\right]$$

In functional notation, the uniform gradient (to) uniform series factor is: $A' = G(A/G, i, n)$.

The uniform gradient uniform series factor may be read from the compound interest tables directly, or computed as

$$(A/G, i, n) = \frac{1 - n(A/F, i, n)}{i}.$$

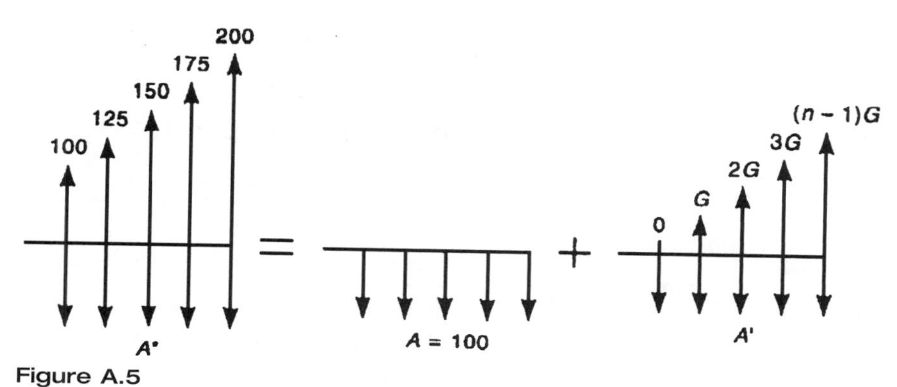

Figure A.5

Note carefully the diagrams for the uniform gradient factors. The first term in the uniform gradient is zero and the last term is $(n - 1)G$. But we use n in the equations and function notation. The derivations (not shown here) were done on this basis, and the uniform gradient compound interest tables are computed this way.

Example **A.11**

For the situation in Example A.10, we wish now to know the uniform annual maintenance cost. Compute an equivalent A for the maintenance costs.

Solution

Refer to Exhibit 2. The equivalent uniform annual maintenance cost is

$$A = 155 + 35(A/G, 6\%, 8) = 155 + 35(3.195) = \$266.83.$$

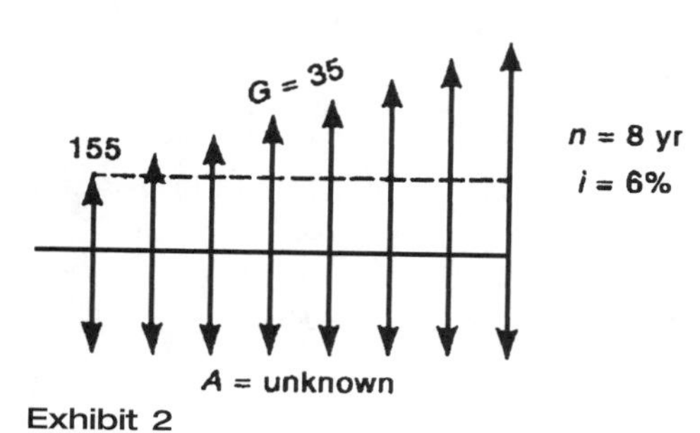

Exhibit 2

Standard compound interest tables give values for eight interest factors: two single payments, four uniform payment series, and two uniform gradients. The tables do *not* give the uniform gradient future worth factor, $(F/G, i, n)$. If it is needed, it may be computed from two tabulated factors:

$$(F/G, i, n) = (P/G, i, n)(F/P, i, n)$$

For example, if $i = 10$ percent and $n = 12$ years, then $(F/G, 10\%, 12) = (P/G, 10\%, 12)\ (F/P, 10\%, 12) = (29.901)(3.138) = 93.83$.

A second method of computing the uniform gradient future worth factor is

$$(F/G, i, n) = \frac{(F/A, i, n) - n}{i}$$

Using this equation for $i = 10$ percent and $n = 12$ years, $(F/G, 10\%, 12) = [(F/A, 10\%, 12) - 12]/0.10 = (21.384 - 12)/0.10 = 93.84$.

Continuous Compounding

Table A.2 Continuous compounding: Functional notation and formulas

Factor	Given	To Find	Functional Notation	Formula
Single payment				
Compound amount factor	P	F	$(F/P, r\%, n)$	$F = P[e^{rn}]$
Present worth factor	F	P	$(P/F, r\%, n)$	$P = F[e^{-rn}]$
Uniform payment series				
Sinking fund factor	F	A	$(A/F, r\%, n)$	$A = F\left[\frac{e^r - 1}{e^{rn} - 1}\right]$
Capital recovery factor	P	A	$(A/P, r\%, n)$	$A = P\left[\frac{e^r - 1}{1 - e^{-rn}}\right]$
Compound amount factor	A	F	$(F/A, r\%, n)$	$F = A\left[\frac{e^{rn} - 1}{e^r - 1}\right]$
Present worth factor	A	P	$(P/A, r\%, n)$	$P = A\left[\frac{1 - e^{-rn}}{e^r - 1}\right]$

r = nominal annual interest rate, n = number of years.

Example A.12

Five hundred dollars is deposited each year into a savings bank account that pays 5 percent nominal interest, compounded continuously. How much will be in the account at the end of five years?

Solution

$A = \$500$, $r = 0.05$, $n = 5$ years.

$$F = A(F/A, r\%, n) = A\left[\frac{e^{rn} - 1}{e^r - 1}\right] = 500\left[\frac{e^{0.05(5)} - 1}{e^{0.05} - 1}\right] = \$2769.84$$

NOMINAL AND EFFECTIVE INTEREST

Nominal interest is the annual interest rate without considering the effect of any compounding. *Effective interest* is the annual interest rate taking into account the effect of any compounding during the year.

Non-Annual Compounding

Frequently an interest rate is described as an annual rate, even though the interest period may be something other than one year. A bank may pay 1 percent interest on the amount in a savings account every three months. The *nominal* interest rate in this situation is $4 \times 1\% = 4\%$. But if you deposited $1000 in such an account, would you have 104%(1000) = $1040 in the account at the end of one year? The answer is no, you would have more. The amount in the account would increase as follows:

Amount in Account

Beginning of year:	1000.00
End of three months:	1000.00 + 1%(1000.00) = 1010.00
End of six months:	1010.00 + 1%(1010.00) = 1020.10
End of nine months:	1020.10 + 1%(1020.10) = 1030.30
End of one year:	1030.30 + 1%(1030.30) = 1040.60

At the end of one year, the interest of $40.60, divided by the original $1000, gives a rate of 4.06 percent. This is the *effective* interest rate.

$$\text{Effective interest rate per year:} \quad i_{\text{eff}} = (1 + r/m)^m - 1$$

where r = nominal annual interest rate
m = number of compound periods per year
r/m = effective interest rate per period.

Example **A.13**

A bank charges 1.5 percent interest per month on the unpaid balance for purchases made on its credit card. What nominal interest rate is it charging? What is the effective interest rate?

Solution

The nominal interest rate is simply the annual interest ignoring compounding, or $12(1.5\%) = 18\%$.

$$\text{Effective interest rate} = (1 + 0.015)^{12} - 1 = 0.1956 = 19.56\%$$

Continuous Compounding

When m, the number of compound periods per year, becomes very large and approaches infinity, the duration of the interest period decreases from Δt to dt. For this condition of *continuous compounding*, the effective interest rate per year is

$$i_{\text{eff}} = e^r - 1$$

where r = nominal annual interest rate.

Example **A.14**

If the bank in Example A.13 changes its policy and charges 1.5 percent per month, compounded continuously, what nominal and what effective interest rate is it charging?

Solution

Nominal annual interest rate, $r = 12 \times 1.5\% = 18\%$

$$\text{Effective interest rate per year, } i_{\text{eff}} = e^{0.18} - 1 = 0.1972 = 19.72\%$$

SOLVING ENGINEERING ECONOMICS PROBLEMS

The techniques presented so far illustrate how to convert single amounts of money, and uniform or gradient series of money, into some equivalent sum at another point in time. These compound interest computations are an essential part of engineering economics problems.

The typical situation is that we have a number of alternatives; the question is, which alternative should we select? The customary method of solution is to express each alternative in some common form and then choose the best, taking both the monetary and intangible factors into account. In most computations an interest rate must be used. It is often called the minimum attractive rate of return (MARR), to indicate that this is the smallest interest rate, or rate of return, at which one is willing to invest money.

Criteria

Engineering economics problems inevitably fall into one of three categories:

1. *Fixed input*. The amount of money or other input resources is fixed. *Example*: A project engineer has a budget of $450,000 to overhaul a plant.
2. *Fixed output*. There is a fixed task or other output to be accomplished. *Example*: A mechanical contractor has been awarded a fixed-price contract to air-condition a building.
3. *Neither input nor output fixed*. This is the general situation, where neither the amount of money (or other inputs) nor the amount of benefits (or other outputs) is fixed. *Example*: A consulting engineering firm has more work available than it can handle. It is considering paying the staff to work evenings to increase the amount of design work it can perform.

There are five major methods of comparing alternatives: present worth, future worth, annual cost, rate of return, and benefit-cost analysis. These are presented in the sections that follow.

PRESENT WORTH

Present worth analysis converts all of the money consequences of an alternative into an equivalent present sum. The criteria are

Category	Present Worth Criterion
Fixed input	Maximize the present worth of benefits or other outputs
Fixed output	Minimize the present worth of costs or other inputs
Neither input nor output fixed	Maximize present worth of benefits minus present worth of costs, or maximize net present worth

Appropriate Problems

Present worth analysis is most frequently used to determine the present value of future money receipts and disbursements. We might want to know, for example, the present worth of an income-producing property, such as an oil well. This should provide an estimate of the price at which the property could be bought or sold.

An important restriction in the use of present worth calculation is that there must be a common analysis period for comparing alternatives. It would be incorrect, for example, to compare the present worth (PW) of cost of pump *A*, expected to last 6 years, with the PW of cost of pump *B*, expected to last 12 years (Fig. A.6). In situations like this, the solution is either to use some other analysis technique (generally, the annual cost method is suitable in these situations) or to restructure the problem so that there is a common analysis period.

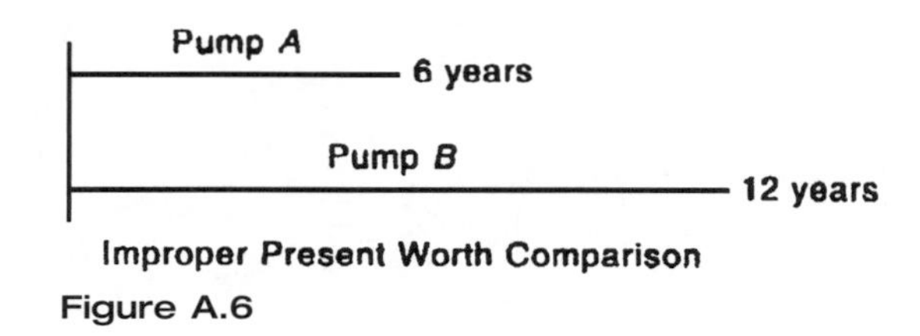

Improper Present Worth Comparison

Figure A.6

In this example, a customary assumption would be that a pump is needed for 12 years and that pump *A* will be replaced by an identical pump *A* at the end of 6 years. This gives a 12-year common analysis period (Fig. A.7). This approach is easy to use when the different lives of the alternatives have a practical least-common-multiple life. When this is not true (for example, the life of *J* equals 7 years and the life of *K* equals 11 years), some assumptions must be made to select a suitable common analysis period, or the present worth method should not be used.

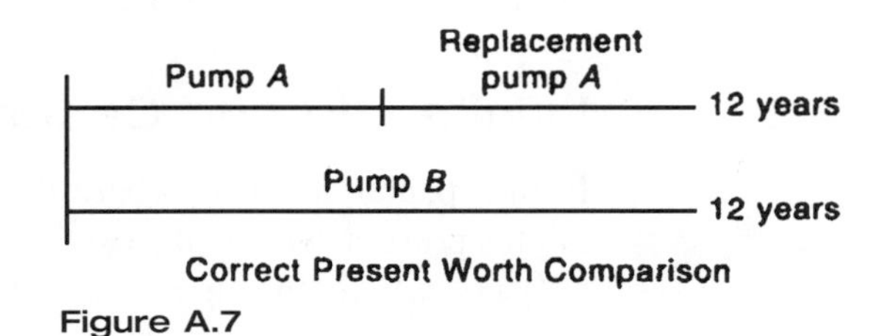

Correct Present Worth Comparison

Figure A.7

Example A.15

Machine *X* has an initial cost of \$10,000, an annual maintenance cost of \$500 per year, and no salvage value at the end of its 4-year useful life. Machine *Y* costs \$20,000, and the first year there is no maintenance cost. Maintenance is \$100 the second year, and it increases \$100 per year thereafter. The machine has an anticipated \$5000 salvage value at the end of its 12-year useful life. If the minimum attractive rate of return (MARR) is 8 percent, which machine should be selected?

Solution

The analysis period is not stated in the problem. Therefore, we select the least common multiple of the lives, or 12 years, as the analysis period.

Present worth of cost of 12 years of machine X:

$$\text{PW} = 10{,}000 + 10{,}000(P/F,\ 8\%,\ 4) + 10{,}000(P/F,\ 8\%,\ 8) + 500(P/A,\ 8\%,\ 12)$$
$$= 10{,}000 + 10{,}000(0.7350) + 10{,}000(0.5403) + 500(7.536) = \$26{,}521$$

Present worth of cost of 12 years of machine Y:

$$\text{PW} = 20{,}000 + 100(P/G,\ 8\%,\ 12) - 5000(P/F,\ 8\%,\ 12)$$
$$= 20{,}000 + 100(34.634) - 5000(0.3971) = \$21{,}478$$

Choose machine Y, with its smaller PW of cost.

Example A.16

Two alternatives have the following cash flows:

	Alternative	
Year	***A***	***B***
0	−\$2000	−\$2800
1	+800	+1100
2	+800	+1100
3	+800	+1100

At a 4 percent interest rate, which alternative should be selected?

Solution

The net present worth of each alternative is computed:

$$\text{Net present worth (NPW)} = \text{PW of benefit} - \text{PW of cost}$$
$$\text{NPW}_A = 800(P/A,\ 4\%,\ 3) - 2000 = 800(2.775) - 2000 = \$220.00$$
$$\text{NPW}_B = 1100(P/A,\ 4\%,\ 3) - 2800 = 1100(2.775) - 2800 = \$252.50$$

To maximize NPW, choose alternative B.

Infinite Life and Capitalized Cost

In the special situation where the analysis period is infinite ($n = \infty$), an analysis of the present worth of cost is called *capitalized cost.* There are a few public projects where the analysis period is infinity. Other examples are permanent endowments and cemetery perpetual care.

When n equals infinity, a present sum P will accrue interest of Pi for every future interest period. For the principal sum P to continue undiminished (an essential requirement for n equal to infinity), the end-of-period sum A that can be disbursed is Pi (Fig. A.8). When $n = \infty$, the fundamental relationship is

$$A = Pi.$$

Some form of this equation is used whenever there is a problem involving an infinite analysis period.

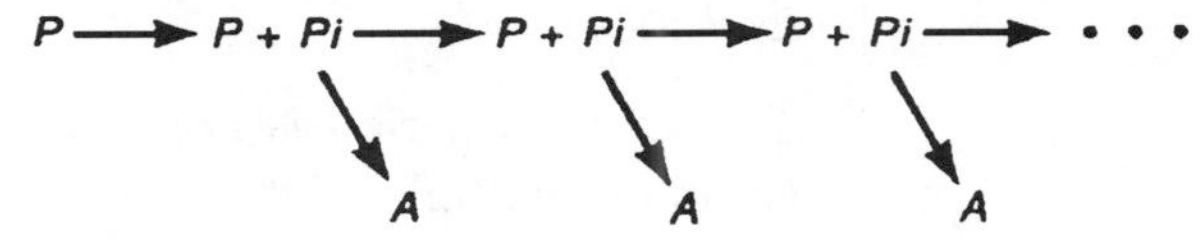

Figure A.8

Example A.17

In his will, a man wishes to establish a perpetual trust to provide for the maintenance of a small local park. If the annual maintenance is \$7500 per year and the trust account can earn 5 percent interest, how much money must be set aside in the trust?

Solution

When $n = \infty$, $A = Pi$ or $P = A/i$. The capitalized cost is $P = A/i = \$7500/0.05 = \$150{,}000$.

FUTURE WORTH OR VALUE

In present worth analysis, the comparison is made in terms of the equivalent present costs and benefits. But the analysis need not be made in terms of the present—it can be made in terms of a past, present, or future time. Although the numerical calculations may look different, the decision is unaffected by the selected point in time. Often we do want to know what the future situation will be if we take some particular couse of action now. An analysis based on some future point in time is called *future worth analysis.*

Category	Future Worth Criterion
Fixed input	Maximize the future worth of benefits or other outputs
Fixed output	Minimize the future worth of costs or other inputs
Neither input nor output fixed	Maximize future worth of benefits minus future worth of costs, or maximize net future worth

Example A.18

Two alternatives have the following cash flows:

Year	Alternative A	Alternative B
0	–\$2000	–\$2800
1	+800	+1100
2	+800	+1100
3	+800	+1100

At a 4 percent interest rate, which alternative should be selected?

Solution

In Example A.16, this problem was solved by present worth analysis at year 0. Here it will be solved by future worth analysis at the end of year 3.
Net future worth (NFW) = FW of benefits – FW of cost

$$\begin{aligned} \text{NFW}_A &= 800(F/A, 4\%, 3) - 2000(F/P, 4\%, 3) \\ &= 800(3.122) - 2000(1.125) = +\$247.60 \end{aligned}$$

$$\begin{aligned} \text{NFW}_B &= 1100(F/A, 4\%, 3) - 2800(F/P, 4\%, 3) \\ &= 1100(3.122) - 2800(1.125) = +\$284.20 \end{aligned}$$

To maximize NFW, choose alternative *B*.

ANNUAL COST

The annual cost method is more accurately described as the method of equivalent uniform annual cost (EUAC). Where the computation is of benefits, it is called the method of equivalent uniform annual benefits (EUAB).

Criteria

For each of the three possible categories of problems, there is an annual cost criterion for economic efficiency.

Category	Annual Cost Criterion
Fixed input	Maximize the equivalent uniform annual benefits (EUAB)
Fixed output	Minimize the equivalent uniform annual cost (EUAC)
Neither input nor output fixed	Maximize EUAB – EUAC

Application of Annual Cost Analysis

In the section on present worth, we pointed out that the present worth method requires a common analysis period for all alternatives. This restriction does not apply in all annual cost calculations, but it is important to understand the circumstances that justify comparing alternatives with different service lives.

Frequently, an analysis is done to provide for a more-or-less continuing requirement. For example, one might need to pump water from a well on a continuing basis. Regardless of whether each of two pumps has a useful service life of 6 years or 12 years, we would select the alternative whose annual cost is a minimum. And this still would be the case if the pumps' useful lives were the more troublesome 7 and 11 years. Thus, if we can assume a continuing need for an item, an annual cost comparison among alternatives of differing service lives is valid. The underlying assumption in these situations is that the shorter-lived alternative can be replaced with an identical item with identical costs, when it has reached the end of its useful life. This means that the EUAC of the initial alternative is equal to the EUAC for the continuing series of replacements.

On the other hand, if there is a specific requirement to pump water for 10 years, then each pump must be evaluated to see what costs will be incurred during the analysis period and what salvage value, if any, may be recovered at the end of the analysis period. The annual cost comparison needs to consider the actual circumstances of the situation.

Examination problems are often readily solved using the annual cost method. And the underlying "continuing requirement" is usually present, so an annual cost comparison of unequal-lived alternatives is an appropriate method of analysis.

Example A.19

Consider the following alternatives:

	A	*B*
First cost	$5000	$10,000
Annual maintenance	500	200
End-of-useful-life salvage value	600	1000
Useful life	5 years	15 years

Based on an 8 percent interest rate, which alternative should be selected?

Solution

Assuming both alternatives perform the same task and there is a continuing requirement, the goal is to minimize EUAC.

Alternative *A:*

$$\begin{aligned} \text{EUAC} &= 5000(A/P, 8\%, 5) + 500 - 600(A/F, 8\%, 5) \\ &= 5000(0.2505) + 500 - 600(0.1705) = \$1650 \end{aligned}$$

Alternative *B:*

$$\begin{aligned} \text{EUAC} &= 10{,}000(A/P, 8\%, 15) + 200 - 1000(A/F, 8\%, 15) \\ &= 10{,}000(0.1168) + 200 - 1000(0.0368) = \$1331 \end{aligned}$$

To minimize EUAC, select alternative *B*.

RATE OF RETURN ANALYSIS

A typical situation is a cash flow representing the costs and benefits. The rate of return may be defined as the interest rate where PW of cost = PW of benefits, EUAC = EUAB, or PW of cost – PW of benefits = 0.

Example A.20

Compute the rate of return for the investment represented by the following cash flow table.

Year:	0	1	2	3	4	5
Cash flow:	–$595	+250	+200	+150	+100	+50

Solution

This declining uniform gradient series may be separated into two cash flows (Exhibit 3) for which compound interest factors are available.

Note that the gradient series factors are based on an *increasing* gradient. Here the declining cash flow is solved by subtracting an increasing uniform gradient, as indicated in the figure.

PW of cost – PW of benefits = 0

$$595 - [250(P/A, i, 5) - 50(P/G, i, 5) = 0$$

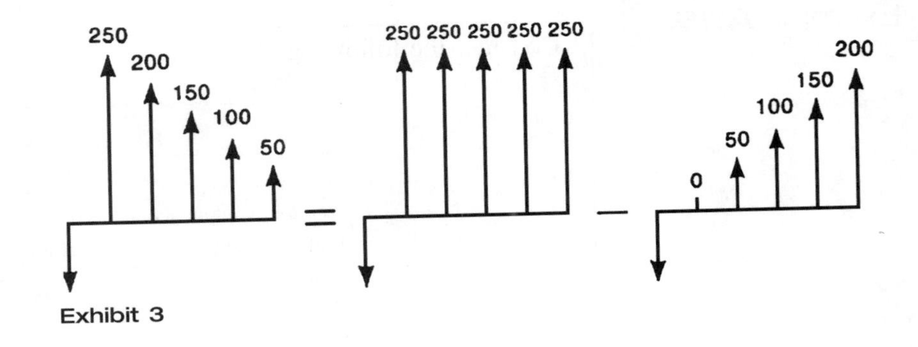

Exhibit 3

Try $i = 10\%$:

$$595 - [250(3.791) - 50(6.862)] = -9.65$$

Try $i = 12\%$:

$$595 - [250(3.605) - 50(6.397)] = +13.60$$

The rate of return is between 10 percent and 12 percent. It may be computed more accurately by linear interpolation:

$$\text{Rate of return} = 10\% + (2\%)\left(\frac{9.65 - 0}{13.60 + 9.65}\right) = 10.83\%.$$

Two Alternatives

Compute the incremental rate of return on the cash flow representing the difference between the two alternatives. Since we want to look at increments of *investment*, the cash flow for the difference between the alternatives is computed by taking the higher initial-cost alternative minus the lower initial-cost alternative. If the incremental rate of return is greater than or equal to the predetermined minimum attractive rate of return (MARR), choose the higher-cost alternative; otherwise, choose the lower-cost alternative.

Example **A.21**

Two alternatives have the following cash flows:

	Alternative	
Year	***A***	***B***
0	–\$2000	–\$2800
1	+800	+1100
2	+800	+1100
3	+800	+1100

If 4 percent is considered the minimum attractive rate of return (MARR), which alternative should be selected?

Solution

These two alternatives were previously examined in Examples A.16 and A.18 by present worth and future worth analysis. This time, the alternatives will be resolved using a rate-of-return analysis.

Note that the problem statement specifies a 4 percent MARR, whereas Examples A.16 and A.18 referred to a 4 percent interest rate. These are really two different ways of saying the same thing: The minimum acceptable time value of money is 4 percent.

First, tabulate the cash flow that represents the increment of investment between the alternatives. This is done by taking the higher initial-cost alternative minus the lower initial-cost alternative:

	Alternative		**Difference Between Alternatives**
Year	***A***	***B***	***B* – *A***
0	–$2000	–$2800	–$800
1	+800	+1100	+300
2	+800	+1100	+300
3	+800	+1100	+300

Then compute the rate of return on the increment of investment represented by the difference between the alternatives:

$$\text{PW of cost} = \text{PW of benefits}$$
$$800 = 300(P/A, i, 3)$$
$$(P/A, i, 3) = 800/300 = 2.67$$
$$i = 6.1\%$$

Since the incremental rate of return exceeds the 4 percent MARR, the increment of investment is desirable. Choose the higher-cost alternative *B*.

Before leaving this example, one should note something that relates to the rates of return on alternative *A* and on alternative *B*. These rates of return, if calculated, are

	Rate of Return
Alternative *A*	9.7%
Alternative *B*	8.7%

The correct answer to this problem has been shown to be alternative *B*, even though alternative *A* has a higher rate of return. The higher-cost alternative may be thought of as the lower-cost alternative plus the increment of investment between them. Viewed this way, the higher-cost alternative *B* is equal to the desirable lower-cost alternative *A* plus the difference between the alternatives.

The important conclusion is that computing the rate of return for each alternative does *not* provide the basis for choosing between alternatives. Instead, incremental analysis is required.

Example A.22

Consider the following:

	Alternative	
Year	*A*	*B*
0	–$200.0	–$131.0
1	+77.6	+48.1
2	+77.6	+48.1
3	+77.6	+48.1

If the MARR is 10 percent, which alternative should be selected?

Solution

To examine the increment of investment between the alternatives, we will examine the higher initial-cost alternative minus the lower initial-cost alternative, or $A - B$.

	Alternative		Increment
Year	*A*	*B*	$A - B$
0	–$200.0	–$131.0	–$69.0
1	+77.6	+48.1	+29.5
2	+77.6	+48.1	+29.5
3	+77.6	+48.1	+29.5

Solve for the incremental rate of return:

$$\text{PW of cost} = \text{PW of benefits}$$
$$69.0 = 29.5(P/A, i, 3)$$
$$(P/A, i, 3) = 69.0/29.5 = 2.339$$

From compound interest tables, the incremental rate of return is between 12 percent and 18 percent. This is a desirable increment of investment; hence we select the higher-initial-cost alternative *A*.

Three or More Alternatives

When there are three or more mutually exclusive alternatives, proceed with the same logic presented for two alternatives. The components of incremental analysis are listed below.

Step 1. Compute the rate of return for each alternative. Reject any alternative where the rate of return is less than the desired MARR. (This step is not essential, but helps to immediately identify unacceptable alternatives.)

Step 2. Rank the remaining alternatives in order of increasing initial cost.

Step 3. Examine the increment of investment between the two lowest-cost alternatives as described for the two-alternative problem. Select the better of the two alternatives and reject the other one.

Step 4. Take the preferred alternative from step 3. Consider the next higher initial-cost alternative and proceed with another two-alternative comparison.

Step 5. Continue until all alternatives have been examined and the best of the multiple alternatives has been identified.

Example A.23

Consider the following:

	Alternative	
Year	*A*	*B*
0	−$200.0	−$131.0
1	+77.6	+48.1
2	+77.6	+48.1
3	+77.6	+48.1

If the MARR is 10 percent, which alternative, if any, should be selected?

Solution

One should carefully note that this is a *three-alternative* problem, where the alternatives are *A*, *B*, and *Do nothing*. In this solution we will skip step 1. Reorganize the problem by placing the alternatives in order of increasing initial cost:

	Alternative		
Year	*Do Nothing*	*B*	*A*
0	0	−$131.0	−$200.0
1	0	+48.1	+77.6
2	0	+48.1	+77.6
3	0	+48.1	+77.6

Examine the *B* – *Do nothing* increment of investment:

Year	*B* 2 *Do Nothing*
0	−$131.0 − 0 = −$131.0
1	+48.1 − 0 = +48.1
2	+48.1 − 0 = +48.1
3	+48.1 − 0 = +48.1

Solve for the incremental rate of return:

$$\text{PW of cost} = \text{PW of benefits}$$
$$131.0 = 48.1(P/A, i, 3)$$
$$(P/A, i, 3) = 131.0/48.1 = 2.723$$

From compound interest tables, the incremental rate of return is about 5 percent. Since the incremental rate of return is less than 10 percent, the *B* – *Do nothing* increment is not desirable. Reject alternative *B*.

Year	*A* – *Do Nothing*
0	−$200.0 − 0 = −$200.0
1	+77.6 − 0 = +77.6
2	+77.6 − 0 = +77.6
3	+77.6 − 0 = +77.6

Next, consider the increment of investment between the two remaining alternatives. Solve for the incremental rate of return:

$$\text{PW of cost} = \text{PW of benefits}$$
$$200.0 = 77.6(P/A, i, 3)$$
$$(P/A, i, 3) = 200.0/77.6 = 2.577$$

The incremental rate of return is 8 percent, less than the desired 10 percent. Reject the increment and select the remaining alternative: *Do nothing*.

If you have not already done so, you should go back to Example A.22 and see how the slightly changed wording of the problem has radically altered it. Example A.22 required a choice between two undesirable alternatives. This example adds the *Do nothing* alternative, which is superior to *A* and *B*.

BENEFIT-COST ANALYSIS

Generally, in public works and governmental economic analyses, the dominant method of analysis is the *benefit-cost ratio*. It is simply the ratio of benefits divided by costs, taking into account the time value of money.

$$B/C = \frac{\text{PW of benefits}}{\text{PW of cost}} = \frac{\text{Equivalent uniform annual benefits}}{\text{Equivalent uniform annual cost}}$$

For a given interest rate, a B/C ratio ≥ 1 reflects an acceptable project. The B/C analysis method is parallel to rate-of-return analysis. The same kind of incremental analysis is required.

Example A.24

Solve Example A.22 by benefit-cost analysis.

Solution

	Alternative		Increment
Year	***A***	***B***	***A* − *B***
0	−$200.0	−$131.0	−$69.0
1	+77.6	+48.1	+29.5
2	+77.6	+48.1	+29.5
3	+77.6	+48.1	+29.5

The benefit-cost ratio for the *A* − *B* increment is

$$B/C = \frac{\text{PW of benefits}}{\text{PW of cost}} = \frac{29.5(P/A, 10\%, 3)}{69.0} = \frac{73.37}{69.0} = 1.06.$$

Since the B/C ratio exceeds 1, the increment of investment is desirable. Select the higher-cost alternative *A*.

BREAKEVEN ANALYSIS

In business, "breakeven" is defined as the point where income just covers costs. In engineering economics, the breakeven point is defined as the point where two alternatives are equivalent.

Example A.25

A city is considering a new \$50,000 snowplow. The new machine will operate at a savings of \$600 per day compared with the present equipment. Assume that the MARR is 12 percent, and the machine's life is 10 years with zero resale value at that time. How many days per year must the machine be used to justify the investment?

Solution

This breakeven problem may be readily solved by annual cost computations. We will set the equivalent uniform annual cost (EUAC) of the snowplow equal to its annual benefit and solve for the required annual utilization. Let X = breakeven point = days of operation per year.

$$\text{EUAC} = \text{EUAB}$$
$$50{,}000(A/P, 12\%, 10) = 600X$$
$$X = 50{,}000(0.1770)/600 = 14.8 \text{ days/year}$$

OPTIMIZATION

Optimization is the determination of the best or most favorable situation.

Minima-Maxima

In problems where the situation can be represented by a function, the customary approach is to set the first derivative of the function to zero and solve for the root(s) of this equation. If the second derivative is *positive*, the function is a minimum for the critical value; if it is *negative*, the function is a maximum.

Example A.26

A consulting engineering firm estimates that their net profit is given by the equation

$$P(x) = -0.03x^3 + 36x + 500 \quad x \geq 0$$

where x = number of employees and $P(x)$ = net profit. What is the optimal number of employees?

Solution

$$P'(x) = -0.09x^2 + 36 = 0 \quad P''(x) = -0.18x$$
$$x^2 = 36/0.09 = 400$$
$$x = 20 \text{ employees.}$$
$$P''(20) = -0.18(20) = -3.6$$

Since $P''(20) < 0$, the net profit is maximized for 20 employees.

Economic Problem—Best Alternative

Since engineering economics problems seek to identify the best or most favorable situation, they are by definition optimization problems. Most use compound interest computations in their solution, but some do not. Consider the following example.

Example A.27

A firm must decide which of three alternatives to adopt to expand its capacity. It wants a minimum annual profit of 20 percent of the initial cost of each increment of investment. Any money not invested in capacity expansion can be invested elsewhere for an annual yield of 20 percent of the initial cost.

Alternative	Initial Cost	Annual Profit	Profit Rate
A	$100,000	$30,000	30%
B	300,000	66,00	22
C	500,000	80,000	16

Which alternative should be selected?

Solution

Since alternative *C* fails to produce the 20 percent minimum annual profit, it is rejected. To decide between alternatives *A* and *B*, examine the profit rate for the $B - A$ increment.

Alternative	Initial Cost	Annual Profit	Incremental Cost	Incremental Profit	Incremental Profit Rate
A	$100,000	$30,000			
			$200,000	$36,000	18%
B	300,000	66,000			

The $B - A$ incremental profit rate is less than the minimum 20 percent, so alternative *B* should be rejected. Thus the best investment of $300,000, for example, would be alternative *A* (annual profit = $30,000) plus $200,000 invested elsewhere at 20 percent (annual profit = $40,000). This combination would yield a $70,000 annual profit, which is better than the alternative *B* profit of $66,000. Select *A*.

Economic Order Quantity

One special case of optimization occurs when an item is used continuously and is periodically purchased. Thus the inventory of the item fluctuates from zero (just prior to the receipt of the purchased quantity) to the purchased quantity (just after receipt). The simplest model for the economic order quantity (EOQ) is

$$\text{EOQ} = \sqrt{\frac{2BD}{E}}$$

where

B = ordering cost, \$/order
D = demand per period, units
E = inventory holding cost, \$/unit/period
EOC = economic order quantity, units

Example A.28

A company uses 8000 wheels per year in its manufacture of golf carts. The wheels cost \$15 each and are purchased from an outside supplier. The money invested in the inventory costs 10 percent per year, and the warehousing cost amounts to an additional 2 percent per year. It costs \$150 to process each purchase order. When an order is placed, how many wheels should be ordered?

Solution

$$\text{EOQ} = \sqrt{\frac{2 \times \$150 \times 8000}{(10\% + 2\%)(15.00)}} = 1155 \text{ wheels}$$

VALUATION AND DEPRECIATION

Depreciation of capital equipment is an important component of many after-tax economic analyses. For this reason, one must understand the fundamentals of depreciation accounting.

Notation

BV = book value
C = cost of the property (basis)
D_j = depreciation in year j
S_n = salvage value in year n

Depreciation is the systematic allocation of the cost of a capital asset over its useful life. *Book value* is the original cost of an asset, minus the accumulated depreciation of the asset.

$$\text{BV} = C - \Sigma(D_j)$$

In computing a schedule of depreciation charges, four items are considered.

1. Cost of the property, C (called the *basis* in tax law).
2. Type of property. Property is classified as either *tangible* (such as machinery) or *intangible* (such as a franchise or a copyright), and as either *real property* (real estate) or *personal property* (everything that is not real property).
3. Depreciable life in years, n.
4. Salvage value of the property at the end of its depreciable (useful) life, S_n.

Straight-Line Depreciation

The depreciation charge in any year is

$$D_j = \frac{C - S_n}{n}.$$

An alternative computation is

$$\text{Depreciation charge in any year, } D_j = \frac{C - \text{ depreciation taken to beginning of year } j - S_n}{\text{Remaining useful life at beginning of year } j}.$$

Sum-of-Years'-Digits Depreciation

$$\text{Depreciation charge in any year, } D_j = \frac{\text{Remaining depreciable life at beginning of year}}{\text{Sum of years' digits for total useful life}} \times (C - S_n)$$

Declining-Balance Depreciation

$$\text{Double declining-balance depreciation charge in any year, } D_j = \frac{2C}{m}\left(1 - \frac{2}{n}\right)^{j-1}$$

$$\text{Total depreciation at the end of } n \text{ years, } C = \left[1 - \left(1 - \frac{2}{n}\right)^n\right]$$

$$\text{Book value at the end of } j \text{ years, } BV_j = C\left(1 - \frac{2}{n}\right)^j$$

For 150 percent declining-balance depreciation, replace the 2 in the three equations above with 1.5.

Sinking-Fund Depreciation

$$\text{Depreciation charge in any year, } D_j = (C - S_n)(A/F, i\%, n)(F/P, i\%, j - 1)$$

Modified Accelerated Cost Recovery System Depreciation

The modified accelerated cost recovery system (MACRS) depreciation method generally applies to property placed in service after 1986. To compute the MACRS depreciation for an item, one must know

1. Cost (basis) of the item.
2. Property class. All tangible property is classified in one of six classes (3, 5, 7, 10, 15, and 20 years), which is the life over which it is depreciated (see Table A.3). Residential real estate and nonresidential real estate are in two separate real property classes of 27.5 years and 39 years, respectively.
3. Depreciation computation.

Table A.3 MACRS classes of depreciable property

Property Class	Personal Property (All Property Except Real Estate)
3-year property	Special handling devices for food and beverage manufacture Special tools for the manufacture of finished plastic products, fabricated metal products, and motor vehicles Property with an asset depreciation range (ADR) midpoint life of 4 years or less
5-year property	Automobiles* and trucks Aircraft (of non–air-transport companies) Equipment used in research and experimentation Computers Petroleum drilling equipment Property with an ADR midpoint life of more than 4 years and less than 10 years
7-year property	All other property not assigned to another class Office furniture, fixtures, and equipment Property with an ADR midpoint life of 10 years or more, and less than 16 years
10-year property	Assets used in petroleum refining and preparation of certain food products Vessels and water transportation equipment Property with an ADR midpoint life of 16 years or more, and less than 20 years
15-year property	Telephone distribution plants Municipal sewage treatment plants Property with an ADR midpoint life of 20 years or more, and less than 25 years
20-year property	Municipal sewers Property with an ADR midpoint life of 25 years or more
Property Class	**Real Property (Real Estate)**
27.5 years	Residential rental property (does not include hotels and motels)
39 years	Nonresidential real property

*The depreciation deduction for automobiles is limited to $2860 in the first tax year and is reduced in subsequent years.

- Use double-declining-balance depreciation for 3-, 5-, 7-, and 10-year property classes with conversion to straight-line depreciation in the year that increases the deduction.
- Use 150%-declining-balance depreciation for 15- and 20-year property classes with conversion to straight-line depreciation in the year that increases the deduction.
- In MACRS, the salvage value is assumed to be zero.

Half-Year Convention

Except for real property, a half-year convention is used. Under this convention all property is considered to be placed in service in the middle of the tax year, and a half-year of depreciation is allowed in the first year. For each of the remaining years, one is allowed a full year of depreciation. If the property is disposed of

Table A.4 MACRS* depreciation for personal property—half-year convention

If the Recovery Year Is	The Applicable Percentage for the Class of Property Is			
	3-Year Class	5-Year Class	7-Year Class	10-Year Class
1	33.33	20.00	14.29	10.00
2	44.45	32.00	24.49	18.00
3	14.81†	19.20	17.49	14.40
4	7.41	11.52†	12.49	11.52
5		11.52	8.93†	9.22
6		5.76	8.92	7.37
7			8.93	6.55†
8			4.46	6.55
9				6.56
10				6.55
11				3.28

*In the *Fundamentals of Engineering Reference Handbook*, this table is called "Modified ACRS Factors."

†Use straight-line depreciation for the year marked and all subsequent years.

prior to the end of the recovery period (property class life), a half-year of depreciation is allowed in that year. If the property is held for the entire recovery period, a half-year of depreciation is allowed for the year following the end of the recovery period (see Table A.4). Owing to the half-year convention, a general form of the double-declining-balance computation must be used to compute the year-by-year depreciation.

$$\text{DDB depreciation in any year, } D_j = \frac{2}{n}(C - \text{depreciation in years prior to } j)$$

Example A.29

A \$5000 computer has an anticipated \$500 salvage value at the end of its five-year depreciable life. Compute the depreciation schedule for the machinery by (a) sum-of-years'-digits depreciation and (b) MACRS depreciation. Do the MACRS computation by hand, and then compare the results with the values from Table A.4.

Solution

(a) Sum-of-years'-digits depreciation:

$$D_j = \frac{n - j + 1}{\frac{n}{2}(n+1)}(C - S_n)$$

$$D_1 = \frac{5 - 1 + 1}{\frac{5}{2}(5+1)}(5000 - 500) = \$1500$$

$$D_2 = \frac{5-2+1}{\frac{5}{2}(5+1)}(5000-500) = \$1200$$

$$D_3 = \frac{5-3+1}{\frac{5}{2}(5+1)}(5000-500) = 900$$

$$D_4 = \frac{5-4+1}{\frac{5}{2}(5+1)}(5000-500) = 600$$

$$D_5 = \frac{5-5+1}{\frac{5}{2}(5+1)}(5000-500) = 300$$

$$\overline{\$4500}$$

(b) MACRS depreciation. Double-declining-balance with conversion to straight-line. Five-year property class. Half-year convention. Salvage value S_n is assumed to be zero for MACRS. Using the general DDB computation,

Year

Year	Computation	Amount
1 (half-year)	$D_1 = \frac{1}{2} \times \frac{2}{5}(5000-0)$	= \$1000
2	$D_2 = \frac{2}{5}(5000-1000)$	= 1600
3	$D_3 = \frac{2}{5}(5000-2600)$	= 960
4	$D_4 = \frac{2}{5}(5000-3560)$	= 576
5	$D_5 = \frac{2}{5}(5000-4136)$	= 346
6 (half-year)	$D_6 = \frac{1}{2} \times \frac{2}{5}(5000-4482)$	= 104
		\$4586

The computation must now be modified to convert to straight-line depreciation at the point where the straight-line depreciation will be larger. Using the alternative straight-line computation,

$$D_5 = \frac{5000-4136-0}{1.5 \text{ years remaining}} = \$576.$$

This is more than the \$346 computed using DDB, hence switch to straight-line for year 5 and beyond.

$$D_6 \text{ (half-year)} = \frac{1}{2}(576) = \$288$$

Answers:

Year	Depreciation SOYD	MACRS
1	$1500	$1000
2	1200	1600
3	900	960
4	600	576
5	300	576
6	0	288
	$4500	$5000

The computed MACRS depreciation is identical to the result obtained from Table A.4.

TAX CONSEQUENCES

Income taxes represent another of the various kinds of disbursements encountered in an economic analysis. The starting point in an after-tax computation is the before-tax cash flow. Generally, the before-tax cash flow contains three types of entries:

1. Disbursements of money to purchase capital assets. These expenditures create no direct tax consequence, for they are the exchange of one asset (money) for another (capital equipment).
2. Periodic receipts and/or disbursements representing operating income and/or expenses. These increase or decrease the year-by-year tax liability of the firm.
3. Receipts of money from the sale of capital assets, usually in the form of a salvage value when the equipment is removed. The tax consequences depend on the relationship between the book value (cost – depreciation taken) of the asset and its salvage value.

Situation	Tax Consequence
Salvage value > Book value	Capital gain on differences
Salvage value = Book value	No tax consequence
Salvage value < Book value	Capital loss on difference

After determining the before-tax cash flow, compute the depreciation schedule for any capital assets. Next, compute taxable income, the taxable component of the before-tax cash flow minus the depreciation. The income tax is the taxable income times the appropriate tax rate. Finally, the after-tax cash flow is the before-tax cash flow adjusted for income taxes.

To organize these data, it is customary to arrange them in the form of a cash flow table, as follows:

Year	Before-Tax Cash Flow	Depreciation	Taxable Income	Income Taxes	After-Tax Cash Flow
0	•				•
1	•	•	•	•	•

Example A.30

A corporation expects to receive $32,000 each year for 15 years from the sale of a product. There will be an initial investment of $150,000. Manufacturing and sales expenses will be $8067 per year. Assume straight-line depreciation, a 15-year useful life, and no salvage value. Use a 46 percent income tax rate. Determine the projected after-tax rate of return.

Solution

Straight-line depreciation, $D_j = \dfrac{C - S_n}{n} = \dfrac{\$150,000 - 0}{15} = \$10,000$ per year

Year	Before-Tax Cash Flow	Depreciation	Taxable Income	Income Taxes	After-Tax Cash Flow
0	−150,000				−150,000
1	+23,933	10,000	13,933	−6,409	+17,524
2	+23,933	10,000	13,933	−6,409	+17,524
•	•	•	•	•	•
•	•	•	•	•	•
•	•	•	•	•	•
15	+23,933	10,000	13,933	−6,409	+17,524

Take the after-tax cash flow and compute the rate of return at which the PW of cost equals the PW of benefits.

$$150,000 = 17,524(P/A, i\%, 15)$$

$$(P/A, i\%, 15) = \frac{150,000}{17,524} = 8.559$$

From the compound interest tables, the after-tax rate of return is $i = 8\%$.

INFLATION

Inflation is characterized by rising prices for goods and services, whereas deflation produces a fall in prices. An inflationary trend makes future dollars have less purchasing power than present dollars. This helps long-term borrowers of money, for they may repay a loan of present dollars in the future with dollars of reduced buying power. The help to borrowers is at the expense of lenders. Deflation has the opposite effect. Money borrowed at one point in time, followed by a deflationary period, subjects the borrower to loan repayment with dollars of greater purchasing power than those borrowed. This is to the lenders' advantage at the expense of borrowers.

Price changes occur in a variety of ways. One method of stating a price change is as a uniform rate of price change per year.

f = General inflation rate per interest period
i = Effective interest rate per interest period

The following situation will illustrate the computations. A mortgage is to be repaid in three equal payments of $5000 at the end of years 1, 2, and 3. If the annual inflation rate, f, is 8% during this period, and a 12% annual interest rate

(i) is desired, what is the maximum amount the investor would be willing to pay for the mortgage?

The computation is a two-step process. First, the three future payments must be converted to dollars with the same purchasing power as today's (year 0) dollars.

Year	Actual Cash Flow		Multiplied by		Cash Flow Adjusted to Today's (yr. 0) Dollars
0	—		—		—
1	+5000	×	$(1 + 0.08)^{-1}$	=	+4630
2	+5000	×	$(1 + 0.08)^{-2}$	=	+4286
3	+5000	×	$(1 + 0.08)^{-3}$	=	+3969

The general form of the adjusting multiplier is

$$(1 + f)^{-n} = (P/F, f, n).$$

Now that the problem has been converted to dollars of the same purchasing power (today's dollars, in this example), we can proceed to compute the present worth of the future payments.

Year	Adjusted Cash Flow		Multiplied by		Present Worth
0	—		—		—
1	+4630	×	$(1 + 0.12)^{-1}$	=	+4134
2	+4286	×	$(1 + 0.12)^{-2}$	=	+3417
3	+3969	×	$(1 + 0.12)^{-3}$	=	+2825
					$10,376

The general form of the discounting multiplier is

$$(1 + i)^{-n} = (P/F, i\%, n).$$

Alternative Solution

Instead of doing the inflation and interest rate computations separately, one can compute a combined equivalent interest rate, d.

$$d = (1 + f)(1 + i) - 1 = i + f + i(f).$$

For this cash flow, $d = 0.12 + 0.08 + 0.12(0.08) = 0.2096$. Since we do not have 20.96 percent interest tables, the problem has to be calculated using present worth equations.

$$\begin{aligned} PW &= 5000(1 + 0.2096)^{-1} + 5000(1 + 0.2096)^{-2} + 5000(1 + 0.2096)^{-3} \\ &= 4134 + 3417 + 2825 = \$10,376 \end{aligned}$$

Example A.31

One economist has predicted that there will be 7 percent per year inflation of prices during the next 10 years. If this proves to be correct, an item that presently sells for $10 would sell for what price 10 years hence?

Solution

$$f = 7\%,\ P = \$10$$
$$F = ?,\ n = 10 \text{ years}$$

Here the computation is to find the future worth *F*, rather than the present worth, *P*.

$$F = P(1 + f)^{10} = 10(1 + 0.07)^{10} = \$19.67$$

Effect of Inflation on Rate of Return

The effect of inflation on the computed rate of return for an investment depends on how future benefits respond to the inflation. If benefits produce constant dollars, which are not increased by inflation, the effect of inflation is to reduce the before-tax rate of return on the investment. If, on the other hand, the dollar benefits increase to keep up with the inflation, the before-tax rate of return will not be adversely affected by the inflation.

This is not true when an after-tax analysis is made. Even if the future benefits increase to match the inflation rate, the allowable depreciation schedule does not increase. The result will be increased taxable income and income tax payments. This reduces the available after-tax benefits and, therefore, the after-tax rate of return.

Example **A.32**

A man bought a 5 percent tax-free municipal bond. It cost $1000 and will pay $50 interest each year for 20 years. The bond will mature at the end of 20 years and return the original $1000. If there is 2% annual inflation during this period, what rate of return will the investor receive after considering the effect of inflation?

Solution

$$d = 0.05,\ i = \text{unknown},\ j = 0.02$$
$$d = i + j + i(j)$$
$$0.05 = i + 0.02 + 0.02i$$
$$1.02i = 0.03,\ i = 0.294 = 2.94\%$$

RISK ANALYSIS

Probability

Probability can be considered to be the long-run relative frequency of occurrence of an outcome. There are two possible outcomes from flipping a coin (a head or a tail). If, for example, a coin is flipped over and over, we can expect in the long run that half the time heads will appear and half the time tails will appear. We would say the probability of flipping a head is 0.50 and of flipping a tail is 0.50. Since the probabilities are defined so that the sum of probabilities for all possible outcomes is 1, the situation is

$$\text{Probability of flipping a head} = 0.50$$
$$\text{Probability of flipping a tail} = \underline{0.50}$$
$$\text{Sum of all possible outcomes} = 1.00.$$

Example A.33

If one were to roll one die (that is, one-half of a pair of dice), what is the probability that either a 1 or a 6 would result?

Solution

Since a die is a perfect six-sided cube, the probability of any side appearing is 1/6.

$$\begin{aligned} \text{Probability of rolling a } 1 &= P(1) = 1/6 \\ 2 &= P(2) = 1/6 \\ 3 &= P(3) = 1/6 \\ 4 &= P(4) = 1/6 \\ 5 &= P(5) = 1/6 \\ 6 &= P(6) = 1/6 \end{aligned}$$

Sum of all possible outcomes = 6/6 = 1. The probability of rolling a 1 or a 6 = 1/6 + 1/6 = 1/3.

In the preceding examples, the probability of each outcome was the same. This need not be the case.

Example A.34

In the game of blackjack, a perfect hand is a 10 or a face card plus an ace. What is the probability of being dealt a 10 or a face card from a newly shuffled deck of 52 cards? What is the probability of being dealt an ace in this same situation?

Solution

The three outcomes examined are to be dealt a 10 or a face card, an ace, or some other card. Every card in the deck represents one of these three possible outcomes. There are 4 aces; 16 10s, jacks, queens, and kings; and 32 other cards.

$$\begin{aligned} \text{Probability of being dealt a 10 or a face card} &= 16/52 = 0.31 \\ \text{Probability of being dealt an ace} &= 4/52 = 0.08 \\ \text{Probability of being dealt some other card} &= 32/52 = \underline{0.61} \\ & 1.00 \end{aligned}$$

Risk

The term *risk* has a special meaning in statistics. It is defined as a situation where there are two or more possible outcomes and the probability associated with each outcome is known. In each of the two previous examples there is a risk situation. We could not know in advance what playing card would be dealt or what number would be rolled by the die. However, since the various probabilities could be computed, our definition of risk has been satisfied. Probability and risk are not restricted to gambling games. For example, in a particular engineering course, a student has computed the probability for each of the letter grades he might receive as follows:

Grade	Grade Point	Probability P(Grade)
A	4.0	0.10
B	3.0	0.30
C	2.0	0.25
D	1.0	0.20
F	0	0.15
		1.00

From the table we see that the grade with the highest probability is a B. This, therefore, is the most likely grade. We also see that there is a substantial probability that some grade other than a B will be received. And the probabilities indicate that if a B is not received, the grade will probably be something less than a B. But in saying that the most likely grade is a B, other outcomes are ignored. In the next section we will show that a composite statistic may be computed using all the data.

Expected Value

In the last example the most likely grade of B in an engineering class had a probability of 0.30. That is not a very high probability. In some other course, say a math class, we might estimate a probability of 0.65 of obtaining a B, again making the B the most likely grade. While a B is most likely in both classes, it is more certain in the math class.

We can compute a weighted mean to give a better understanding of the total situation as represented by various possible outcomes. When the probabilities are used as the weighting factors, the result is called the *expected value* and is written

$$\text{Expected value} = \text{Outcome}_A \times P(A) + \text{Outcome}_B \times P(B) + \cdots$$

Example A.35

An engineer wishes to determine the risk of fire loss for her $200,000 home. From a fire rating bureau she obtains the following data:

Outcome	Probability
No fire loss	0.986 in any year
$10,000 fire loss	0.010
40,000 fire loss	0.003
200,000 fire loss	0.001

Compute the expected fire loss in any year.

Solution

$$\text{Expected fire loss} = 10{,}000(0.010) + 40{,}000(0.003) + 200{,}000(0.001) = \$420$$

REFERENCE

Newnan, Donald G. *Engineering Economic Analysis*, 5th ed. Engineering Press, San Jose, CA, 1995.

APPENDIX B

Suggested Study Reference List

Mechanics and Machine Design

Church, A. H., *Mechanical Vibrations,* 2nd ed., Wiley, New York, 1963.
Creamer, R. H., *Machine Design,* 3rd ed., Addison-Wesley, Reading, Mass., 1984.
Dudley, D. W., *Gear Handbook,* 2nd ed., McGraw-Hill, New York, 1991.
Faires, V.M., *Design of Machine Elements,* 4th ed., Macmillan, New York, 1965.
Ford, H., and J. M. Alexander, *Advanced Mechanics of Materials,* Wiley, New York, 1977.
Haberman, *Vibration Analysis,* Merrill, Chicago, 1968.
Hibbeler, R. C., *Engineering Mechanics,* 10th ed., Prentice-Hall, Englewood Cliffs, N.J., 2004.
Hinkle, R. T., *Kinematics of Machines,* 2nd ed., Prentice-Hall, Englewood Cliffs, N.J., 1960.
Juvinall, R. C., and K. M. Marshek, *Fundamentals of Machine Component Design,* 3rd ed., Wiley, New York, 2003.
Oberg, E., et al., *Machinery's Handbook,* 27th ed., Industrial Press, New York, 2004.
Olsen, G. A., *Elements of Mechanics of Materials,* 3rd ed., Prentice-Hall, Englewood Cliffs, N.J., 1974.
Riley et al., *Mechanics of Materials,* 5th ed., Wiley, 1999.
Spotts, N. F., *Design of Machine Elements,* 8th ed., Prentice-Hall, Englewood Cliffs, N.J., 2003.
Timoshenko, S., *Vibration Problems in Engineering,* 5th ed., Wiley, New York, 1990.
Willems, N., *Strength of Materials,* McGraw-Hill, New York, 1981.

Fluid Mechanics and Hydraulics

Binder, R. C., *Fluid Mechanics,* 5th ed., Prentice-Hall, Englewood Cliffs, N.J., 1973.
Brater, E. F., and H. W. King, *Handbook of Hydraulics,* 6th ed., McGraw-Hill, New York, 1976.
Centrifugal Pump Application Manual, Buffalo Forge Co., 1959.
Davis, C. V., and K. E. Sorenson, *Handbook of Applied Hydraulics,* 3rd ed., McGraw-Hill, New York, 1969.

Finnemore, E., and J. B. Franzinia, *Fluid Mechanics with Engineering Applications,* 10th ed., McGraw-Hill, New York, 2001.
Hicks, T. G., and T. Edwards, *Pump Application Engineering,* McGraw-Hill, New York, 1971.
Streeter, V. L., and E. B. Wylie, *Fluid Mechanics,* 8th ed., McGraw-Hill, New York, 1985.
Thompson, P. A., *Compressible-Fluid Dynamics,* McGraw-Hill, New York, 1971.

Thermodynamics, Heat and Power

ASME *Power Test Codes,* American Society of Mechanical Engineers.
Cengel, Y. A., and M.A. Boles, *Thermodyanmics: An Engineering Approach,* 4th ed., McGraw-Hill, New York, 2002.
Jones, J. B., and G. A. Hawkins, *Engineering Thermodynamics,* 2nd ed., Wiley, New York, 1986.
Moran, M. J., and H. N. Shapiro, *Fundamentals of Thermodynamics,* 5th ed., Wiley, New York, 2003.
Potter, P. J., *Power Plant Theory and Design,* 2nd ed., Krieger, New York, 1988.
Zemansky, M. W., *Heat and Thermodynamics,* 6th ed., McGraw-Hill, New York, 1981.

Engines and Turbines

El-Wakil, M. M., *Powerplant Technology,* McGraw-Hill, New York, 2002.
Lichty, L. C., *Internal Combustion Engines,* McGraw-Hill, New York, 1951.
Shepherd, D. G., *Principles of Turbomachinery,* Macmillan, New York, 1961.

Heat Transfer

Holman, J. P., *Heat Transfer,* 9th ed., McGraw-Hill, New York, 2002.
Incropera, F. P., and D. P. Dewitt, *Fundamentals of Heat and Mass Transfer,* 5th ed., Wiley, New York, 2001.
Welty, J. R., *Engineering Heat Transfer,* Wiley, New York, 1978.

Refrigeration and Air Conditioning

ASHRAE *Guide and Data Book Applications,* 1980.
ASHRAE *Guide and Data Book Systems,* 1980.
ASHRAE *Handbook of Fundamentals,* 2005.
Carrier Air Conditioning Company, *Handbook of Air Conditioning System Design,* McGraw-Hill, New York, 1966.
Dossat, R. J., *Principles of Refrigeration,* 5th ed., Prentice-Hall, New York, 2002.
Fan Engineering, 7th ed., Buffalo Forge, 1970.
Hemeon, W., *Plant and Process Ventilation,* 3rd ed., CRC Press, 1998.
McQuiston, F. C. and J. D. Parker, *Heating, Ventilating and Air Conditioning,* 6th ed., Wiley, New York, 2005.
Stoecker, W. F., *Design of Thermal Systems,* 3rd ed., McGraw-Hill, New York, 1989.

Zimmerman and Levine, *Psychrometric Tables and Charts,* Industrial Research Service, 1954.

Environmental and Sound Control

Austin, P. R., *Design and Operation of Clean Rooms,* rev. ed., Business News, Birmingham, Mich. 1965.

Clayton, G. D., and F. E. Clayton, *Patty's Industrial Hygiene and Toxicology,* 5th ed., rev.; vol. 1: *General Principles,* Wiley Interscience, New York, 2000.

Clayton, G. D., and F. E. Clayton, *Patty's Industrial Hygiene and Toxicology,* 5th ed., rev.; vols. 2A, 2B, and 2C: *Toxocology,* Wiley Interscience, New York, 2000.

Constance, J. A., "Why Some Dust Control Systems Don't Work," *Pharmaceutical Engineering,* vol. 3, no. 1, January-February 1983).

Constance, J. A., "Reverse Contamination Caused by Dust Control Exhaust Systems," *Pharmaceutical Engineering,* vol. 3, no. 2, March–April 1983.

Constance, J. D., *Controlling In-Plant Airborne Contaminants: Systems Design and Calculations,* Marcel Dekker, New York, 1983.

Industrial Ventilation: A Manual of Recommended Practice, 25th ed., American Conference of Governmental Industrial Hygienists, 2004.

Jennings, B. H., *Environmental Engineering,* IEP, New York, 1970.

Lewis, R. J., *Sax's Dangerous Properties of Industrial Materials,* 11th ed., John Wiley, New York, 2004.

The Occupational Environment—Its Evaluation and Control, AIHA Press, Fairfax, VA, 1997.

Threlkeld, J. L., *Thermal Environmental Engineering,* 3rd ed., Prentice-Hall, Englewood Cliffs, N.J., 1998.

Nuclear Power

Etherington, H., *Nuclear Engineering Handbook,* McGraw-Hill, New York, 1958.

Glasstone, S., and A. Sesonske, *Nuclear Reactor Engineering,* Springer Verlag, New York, 1995.

Plant and Production Engineering

Apple, J. M., *Material Handling Systems Design,* Wiley, New York, 1972.

Bakerjian, R., and W. Cubberly (eds.), *Tool and Manufacturing Engineers Handbook Desk Edition,* Society of Manufacturing Engineers, Dearborn, MI, 1988.

Bolz, H. A., and G. E. Hagemann, *Materials Handling Handbook,* Wiley, New York, 1958.

Higgins, L., and K. Mobley, *Maintenance Engineering Handbook,* 6th ed., McGraw-Hill, New York, 2002.

Kraus, Milton N. and staff of Chemical Engineering Magazine, *Pneumatic Conveying of Bulk Materials,* 2nd ed., McGraw-Hill, New York, 1980.

Society of Tool and Manufacturing Engineers, *Handbook of Fixture Design,* McGraw-Hill, New York, 1962.

Stanier, W., *Plant Engineering Handbook,* 2nd ed., McGraw-Hill, New York, 1959.

Metrication

Adams, H. F. R., *SI Metric Units—An Introduction,* McGraw-Hill Paperbacks, 1974.

Conversion Factors, Englehard Industries Division of Englehard Minerals and Chemicals Corp., Murray Hill, N.J. 07974.

Feirer, J. L., *SI Metric Handbook,* The Metric Company, 1985.

Gaspe, Marc S., and Nicola Scianna, "The Problems of Dual Dimensioning," *Consulting Engineer,* June 1977.

Manuals and Handbooks—General

Avallone, E. A., and T. Baumeister, *Marks Standard Handbook for Mechanical Engineers,* 10th ed., McGraw-Hill, New York, 1996.

Heald, C. C., *Cameron Hydraulic Data,* 18th ed., Ingersoll-Rand, 1994.

Keenan, J. H., F. G. Keyes, P. G. Hill, and J. G. Moore, *Steam Tables: Thermodynamic Properties of Water, Including Vapor, Liquid and Solid,* Wiley, New York, 1969 (English units), 1978 (S.I. units).

Kent, R. T., *Mechanical Engineer's Handbook,* 12th ed., Wiley, New York, 1950.

In addition to the above references, much help can be obtained from the various technical publications: *Chemical Engineering; Heating, Piping and Air Conditioning; Machine Design; Nucleonics; Plant Engineering; Power; Power Engineering; Product Engineering; Research and Development; Specifying Engineer.*

INDEX